Conservation and Environmental Management in Madagascar

Madagascar is one of the most biologically diverse places on the planet, the result of 160 million years of isolation from the African mainland. More than 80 per cent of its species are not found anywhere else on Earth. However, this highly diverse flora and fauna is threatened by habitat loss and fragmentation and the island has been classified one of the world's highest conservation priorities.

Drawing on insights from geography, anthropology, sustainable development, political science and ecology, this book provides a comprehensive assessment of the status of conservation and environmental management in Madagascar. It describes how conservation organisations have been experimenting with new forms of protected areas, community-based resource management, ecotourism and payments for ecosystem services. But the country must also deal with pressing human needs. The problems of poverty, development, environmental justice, natural resource use and biodiversity conservation are shown to be interlinked in complex ways. Authors address key questions, such as who are the winners and losers in attempts to conserve biodiversity? And what are the implications of new forms of conservation for rural livelihoods and environmental justice?

Ivan R. Scales is McGrath Lecturer in Human Geography, St Catharine's College, University of Cambridge, UK.

Earthscan Conservation and Development series

Series Editor: W.M. Adams, Moran Professor of Conservation and Development, Department of Geography, University of Cambridge, UK

Conservation and Sustainable Development
Linking practice and policy in Eastern Africa
Edited by Jon Davies

Conservation and Environmental Management in Madagascar
Edited by Ivan R. Scales

Forthcoming titles:

Conservation and Development in Cambodia
New frontiers in nature, society and community
Edited by Sarah Milne and Sanghamitra Mahanty

For further information please visit the series page on the Routledge website: http://www.routledge.com/books/series/ECCAD

Conservation and Environmental Management in Madagascar

Edited by Ivan R. Scales

Routledge
Taylor & Francis Group
LONDON AND NEW YORK

First published 2014
by Routledge
2 Park Square, Milton Park, Abingdon, Oxon OX14 4RN

and by Routledge
711 Third Avenue, New York, NY 10017

Routledge is an imprint of the Taylor & Francis Group, an informa business

© 2014 Ivan R. Scales, selection and editorial material; individual chapters, the contributors

The right of the editor to be identified as the author of the editorial material, and of the authors for their individual chapters, has been asserted in accordance with sections 77 and 78 of the Copyright, Designs and Patents Act 1988.

All rights reserved. No part of this book may be reprinted or reproduced or utilised in any form or by any electronic, mechanical, or other means, now known or hereafter invented, including photocopying and recording, or in any information storage or retrieval system, without permission in writing from the publishers.

Trademark notice: Product or corporate names may be trademarks or registered trademarks, and are used only for identification and explanation without intent to infringe.

British Library Cataloguing-in-Publication Data
A catalogue record for this book is available from the British Library

Library of Congress Cataloging-in-Publication Data
Conservation and environmental management in Madagascar / edited by Ivan R. Scales.
pages cm
Includes bibliographical references and index.
1. Conservation of natural resources–Madagascar. 2. Biodiversity conservation–Madagascar. 3. Natural resources–Madagascar–Management. 4. Environmental policy–Political aspects–Madagascar. I. Scales, Ivan R., author, editor of compilation.
S934.M28.C65 2014
333.7209691–dc23
2013030323

ISBN: 978-0-415-52877-1 (hbk)
ISBN: 978-0-203-11831-3 (ebk)

Typeset in Baskerville
by GreenGate Publishing Services, Tonbridge, Kent

Printed and bound by CPI Group (UK) Ltd, Croydon, CR0 4YY

In memory of Professor Robert Dewar (1949–2013):
anthropologist known for his pioneering work on the
prehistory and palaeoecology of Madagascar

Contents

	List of illustrations	x
	List of contributors	xiii
	Preface	xix
	Acronyms and abbreviations	xxii
1	Introduction: conservation at the crossroads: biological diversity, environmental change and natural resource use in Madagascar IVAN R. SCALES	1

PART 1
Madagascar's biological diversity: from deep time to the arrival of humans 15

2	Explaining Madagascar's biodiversity JÖRG U. GANZHORN, LUCIENNE WILMÉ AND JEAN-LUC MERCIER	17
3	Early human settlers and their impact on Madagascar's landscapes ROBERT E. DEWAR	44

PART 2
Paradise lost? The myths, narratives and received wisdoms at the heart of conservation research and policy 65

4	Deforestation in Madagascar: debates over the island's forest cover and challenges of measuring forest change WILLIAM J. MCCONNELL AND CHRISTIAN A. KULL	67
5	The drivers of deforestation and the complexity of land use in Madagascar IVAN R. SCALES	105

PART 3
The politics of biodiversity conservation and environmental management   127

6 A brief history of the state and the politics of natural resource use in Madagascar   129
   IVAN R. SCALES

7 The roots, persistence, and character of Madagascar's conservation boom   146
   CHRISTIAN A. KULL

8 The transfer of natural resource management rights to local communities   172
   JACQUES POLLINI, NEAL HOCKLEY, FRANK D. MUTTENZER AND BRUNO S. RAMAMONJISOA

9 Conservation politics in Madagascar: the expansion of protected areas   193
   CATHERINE CORSON

10 The Durban Vision in practice: experiences in the participatory governance of Madagascar's new protected areas   216
   MALIKA VIRAH-SAWMY, CHARLIE J. GARDNER AND ANITRY N. RATSIFANDRIHAMANANA

PART 4
Making conservation pay? Incentive-based conservation, the commodification of Madagascar's nature and conflicting views of landscape and nature   253

11 Tourism, conservation and development in Madagascar: moving beyond panaceas?   255
   IVAN R. SCALES

12 Bioprospecting a biodiversity hotspot: the political economy of natural products drug discovery for conservation goals in Madagascar   271
   BENJAMIN D. NEIMARK AND LAURA M. TILGHMAN

13 Incentivising forest conservation: payments for environmental services and reducing carbon emissions from deforestation   299
   LAURA BRIMONT AND CÉCILE BIDAUD

14 Contrasting visions of nature and landscapes   320
    JEFFREY C. KAUFMANN

15 Conclusion: the future of biodiversity conservation and
    environmental management in Madagascar: lessons from
    the past and challenges ahead   342
    IVAN R. SCALES

    Index   361

Illustrations

Figures

1.1	Map of Madagascar showing the location of key case studies discussed in the book	5
1.2	Timeline of significant political and economic events in Madagascar	8
2.1	Species–area relationship for bats (including flying foxes) on islands of different size	19
2.2	Number of species described for Madagascar (since Linnaeus in 1758) in a range of vertebrate and invertebrate taxonomic groups	20
2.3	Geological time scales, phases of continental drift and colonization of Madagascar by selected vertebrate groups	23
2.4	Bathymetric profile (underwater topography) between Antarctica and Madagascar	25
2.5	Vegetation formations and climatic conditions forming distinct ecotypes of Madagascar	29
2.6	Left: schematic sequence of the retreat of forests during cold dry periods and forest expansion during warm, humid episodes. Right: resulting centers of endemism and retreat-dispersion watershed	32
3.1	Location of key palaeontological and archaeological sites	46
4.1	Forest patches in the highlands landscape. Are these images of deforestation or reforestation?	68
4.2	Current vegetation zones of Madagascar	69
4.3	James Sibree's (1879) Physical Map of Madagascar	72
4.4	Key steps in the classification of satellite imagery	80
5.1	Forest recently cleared for *hatsake* (swidden cultivation) in western Madagascar	111
5.2	First year of cultivation following forest clearance	112

7.1	Annual expenditures by WWF in Madagascar – currently celebrating 50 years of work on the island – are illustrative of the conservation boom and its persistence	147
7.2	Bilateral aid from Switzerland to Madagascar, 1963–2011	148
7.3	Timeline of key events in Madagascar politics and conservation, 1970–2012	150
8.1	Key stakeholders in early stages of GELOSE management transfers	175
9.1	Map showing the expansion of protected areas and location of case studies discussed in the chapter	206
10.1	A view of the sacred forest on a hill in Ankodida protected area	230
10.2	A local meeting with staff from the WWF for the establishment of Ankodida protected area	230
10.3	Protected area zoning of Ankodida	231
10.4	Governance structure of Ankodida protected area as a hybrid between traditional and modern forms of governance	232
10.5	Protected area zoning of Ranobe PK32	233
10.6	A view of the extent of *hatsake* (swidden agriculture for maize cultivation) in the forest frontier of Ranobe PK32 from the air	234
10.7	View of on-the-ground deforestation for maize cultivation, also showing migrant shelter	235
10.8	Governance structure of Ranobe PK32 in 2006 based mainly on municipal governance	235
10.9	A new structure in 2012 to ensure more representation of local actors as a hybrid between local and municipal governance	237
11.1	Ring-tailed lemurs (*Lemur catta*) in Berenty Nature Reserve, one of Madagascar's biggest tourist draws	257
11.2	Baobab Alley: rows of Grandidier's baobab (*Adansonia grandidieri*) near Morondava. Tourist curiosity or conservation icon?	261
12.1	Rosy periwinkle: medicinal wonder plant or the global case of 'biopiracy'?	277
12.2	ICBG-Madagascar project plant collection locations	284
13.1	The distribution of carbon revenues from the Makira REDD+ project	309
13.2	The share of conservation agreement's budget in US$ (VOI Taratra, Didy municipality, CAZ, 2011)	312

Tables

2.1	Endemic plant and vertebrate families, genera and species in some global biodiversity hotspots in the tropics and subtropics	21
3.1	Youngest radiocarbon determination for extinct species in Madagascar	59
4.1	Colonial period estimates of Madagascar's forest cover	73
4.2	Inconsistent re-interpretations of total forest cover according to the 1949–1957 air photos (and the topographic maps derived from these)	74
4.3	Analysis of articles in scholarly publications making the 90 per cent claim	86
4.4	Description of the sources cited by articles in Table 4.3 to justify their 90 per cent claims	90
10.1	IUCN protected area categories and how they are being applied in Madagascar	218
10.2	IUCN protected area governance types and how they are being applied in Madagascar	221
10.3	New protected areas within the Durban Vision and the delegated institutions supporting them	223
12.1	Role and function of US- and Madagascar-based funding institutions and research organizations in the ICBG-Madagascar	285
12.2	Malagasy terms used to describe upfront compensation development activities (Phase I)	287

Boxes

2.1	Biodiversity	18
2.2	Island biogeography and speciation	18
4.1	Using satellite imagery to classify land cover	80
6.1	Environmental politics and different forms of power	130
6.2	'Tribes', ethnicity and the politics of identity	138
10.1	Definition of terms concerning protected areas	226
12.1	Madagascar's rosy periwinkle: fodder for the debate	277

Contributors

Cécile Bidaud has a PhD from the Graduate Institute of International and Development Studies in Geneva. Her thesis was on the science and politics of carbon storage in Malagasy forests. She is currently working for the French Institute of Research for Development and supported by the Malagasy Center of Economy, Environment and Equity for the Development of Madagascar (C3EDM) at the University of Antananarivo, where she is analysing the use of ecosystem services in environmental programmes and policies with the SERENA research programme. She also works on the scientific basis and governance of mining company offsets and their influence on national policies.

Laura Brimont is finishing a PhD in economics at the Agricultural Research Center for Development (CIRAD) in Montpellier, France. Her thesis is evaluating the impact of Reducing Emissions from Deforestation and Forest Degradation (REDD+) projects and various incentive-based tools for tropical forests conservation in Madagascar. She is using a multi-disciplinary approach combining economics, political science and management, and drawing on both qualitative and quantitative methods. She has a Masters degree in the economics of international relations from Sciences Po (Paris) and has also carried out fieldwork in Cameroon.

Catherine Corson is Assistant Professor of Environmental Studies at Mount Holyoke College, Massachusetts. As a political ecologist, her current research uses ethnography to explore questions of power, knowledge and justice in environmental governance in case studies that stretch from rural villages to international policy arenas. She has conducted ethnographic field research in Zimbabwe, Australia, Madagascar and the United States on rural women's microenterprises and on local and indigenous peoples' access to resources in protected areas. Her current intellectual interest in the use of ethnography to study global environmental policy-making processes is informed by a decade of prior professional experience in international environment and development policy, research and consulting. She received her PhD from the

University of California at Berkeley and she has Masters degrees from Cornell University and University College London.

Robert E. Dewar's primary research interests were in prehistoric settlement patterns, and particularly how people transform the landscapes they occupy and exploit. These interests led to his initial PhD research on early agriculture in Taiwan, and subsequent work on Connecticut prehistory and population trends in pre-Columbian central Mexico. He is most widely known, however, for his deep engagement with the prehistory and palaeoecology of Madagascar, where he worked from 1976 until the time of his death in 2013. He majored in Anthropology at Brown University and received his PhD in Anthropology from Yale University. After many years as a Professor of Anthropology at the University of Connecticut, he became a Fellow of the McDonald Institute for Archaeological Research, University of Cambridge. At the time of this death, he was a Senior Research Scientist at Yale University.

Jörg U. Ganzhorn is Professor of Zoology at the Department of Animal Ecology and Conservation at Hamburg University. He started working in Madagascar in 1984, focusing on the ecology of forest ecosystems, lemurs and mammals in general. He is interested in identifying ways to reconcile conservation and the survival of Madagascar's native ecosystems with the economic development of the country.

Charlie J. Gardner is a doctoral student at the Durrell Institute of Conservation and Ecology, University of Kent. His research focuses on reconciling conservation and development in Madagascar's rapidly expanding protected area system. Having previously worked in conservation in Kenya, Mauritius and the UK, he has been living in Toliara since 2006, and since 2010 has been a scientific advisor to WWF's Ala Maiky Programme. He holds a BSc in Zoology from the University of Leeds and an MSc in Conservation Biology from the University of Kent.

Neil Hockley is Research Lecturer in Economics and Policy at Bangor University's School of Environment, Natural Resources and Geography. He first lived and worked in Madagascar in 2001, carrying out fieldwork for his PhD on the welfare impacts of protected areas and working as a consultant to USAID and Conservation International. His research interests include: the role of individuals, communities and the state in environmental management; the use and abuse of economic evidence in decision-making; and measuring the welfare impacts of environmental policies. He holds degrees in Natural Sciences and Economics from the Universities of Cambridge, Edinburgh and Wales.

Jeffrey C. Kaufmann is Professor of Anthropology in the Department of Anthropology and Sociology at the University of Southern Mississippi. He holds a PhD in Anthropology from the University of Wisconsin–Madison with concentrations in cultural anthropology and African Studies. He also did graduate studies in philosophy at Arizona State University. He received a BS in sociology and a BA in philosophy from Montana State University. He is the author of *Greening the Great Red Island: Madagascar in Nature and Culture* (Africa Institute of South Africa, 2008), guest editor of New Perspectives on Conservation in Madagascar for *Conservation & Society* and guest editor of a volume on Emerging Histories in Madagascar for *Ethnohistory*. He was honored with the Heizer Prize from the American Ethnohistorical Society for best article on the methodology of ethnohistory, with the Mississippi Humanities Teacher Award and USM's Community Service Award.

Christian A. Kull is a geographer with interests in the human dimensions of environmental change. He investigates the political and social aspects of resource management issues such as fire, invasive species, small-scale farming, deforestation, tree planting and protected areas conservation, particularly in developing country contexts. He is also interested in the history of human-caused land-use change and plant movements. He has researched Madagascar for two decades and is author of *Isle of Fire: The Political Ecology of Landscape Burning in Madagascar* (University of Chicago Press, 2004). Trained in the United States (BA Dartmouth, MA Colorado, MSc Yale, PhD Berkeley), he has taught at McGill University (Canada) and is currently Associate Professor at Monash University in Melbourne, Australia.

William J. McConnell is Associate Director of the Center for Systems Integration and Sustainability and Associate Professor in the Department of Fisheries and Wildlife at Michigan State University. He holds a BA in the Political Economy of Natural Resources from the University of California, Berkeley; an MA in International Development and Social Change; and a PhD in Geography from Clark University. He served as Science Officer for the Land Use and Cover Change (LUCC) project after completing post-doctoral work at the Anthropological Center for Training at Indiana University. He has been working in West and East Africa since 1983 and, since 1994, in Madagascar, where he has participated in agricultural development and biodiversity conservation work. His main expertise is in land change science and particularly in collaborative land-use planning.

Jean-Luc Mercier (PhD 1972, Dr. Sc 1976) is Emeritus Professor at the University of Strasbourg, where he taught dynamic geomorphology. He is interested in weathering and soils, processes in geomorphology, heat and mass transfer in watersheds, measures and modelling.

Frank D. Muttenzer is Postdoctoral Fellow in Anthropology at the University of Toronto, and Habile candidate in ethnology at the University of Luzern since 2010. His current ethnographic fieldwork is about the representations of livelihood and ecology in the rituals of the Malagasy fishing people. He has worked in Madagascar since 2003, carrying out field research on forest policy, land-use management and legal pluralism, while implementing three research partnerships between the Graduate Institute of International and Development Studies and ESSA Forêts. The research has been published in the book *Déforestation et droit coutumier à Madagascar* (Karthala, 2010). He received his first degree in Law from the University of Basel, followed by an MA in Legal Theory from the Catholic University of Brussels, an MA in Development Studies and a PhD in Development Studies, both from the University of Geneva.

Benjamin D. Neimark is currently a Lecturer in the Lancaster Environmental Centre, Lancaster University, UK. He has worked in Madagascar since 1999. His Masters of Science degree, completed in 2001 at Cornell University, investigated strategies to improve small-scale agriculture using innovative agroforestry methods. In collaboration with the Cornell International Institute of Food Agriculture and Development (CIIFAD), he developed techniques to improve the direction and speed of domestication of threatened Malagasy forest and fruit species, in order to increase food security and provide added income. His recent research focuses on the political economy of biological prospecting and agrofuel production. His work has been funded by the Fulbright IIE, *National Geographic* and the Heinz Foundation and has been published in *Geoforum, Journal of Peasant Studies* and *Development and Change.*

Jacques Pollini holds an MA in Ecology and Agronomy and a PhD in Natural Resources. He has worked in international development and conservation programmes for over ten years in South East Asia and Africa. He is currently conducting research for the Responsive Forest Governance Initiative of the University of Illinois at Urbana Champaign. His main areas of expertise are swidden cultivation and land-use changes on forest frontiers. His research concerns the resilience of peasant societies, decentralisation of natural resources management and REDD+, with an emphasis on Madagascar and the Congo basin.

Bruno S. Ramamonjisoa is Head of the Forest Department at the School of Agronomy, University of Antananarivo. He trained in Madagascar and Nancy (France) in forest economics and policy, and has worked on the social aspects of forestry management (especially deforestation) in Madagascar since 1988. He supervised research on community management of forest and marine resources in the region of Toliara and Nosy Be between 2000 and 2006.

Anitry N. Ratsifandrihamanana is Design and Impact Advisor at World Wide Fund for Nature (WWF) International, based in Madagascar, where she has worked in conservation for the past 22 years, most of which have focused on the conservation of the spiny forests of the South and Southwest of the island. As former Conservation Director for WWF-Madagascar, she took an active part in the designing and implementation of the Durban Vision and co-led the new Malagasy Protected Area Commission in its early stages. She holds an MA in Education from the University of Antananarivo and Cornell University.

Ivan R. Scales is McGrath Lecturer and Director of Studies in Human Geography at St Catharine's College, University of Cambridge. He has worked in Madagascar since 2002, firstly as a Project Officer for the Tropical Biology Association, running its field courses in the tropical dry forests of western Madagascar, and subsequently carrying out research on deforestation, rural livelihoods, environmental values and biodiversity conservation. He received his first degree in Ecology from the University of Durham, followed by an MA in Anthropology from University College London, and a PhD in Geography from the University of Cambridge. He has also carried out fieldwork in Cameroon, French Guiana, The Gambia and Senegal.

Laura M. Tilghman is a PhD candidate in the Department of Anthropology at the University of Georgia. She received a BS in Environmental Biology and a BA in Environmental Studies from the University of Vermont. Besides studying the political ecology of bioprospecting, her research projects in Madagascar have covered a wide range of topics including entomophagy, small-scale sapphire mining and most recently migration and rural–urban linkages.

Malika Virah-Sawmy works as an independent researcher with various universities in South Africa and Australia on sustainable agriculture, responsible mining and conservation governance. She is also a World Wide Fund for Nature Research Associate. Since 2004, she has been working in Madagascar with local communities, businesses, NGOs and government to find shared solutions for building socio-ecological sustainability in various sectors, with a focus on production practices relying on natural resources. She holds a PhD in Natural Resource Management from the University of Oxford for her work on mining–conservation conflicts in Madagascar's critical threatened littoral forest, using the ecosystem's resilience to past climatic changes to forecast its adaptive capacity in the face of large-scale mining and future climate change. From 2009 to 2012 she coordinated WWF's Terrestrial Conservation Programme in Madagascar, where she provided a valuable contribution to improving community engagement practices and sustainable livelihoods.

Lucienne Wilmé is Research Associate with the Madagascar Research and Conservation Program at Missouri Botanical Gardens and Editor in Chief of the peer-reviewed journal *Madagascar Conservation & Development* (www.journalmcd.com). She has been actively involved in diverse aspects of research and conservation in Madagascar for more than 25 years. She went from extensive fieldwork all over Madagascar's forests, to understanding bird communities, to the compilation of a rich database on the vertebrate fauna of the island, into biogeography with the processing of data with GIS technologies. She is mainly self-taught and received her PhD in Geography from the University of Strasbourg in 2012.

Preface

I first travelled to Madagascar in October 2002 as a project officer for the Tropical Biology Association, a non-governmental organisation based in the United Kingdom that works in partnership with African institutions to build expertise in biodiversity conservation and research. I had recently finished a Masters degree in the Anthropology and Ecology of Development at University College London following my undergraduate training in the biological sciences.

My job was to oversee a new field course in the dry-deciduous forests of western Madagascar. I had the privilege of working with an inspiring group of young conservationists and getting to see first hand the island's remarkable biodiversity. I also got to witness some of the environmental and social problems Madagascar faces: dramatic grassland fires and deforestation, precarious rural livelihoods and severe poverty. I returned to western Madagascar in 2003 to help run another field course for the Tropical Biology Association. Deforestation in western Madagascar was still progressing at an alarming rate, despite the efforts and interventions of government ministries and international conservation NGOs. It was at this point that I decided to undertake a PhD with Bill Adams to better understand the drivers of forest loss and the factors influencing the livelihood choices of households at the forest frontier. During the course of my fieldwork, I discovered that deforestation in western Madagascar was not simply the outcome of poverty or population growth but the result of a complex interplay between a range of political, economic and cultural factors.

I learned a number of important lessons during the course of my research. The first is that conservation and environmental management policy in Madagascar has often been based on problematic assumptions about the history of landscapes and the factors that shape land use decisions. The second is that a solid understanding of human–environment interactions requires a flexible approach able to deal with insights from different disciplines. This is easier said than done. Finally, if such interdisciplinary insights are to shape policy, more conversations need to happen between researchers and policy-makers and between 'experts' and 'laypeople'. As I discuss in the final chapter of this volume, there are often significant barriers to such interactions.

I am fortunate that my multi-disciplinary training, as well as my experience of both the practical and theoretical side of conservation and environmental management in sub-Saharan Africa, has allowed me to engage with a broad set of individuals, groups and organisations, from farmers at the forest frontier to field researchers, conservationists and government ministers. When Tim Hardwick at Earthscan asked if I had any good ideas for an edited volume on conservation and development, the work of my colleagues and my own experiences of working on environmental issues in Madagascar sprang to mind.

This book brings together a diverse set of perspectives on the issues of conservation, environmental management and poverty alleviation in Madagascar. The island presents policymakers with a classic conservation and environmental management conundrum: how to protect biodiversity at the same time as raising rural households out of poverty. In response to this challenge, Madagascar's policy landscape has changed dramatically over the past 30 years, with a growing recognition that the creation of protected areas has imposed significant costs on the communities living around them. Policymakers have tried to move beyond coercive legislation to the greater involvement of rural households in the management of natural resources. Furthermore, the past 15 years have seen a growing body of research that provides insights into the political, economic and cultural factors that have influenced the success and failure of conservation policy. It is thus a timely opportunity to look back at the lessons that can be learned, and to look forward at the key questions and challenges ahead.

As always with a project of this nature the debts are substantial. First and foremost, I would like to thank all the contributors for their efforts, insights and patience. I have learned a huge amount from their contributions and hope I get the chance to work with them again. Thanks to Barry Ferguson for his help with identifying the key themes and issues at the early stages of this project.

Thanks also to my wife, Helen Scales. When I started working in Madagascar I had just met Helen. By the time I started my PhD fieldwork, Helen had become my spouse and my research a very demanding mistress. Thanks to the Scales family (Micki, Richard, James, Johanna, Keith, Josh and Bella) for their support over the years. Thanks to Rosie Trevelyan at the Tropical Biology Association for giving me my first job in tropical ecology and conservation and the opportunity to observe first hand the realities of environmental management in sub-Saharan Africa. Thanks also to Bill Adams for his continued intellectual guidance and his encouragement to take on this edited volume. I would like to thank St Catharine's College for the extended sabbatical leave that created important writing and editing time for this book.

I would like to dedicate this book to Robert Dewar, who passed away while the book was being finished. I would also like to thank Alison Richard for helping with the final edits on Robert's chapter. As readers will see both from his own chapter in this volume but also from the citations in other chapters, his contribution to our understanding of the island's archaeology and early settlement by humans, as well as its biodiversity and biogeography more generally, was immense. Robert was always generous with his time and very supportive of other researchers, especially young academics. As one of the authors remarked upon hearing the news of his death, it is a privilege to have work published in such esteemed company. I can only echo this sentiment.

Ivan R. Scales
August 2013
St Catharine's College, Cambridge, UK

Acronyms and abbreviations

ABS	Access and Benefit-Sharing
ANAE	*Association Nationale d'Actions Environnementales*
ANGAP	*Association Nationale pour la Gestion des Aires Protégées*
ASB	Alternatives to Slash-and-Burn
AVHRR	Advanced Very High Resolution Radiometer
CAZ	Corridor Ankaniheny-Zahamena
CBD	Convention on Biological Diversity
CBNRM	community-based natural resource management
CI	Conservation International
CIRAD	*Centre de Coopération International en Recherche Agronomique pour le Développement*
CNARP	*Centre National d'Applications des Recherches Pharmaceutiques*
CNRE	*Centre National de Recherches Sur l'Environnement*
COAP	*Code des Aires Protégées*
COBA	*Communauté de Base*
DGEF	*Direction Générale des Eaux et Forêts*
DSAP	*Direction du Système des Aires Protégées*
DWCT	Durrell Wildlife Conservation Trust
FAO	Food and Agriculture Organization of the United Nations
FCPF	Forest Carbon Partnership Facility
GCF	*Gestion Contractualisée des Forêts*
GELOSE	*Gestion Locale Sécurisée*
GIS	Geographic Information Systems
ICBG	International Cooperative Biodiversity Groups program
ICDP	Integrated Conservation and Development Project
IEFN	*Inventaire Ecologique Forestier National*
IMF	International Monetary Fund
INBio	National Biodiversity Institute, Costa Rica
IPR	Intellectual Property Rights
IUCN	International Union for Conservation of Nature
LAC	Local Area Coverage
MBG	Missouri Botanical Gardens
MBI	market-based instrument

MEF	*Ministère des Eaux et Forêts*
METT	Management Effectiveness Tracking Tool
MinEnvEF	*Ministère de l'Environnement, des Eaux et Forêts*
MNP	Madagascar National Parks
MPA	Marine Protected Area
MSS	Multispectral Scanner
NCI	National Cancer Institute
NEAP	National Environmental Action Plan
NGO	non-governmental organisation
NIH	National Institutes of Health
NSF	National Science Foundation
ONE	*Office National pour l'Environnement*
PES	Payments for Environmental Services
PHCF	*Projet Holistique de Conservation de la Forêt*
POSEI	*Programmes d'Option Spécifiques à l'Eloignement et à l'Insularité*
QMM	QIT Madagascar Minerals
REDD+	Reducing Emissions from Deforestation and Forest Degradation
SAPM	*Système des Aires Protégées de Madagascar*
TEK	traditional environmental knowledge
TRIPS	Trade Related Aspects of Intellectual Property Rights
UNCED	United Nations Conference on Environment and Development
UNDP	United Nations Development Programme
UNEP	United Nations Environmental Programme
UNESCO	United Nations Educational, Scientific and Cultural Organization
USAID	United States Agency for International Development
WCS	Wildlife Conservation Society
WWF	World Wide Fund for Nature (formerly the World Wildlife Fund)

1 Introduction

Conservation at the crossroads: biological diversity, environmental change and natural resource use in Madagascar

Ivan R. Scales

In 2007, the Malagasy government published a five-year 'Madagascar Action Plan' for development. As well as setting out targets for economic growth and poverty reduction, it placed strong emphasis on the environment:

> We will become a 'green island' again. Our commitment is to care for, cherish and protect our extraordinary environment. The world looks to us to manage our biodiversity wisely and responsibly – and we will. Local communities will be active participants in environmental conservation under the guidance of bold national policies.
> (MAP, 2007, p97)

As a statement of intent, it certainly ticked all the right boxes, stressing the global significance of the island's biodiversity but also balancing it with the need to involve local communities in the management of natural resources. However, to experienced observers of Madagascar's environmental politics, this was nothing new. Over the past 30 years, the island has been a hotbed of conservation activity and never short of bold plans and policies.

At first glance, the case for urgent action seems clear. Madagascar is one of the most biologically diverse places on the planet, the result of 160 million years of isolation from the African mainland (Krause, 2005). It has over 13,000 species of plants (Phillipson *et al.*, 2006), 700 species of vertebrates and more than 80 per cent of its species are endemic[1] (Ganzhorn *et al.*, 2001; Goodman and Benstead, 2005). This highly diverse flora and fauna is threatened by habitat loss and fragmentation, mostly due to forest clearance. The island has been classified as the world's hottest biodiversity 'hotspot'[2] (Ganzhorn *et al.*, 2001) and one of the world's highest conservation priorities (Myers *et al.*, 2000). As William McConnell (2002, p10) writes: 'Few places on Earth evoke such simultaneous awe and consternation as Madagascar, a country with unique biological riches on a seemingly immutable path of impoverishment.'

As well as these environmental challenges, Madagascar must also deal with considerable and pressing human needs. In 2010, the island's human population stood at over 20 million, with a growth rate of 2.9 per cent a year and a per capita Gross Domestic Product (GDP) of just US$ 421.[3] Poverty has increased over the past 30 years, with a wide variety of socio-economic indicators declining and seven out of ten people living below the World Bank's poverty line (World Bank, 1996). Over 70 per cent of Madagascar's population is rural, relying mostly on subsistence agriculture or pastoralism and depending directly on the island's ecosystems for a wide range of goods and services (Rasambainarivo and Ranaivoarivelo, 2006).

As if this didn't make environmental management challenging enough, Madagascar also suffers from political instability, most recently in 2002 and 2009. In 2001, presidential elections were held. The then mayor of Antananarivo, Marc Ravalomanana, claimed he had won an outright majority in the first round of voting and alleged the election had been rigged. President Didier Ratsiraka refused to stand down, also claiming victory. This resulted in months of tension, strikes, street protests and violent outbreaks between supporters of the two politicians. President Ratsiraka established a power base in the eastern port city of Toamasina, cutting off vital supply routes to the capital. Through international pressure, the situation was eventually resolved, with the USA recognizing Ravalomanana as the new president in June 2002 and France offering Ratsiraka exile in Paris in July 2002. However, stability was short lived. In 2009, president Marc Ravalomanana was unconstitutionally ousted by a political movement led by the new major of Antananarivo, Andry Rajoelina. As of July 2013, Rajoelina was still in charge of the 'High Transitional Authority', with a timetable for new presidential elections yet to be decided. These problems follow decades of tumultuous national politics.

Madagascar thus presents a classic conservation and environmental management conundrum: how to protect biodiversity at the same time as delivering economic growth and raising people out of poverty in often difficult political circumstances. The challenge is considerable and it is perhaps not surprising that people writing about Madagascar often succumb to hyperbole. Madagascar's environmental discourse is full of dramatic language and tales of impending crisis:

> Twenty years ago, Britain's Prince Philip described Madagascar as a nation committing ecological suicide. It was an apt assessment. The country seemed to be set on transforming its last remaining forests to ash and dumping its fertile but eroding soils into the Indian Ocean.
> (Norris, 2006, p960)

However, the danger with such rhetoric is that it hides the fact that problems of poverty, environmental justice, natural resource use and biodiversity conservation are interlinked in complex ways. As this book shows, policy has often struggled to deal with such complexities.

Madagascar's conservation landscape has changed dramatically over the past 30 years and environmental policy is at a crossroads. There has been a growing recognition that the creation of protected areas has imposed significant costs on rural communities due to loss of access to natural resources (Ferraro, 2002). Conservationists have tried to move beyond coercive legislation and a model of 'fortress conservation' to the greater involvement of communities in the management of natural resources (Pollini and Lassoie, 2011). Conservation organizations and government ministries have experimented with a wide range of community-based schemes, from agro-forestry (Pollini, 2009) to nature tourism and even awarding prizes for conservation competitions (Sommerville et al., 2010). In an attempt to reduce rural poverty and create revenue streams to pay for conservation activities, conservationists have begun to engage with incentive-based mechanisms and payments for ecosystem services. At the same time the government, as part of its 'Durban Vision', has recently tripled the extent of the island's protected areas, creating 125 new protected areas and sustainable forest management sites (see Chapter 9 by Corson). Policy thus continues to reflect tensions between coercion and local participation, as well as between preservationist and utilitarian views of nature. Hundreds of millions of dollars have been spent, with mixed results (Kull, 1996).

It is thus a timely opportunity to ask what lessons can be learned from the experiences of biodiversity conservation and environmental management in Madagascar. How effective have different policies been? Who have the winners and losers been in attempts to conserve biodiversity? What are the implications of emerging forms of conservation for rural livelihoods? This book addresses these questions by pulling together a diverse range of experiences, drawing on insights from different academic disciplines (geography, anthropology, environmental history, political science, archaeology, palaeoecology and biology) and bridging the gap between research, policy and practice.

There has been a wealth of research on conservation-related issues in Madagascar. Publications such as *Natural Change and Human Impact in Madagascar* (Goodman and Patterson, 1997) and *The Natural History of Madagascar* (Goodman et al., 2004) are testament to the depth of scholarship. However, there have been major gaps, both in the academic literature and in our understanding. Perhaps the single biggest limitation to date has been the fact that the majority of academic literature on Madagascar can be classified under the banner of biological science, ranging from taxonomy to applied conservation biology. Without downplaying the importance of a solid understanding of ecological processes and biological diversity, it is clear that conservation and environmental management are as much about the choices that people make as they are about ecosystems or endangered species, and that the biological sciences therefore cannot provide all the answers (Balmford and Cowling, 2006; Mascia et al., 2003).

The past 15 years have seen a growing body of research that provides insights into the political, economic and cultural factors that have influenced the success and failure of conservation policy in Madagascar. The chapters in this book provide a summary of the key insights of this rich literature, as well as the remaining challenges and questions. I hope that this volume, as well as providing an overview and analysis of the major conservation and environmental management issues and dispelling a few myths in the process, stimulates conversation and debate. As readers will see, conservation in Madagascar is highly contested both 'on the ground' and in academic journals. I believe that recognizing this is crucial to the future of conservation on the island (Figure 1.1).

Outline

This book is divided into four parts. Part 1 sets the scene by presenting an overview of Madagascar's biodiversity, long-term environmental changes and the impact of early human settlers. One of the main reasons Madagascar has received so much attention from international conservation organizations is its remarkable level of biological endemism. In Chapter 2, Jörg U. Ganzhorn, Lucienne Wilmé and Jean-Luc Mercier provide an introduction to the island's biological diversity and its evolutionary history. In order to understand the origins of Madagascar's remarkable flora and fauna, we must first look to its geological past. Continental drift over a period of 160 million years led to the island's isolation and the majority of the taxa present on Madagascar evolved from colonization events rather than from any stock present on the landmass at the time of isolation. As well as geological isolation, biogeography and climate change have also played a key role in shaping Madagascar's biodiversity. Madagascar is often described as 'the island continent' due to its wide range of environments – from lowlands to highlands, arid spiny forests to rainforests. The diversity of biomes, along with changes in vegetation in response to variations in temperature and rainfall, have resulted in high levels of micro-endemism.

Madagascar's flora and fauna have undergone extensive changes over the past 10,000 years. The island has experienced a significant number of extinctions, as well as considerable land cover change. This has been the result of a complex set of factors including climate variability and the impacts of human activities. In Chapter 3, Robert E. Dewar focuses on Madagascar's Holocene palaeo-environment, paying particular attention to the evidence for the impact of early human settlers on the island's flora and fauna. His chapter discusses new evidence that pushes back the date of human arrival, with archaeological traces of foragers visiting rock shelters in northern Madagascar from at least 2000BC. The chapter also dispels a few myths about the extent to which early human settlers impacted on the island's flora and fauna, challenging overly simplistic stories and questioning the problematic nature of ideas of the 'original' vegetation of Madagascar, especially given the incompleteness of our evidence of both the present and the past. Until recently, it was supposed that the first people on Madagascar imported fire and the result was a 'giant fire' that was utterly

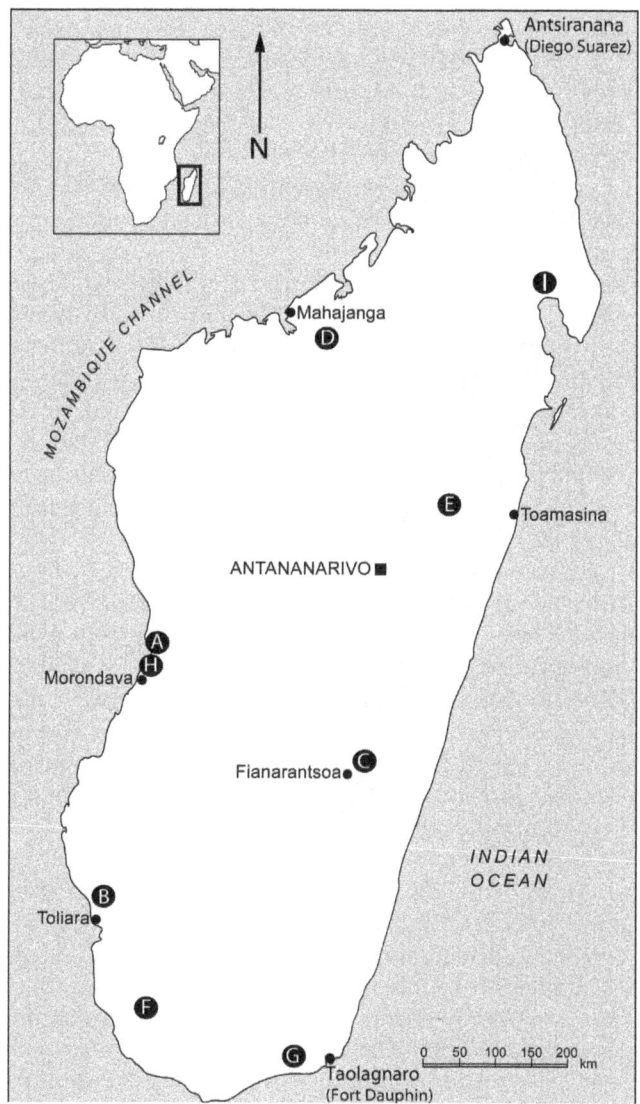

Figure 1.1 Map of Madagascar showing the location of key case studies discussed in the book. A: Dry-deciduous forests of western Madagascar (Chapter 5); B: Spiny forests of southwestern Madagascar (Chapter 5), including Ranobe PK32 protected area (Chapter 10); C: Eastern rainforest including Fandriana-Vondrozo corridor and Ranomafana National Park (Chapters 8 and 9); D: Communities near Ankarafantsika (Chapter 8); E: Eastern rainforests including Ankeniheny-Zahamena corridors (Chapters 8 and 9), the ICBG-Madagascar bioprospecting project (Chapter 12) and Didy village conservation agreements and 'Alternatives to Slash-and-Burn' (Chapter 13); F: Mahafale Plateau (Chapters 8 and 14); G: Ankodida community-managed protected area (Chapter 10); H: Baobab Avenue (Chapter 11); I: Makira REDD+ project (Chapter 13)

destructive to a forested but fragile landscape. However, palaeoecological research shows that periodic fires have been an important element of many Malagasy ecosystems for tens of thousands of years. Dewar persuasively argues the importance of recognizing the complexity of social and environmental systems and acknowledging the limitations of the tools at our disposal for exploring them. More research is required to piece together accurate place-specific accounts of ecological change and human impacts, and major puzzles remain.

Unfortunately, conservation planning in Madagascar has often been based on received wisdoms and untested assumptions. In Part 2, the book explores a set of ecological and social issues that have often been misunderstood and misrepresented. Chapter 4, by William J. McConnell and Christian A. Kull, tackles the debate surrounding the extent of Madagascar's forests and rates of forest loss. Much of the writing on Madagascar's forests is based on a flawed understanding of the island's environmental history (Kull, 2000). Perhaps the best example of this is the 'island–forest' hypothesis. For well over a century, environmental thinking has been dominated by the idea that Madagascar was once entirely covered by forest and that human action has led to the loss of 'over 90 per cent' of forest cover. Such dramatic statistics are often quoted to justify conservation action. Despite a growing body of evidence questioning the validity of such figures, the so-called 'island–forest' hypothesis has remained remarkably persistent. McConnell and Kull explore the origins of the hypothesis and trace its path through the academic literature. Moving to twentieth-century estimates of forest cover and forest loss, McConnell and Kull show how our understanding of contemporary deforestation is little better. Measurements of Madagascar's forest cover have varied hugely. Despite advances in our abilities to monitor forest cover using satellite imagery, forest assessment is marred by poor practices. The authors show how estimates of forest cover and forest loss are highly dependent on a series of decisions made by the researcher. The chapter touches on important questions about how conservation science is practised and how it influences conservation policy. It provides a powerful example of how received wisdom gets constructed and reproduced, and a reminder of how crisis narratives often obscure important issues and questions.

If land cover change in Madagascar has been the source of much misinformation, the land use practices of rural Malagasy are perhaps even more misunderstood. Deforestation has been at the heart of environmental discourse in Madagascar and the blame for forest loss is usually placed on poor rural households, who are portrayed as being caught in a Malthusian spiral of increased population and decreased land productivity. Poverty certainly plays an important role in shaping land use decisions. However, once again the received wisdom on environmental change in Madagascar hides and simplifies more than it reveals. Chapter 5 focuses on the drivers of deforestation, with a particular focus on the swidden farming practices of rural households. This chapter takes an historical approach, looking at the drivers of deforestation and forest degradation during the course of the twentieth century and showing that a wide range of land uses, and not simply household agriculture, have led to changes in forest cover. Furthermore, when rural households clear forest, they do so based on a complex range of environmental, cultural and economic factors. The chapter shows how broader political and economic factors, which are too often ignored by conservation policy, also play a key role in forest loss with changes in the price of commodities stimulating booms in the cultivation of cash crops and associated deforestation.

Having explored the context of conservation in Madagascar, in Part 3 the book moves to a review, analysis and critique of conservation policy (Figure 1.2). Chapter 6 provides an overview of the state's role in environmental management from the pre-colonial state, through the colonial government's efforts to manage the island's forests and natural resources, to independence from France. The state has played and continues to play a major part in shaping human–environment interactions. Not only does it create and control the legal framework for land tenure and resource use, it also holds the power to enforce rules. While there are important changes over time, a common thread has been the desire of the state to regulate particular activities, especially the use of fire. An understanding of the history of state-based politics is important not only because of its role in laying the foundations for contemporary environmental legislation, but also because it raises important questions about the relationship between the state and natural resource use.

In Chapter 7, Christian A. Kull looks at the roots of modern policy in Madagascar and how the island became a poster-child for global biodiversity conservation. Since the late 1970s, a rapid expansion in conservation policy and spending has occurred. This boom is tied to a complex set of interacting factors including environmental change (both real and imagined), national politics, the birth of global environmentalism and geopolitical realities. The chapter also provides an overview of the changes that have occurred in conservation ideology and practice – from strict protected areas to community-based conservation and back again.

During the 1990s, Madagascar moved towards the decentralization of resource management and community involvement in conservation, mirroring global trends in conservation policy. Chapter 8, by Jacques Pollini, Neil Hockley, Frank D. Muttenzer and Bruno S. Ramamonjisoa, provides an overview of community-based conservation and natural resource management in Madagascar, with a particular focus on *Gestion Locale Sécurisée* (GELOSE), which was supposed to address the limitations of both strict 'fortress conservation' and earlier experiments in integrated conservation and development. While GELOSE started off with bold plans for the local management of natural resources, in practice it has tended to follow the familiar 'top-down' politics of conservation. Management transfers

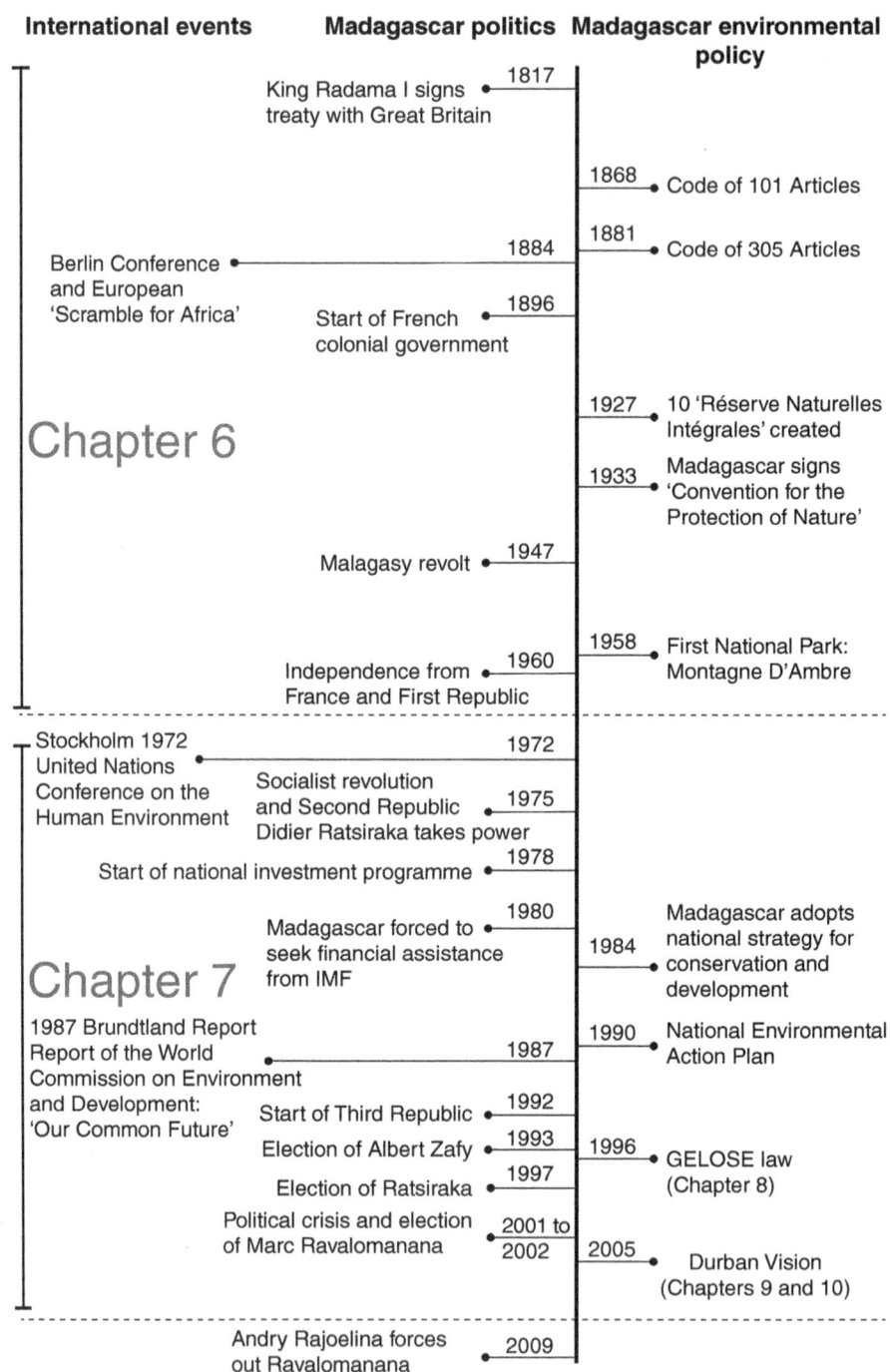

Figure 1.2 Timeline of significant political and economic events in Madagascar

in Madagascar have mostly attempted to involve local communities in achieving natural resource management goals that are not their own, prioritizing biodiversity conservation over livelihoods and poverty alleviation. Participatory processes must do more to consider the likely legitimacy and meaningfulness of newly created resource management institutions, as well as recognize existing institutions and power structures. The chapter shows that there are two fundamental questions when it comes to community involvement in natural resource management – *who* participates and *how* do they participate?

While conservation organizations have attempted to involve communities in the management of natural resources, the most significant recent development in conservation policy has centred around Madagascar's protected areas. In 2003, the president of Madagascar at the time, Marc Ravalomanana, announced the 'Durban Vision' for Madagascar at the 5th International Union for Conservation of Nature (IUCN) World Park Congress. He declared that the country's protected areas would be tripled to cover six million hectares. In Chapter 9, Catherine Corson provides a critique of the 'Durban Vision' process. Her chapter raises important questions about the point at which communities should be consulted in conservation planning. Rather than involve rural Malagasy in environmental decision-making, the Durban Vision has presented communities with a *fait accompli*. In the years following the declaration, it has become evident that many are unaware that they are now living in and around protected areas, with serious implications for natural resource use. Corson argues persuasively that, rather than expanding potentially ineffective conservation territories, conservationists should instead have focused on making the existing parks system effective.

Chapter 10, by Malika Virah-Sawmy, Charlie J. Gardner and Anitry N. Ratsifandrihamanana, offers a different perspective on the Durban Vision. Their chapter provides an overview of the evolution of Madagascar's protected areas. Drawing on a set of case studies, they use the lenses of 'legitimacy', 'inclusiveness', 'fairness', 'accountability and transparency' and 'direction and effectiveness' to explore how the Durban Vision is working in practice and the strengths and weaknesses of attempts to improve local involvement and shared governance.

In the fourth and final part, the book moves to a series of analyses of contemporary attempts to conserve biodiversity. Over the past 30 years, conservation policy has faced the same core problem – how to create mechanisms that might pay for conservation activities as well as providing genuine viable alternatives for rural households that are highly dependent on the natural resources provided by ecosystems such as forests. The struggle to create alternative forms of revenue from natural resources has taken many forms. Chapter 11 discusses the potential of nature-based tourism as a tool for promoting conservation. With Madagascar's unique and charismatic biodiversity, integrating conservation and development through nature tourism seems an obvious solution. However, the reality has so far not matched expectations. The chapter uses a political ecology approach to argue that tourism's lack of success to date has been due to two main factors: i) conflicting perceptions and priorities between different stakeholders; and ii) an uneven distribution of the costs and benefits of managing rainforests for tourism. As a result, expectations that tourism will deliver a 'win–win' solution are often overly ambitious and unrealistic.

Attempts to create financial flows from Madagascar's ecosystems and provide incentives to conserve wildlife have taken many other guises. A common reason given for the need to protect biodiversity is that ecosystems such as rainforests provide humans with a rich store of useful products. Madagascar's incredible biodiversity has already provided numerous medicines now used globally, the most famous cases including extracts of rosy periwinkle (*Catharanthus roseus*) used to treat leukemia. In Chapter 12, Benjamin D. Neimark and Laura M. Tilghman provide an overview and critique of bioprospecting – the process whereby biodiversity is used in the search for new drugs. Bioprospecting is driven largely by Western pharmaceutical companies and aided by governments. Products developed from Madagascar's biodiversity have raised millions of dollars and yet very little of that money has been returned to the island. This raises the classic questions of political economy – who are the winners and losers when it comes to resource use, what are the costs and benefits, and who controls how these are allocated? The chapter touches on questions of intellectual property, indigenous rights and local knowledge and in doing so critically assesses the potential of bioprospecting as a tool for conservation.

The latest attempts to commodify Madagascar's biodiversity involve incentive-based instruments such as Payments for Ecosystem Services (PES). Drawing on ideas from environmental economics – where environmental degradation is seen as the result of market failure – PES emphasizes the various important functions that ecosystems perform, including carbon sequestration and watershed protection. Nature is thus thought of as a flow of services that can be given an economic value. Following on from this, the beneficiaries of the ecosystem services pay those who maintain them. PES schemes thus propose to make direct payments to individuals or groups in exchange for them modifying their land use practices and preserving ecosystem functions. In Chapter 13, Laura Brimont and Cécile Bidaud explore the latest attempts by conservationists in Madagascar to generate income from Madagascar's biodiversity through incentive-based instruments, paying particular attention to schemes based on Reducing Emissions from Deforestation and Forest Degradation (REDD+). Drawing on recent case study evidence, they argue that incentive-based instruments face significant challenges. First, the institutions created to distribute benefits from commodified ecosystem services generally fail to represent the common interests of communities. They are often hampered by corruption, the monopolization of power and bad governance, with some individuals taking advantage to enrich themselves to the detriment of others who are generally the most dependent on natural resources. Second, incentive-based mechanisms have been incapable of generating sufficient income and alternative livelihoods for rural households. For example, income from carbon offset schemes is insufficient to encourage households to switch from forest clearance to more intensified agriculture or move to non-agricultural activities. Finally, the ability of incentive-based instruments such as REDD+ to generate funds to pay for conservation activities is shaped by political commitments and factors such as the price of carbon, which is itself subject to complex political and economic dynamics. Brimont and Bidaud conclude that incentive-based instruments are unlikely to be a miracle solution to Madagascar's conservation problems and must be considered as just one option, taking into account their limitations.

As well as questions of equity and justice, another recurring theme when analysing conservation and environmental management is the difference in perceptions between stakeholders. Chapter 14, by Jeffrey C. Kaufmann, shows the considerable differences between a Western conservation worldview and local beliefs and practices surrounding nature. Conservation policy has tended to have an instrumentalist view of Malagasy culture, seeing it as something to be used and modified for the overarching goal of protecting wildlife. Conservationists have attempted to use local rules and taboos to foster desired behaviour. Kaufmann's chapter reminds us that this can be problematic for a number of reasons. First, the new institutions they create often do not fit well with existing beliefs, institutions and power structures. Second, such conservation interventions raise a strong ethical question: should policy override local practices and undermine social cohesion in the name of conservation?

In the final chapter, I consider the future of biodiversity conservation and environmental management in Madagascar. Pulling together the major insights and arguments from the diverse contributions to the volume, I sketch out the principal lessons that can be taken away from over a century of policy. I also highlight the main areas for future research. I argue that research and policy need to pay more attention to understudied species and ecosystems. Policy must also be more willing to acknowledge the different perceptions and priorities of different stakeholders and be ready to make trade-offs between various environmental and social goals. Human–environment interactions are invariably messy, complicated and contingent on local realities. As well as this shift in expectations, I suggest that there needs to be a re-alignment in the power dynamics of natural resource management, with greater involvement of rural households and communities in environmental research, planning and practice. The key to all the above will be moving beyond traditional boundaries: between academic disciplines; between research and policy; and between 'experts' and 'laypeople'.

Acknowledgements

Thanks to Phil Stickler (Department of Geography, University of Cambridge) for his help with the map.

Notes

1 'Endemic' means 'particular to a defined geographical area'. Biologically, it refers to taxonomic groups of biological organisms that are unique to a defined geographic location. Micro-endemism refers to taxonomic groups that are highly localized – for example, the giant jumping rat (*Hypogeomys antimena*) is found only in small area (less than 200 km^2) of dry-deciduous forest in western Madagascar (Sommer *et al.*, 2002).
2 Biodiversity hotspots are defined as 'areas featuring exceptional concentrations of endemic species and experiencing exceptional loss of habitat' (Myers *et al.*, 2000). Conservation International specifies that hotspots must contain at least 1,500 species of vascular plants (> 0.5 per cent of the world's total) as endemics, and must have lost at least 70 per cent of its original habitat (CI, 1997). It is important to note that the extent of Madagascar's forest loss is a subject of considerable debate (see Chapter 4 by McConnell and Kull). Some would question whether Madagascar has in fact lost 70 per cent of its 'original habitat'.
3 Poverty data for 2010 from The World Bank website (accessed 9 August 2012): http://web.worldbank.org.

References

Balmford, A. and Cowling, R. M. (2006) 'Fusion or failure? The future of conservation biology', *Conservation Biology*, vol 20, pp692–695.
CI (1997) *Global Biodiversity Hotspots*, Conservation International, Washington.
Ferraro, P. J. (2002) 'The local costs of establishing protected areas in low-income nations: Ranomafana National Park, Madagascar', *Ecological Economics*, vol 43, pp261–275.
Ganzhorn, J. U., Lowry II, P. P., Shatz, G. E. and Sommer, S. (2001) 'The biodiversity of Madagascar: one of the world's hottest hotspots on its way out', *Oryx*, vol 35, pp346–348.
Goodman, S. M. and Benstead, J. P. (2005) 'Updated estimates of biotic diversity and endemism for Madagascar', *Oryx*, vol 39, pp73–77.
Goodman, S. M., Benstead, J. P. and Schutz, H. (2004) *The Natural History of Madagascar*, University of Chicago Press, Chicago.
Goodman, S. M. and Patterson, B. D. (1997) *Natural Change and Human Impact in Madagascar*, Smithsonian Institution Press, Washington.
Krause, D. W. (2005) 'Late cretaceous vertebrates of Madagascar: a window into Gondwanan biogeography at the end of the age of dinosaurs', in S. M. Goodman and J. P. Benstead (eds) *The Natural History of Madagascar*, University of Chicago Press, Chicago.
Kull, C. A. (1996) 'The evolution of conservation efforts in Madagascar', *International Environmental Affairs*, vol 8, pp50–86.
Kull, C. A. (2000) 'Deforestation, erosion, and fire: degradation myths in the environmental history of Madagascar', *Environment and History*, vol 6, pp423–450.

MAP (2007) *Madagascar Action Plan 2007–2012: A Bold and Exciting Plan for Rapid Development*, Antananarivo.
Mascia, M. B., Brosius, P. J., Dobson, T. A., Forbes, B. C., Horowitz, L., McKean, M. A. and Turner, N. J. (2003) 'Conservation and the social sciences', *Conservation Biology*, vol 17, pp649–650.
McConnell, W. J. (2002) 'Madagascar: Emerald isle or paradise lost?', *Environment*, vol 44, pp10–22.
Myers, N., Mittermeier, R. A., Mittermeier, C. G., da Fonseca, G. A. B. and Kent, J. (2000) 'Biodiversity hotspots for conservation priorities', *Nature*, vol 403, pp853–858.
Norris, S. (2006) 'Madagascar defiant', *BioScience*, vol 56, pp960–965.
Phillipson, P. B., Schatz, G. E., Lowry II, P. P. and Labat, J. (2006) 'A catalogue of the vascular plants of Madagascar', in S. A. Ghazanfar and H. Beentje (eds) *Taxonomy and Ecology of African Plants, their Conservation and Sustainable Use: Proceedings of the 17th AETFAT Congress Addis Abbaba, Ethiopia*, Kew Publishing Ltd, London.
Pollini, J. (2009) 'Agroforestry and the search for alternatives to slash-and-burn cultivation: from technological optimism to a political economy of deforestation', *Agriculture, Ecosystems and Environment*, vol 133, pp48–60.
Pollini, J. and Lassoie, J. P. (2011) 'Trapping farmer communities within global environmental regimes: the case of the GELOSE legislation in Madagascar', *Society and Natural Resources*, vol 24, pp814–830.
Rasambainarivo, J. H. and Ranaivoarivelo, N. (2006) *Madagascar: Country Pasture/Forage Resource Profiles*, Food and Agriculture Organization of the United Nations, Rome.
Sommer, S., Toto Volahy, A. and Seal, U. S. (2002) 'A population and habitat viability assessment for the highly endangered giant jumping rat (*Hypogeomys antimena*), the largest extant endemic rodent of Madagascar', *Animal Conservation*, vol 5, pp263–273.
Sommerville, M., Milner-Gulland, E. J., Rahajaharison, M. and Jones, J. P. G. (2010) 'Impact of a community-based payment for environmental services intervention on forest use in Menabe, Madagascar', *Conservation Biology*, vol 24, pp1488–1498.
World Bank (1996) *Madagascar Poverty Assessment*, World Bank, Washington. DC.

Part 1
Madagascar's biological diversity
From deep time to the arrival of humans

2 Explaining Madagascar's biodiversity

Jörg U. Ganzhorn, Lucienne Wilmé and Jean-Luc Mercier

Introduction

Madagascar is of special interest to biologists for reasons that also help explain the attention the island receives from international conservation organizations. First and foremost, the majority of Madagascar's plant and animal species are endemic, occurring nowhere else on the planet.[1] This applies to more than 90 per cent of the island's 13,000+ plant species (Phillipson *et al.*, 2006), and the same is true for the various groups of animals, with rates of endemism ranging from 37 per cent in birds to 100 per cent in amphibians and lemurs (Goodman and Benstead, 2003, 2005).

Not only does Madagascar have high levels of endemism, the endemic groups present on the island are ancient, having evolved from their closest relatives many million years ago and therefore forming groups without any close relatives elsewhere (Crottini *et al.*, 2012; Holt *et al.*, 2013). In contrast to other large islands, the vast majority of Madagascar's plants and animals are restricted to one landmass. This has important implications for conservation, as the extinction of a species in Madagascar usually means global extinction.

Madagascar is the Earth's fourth largest island and can be seen as an intermediate between a continent and an island (de Wit, 2003), hence its common label of 'the island continent'. This has played a major evolutionary role – Madagascar is large enough to have a number of very distinct biomes, allowing colonizing species to radiate into distinct species in different environments. It is the combination of high levels of endemism, phylogenetic distinctness[2] and the large number of species at high risk of extinction that makes Madagascar one of the most important sites for biodiversity conservation worldwide (Myers *et al.*, 2000).

A brief overview of Madagascar's biodiversity

Contrary to the impression that might be given by Madagascar's status as one of the world's top biodiversity (Box 2.1) 'hotspots' (Myers *et al.*, 2000), the island is not exceptionally rich in the number of species per unit area. This is particularly evident when comparing Madagascar to other islands. According to the principles of island biogeography (Box 2.2), the species richness found on Madagascar is more or less what would be expected from an island of that size.

Box 2.1 Biodiversity

The term *biodiversity* was originally introduced by Wilson and Peters (1988) in the context of conservation biology. It built upon the notion of biological diversity, or the diversity of biological systems, including the interactions between living organisms. The term can be defined in various ways, ranging from a simple description of species numbers (better defined as species richness) to complex biological systems, their genetic basis and interactions (Magurran and McGill, 2010). Since conservation has integrated economic, political, cultural and even philosophical aspects, the term *biodiversity* or simply *diversity* has been used in a plethora of contexts and meanings (Naeem *et al.*, 2009). In the context of this chapter, *biodiversity* is used in a loosely defined way as a substitute for species richness and the complexity of ecosystems.

Box 2.2 Island biogeography and speciation

The concept of 'island biogeography' was developed by MacArthur and Wilson (1967) to predict the number of species on islands. The number is related to speciation in situ, colonization and extinction in relation to the size and remoteness of the island. Since formulated by MacArthur and Wilson, the concept has been applied to habitats embedded in a matrix of different habitats and to species–area relationships in general (Rosenzweig, 1995). The evolution of distinct species requires some kind of population structure or subdivision that limits gene flow between individuals. The spatial extent of gene flow is correlated with the mobility and dispersal potential of species. Therefore large and mobile species require larger areas for speciation to occur than less mobile species (Kisel and Barraclough, 2010). For example, amphibians have radiated into distinct species on islands as small as Jamaica, while large mammals have not evolved endemic species on islands smaller than Madagascar (Terborgh, 1992).

Looking at bats, for example (Figure 2.1), Madagascar's species richness is consistent with the species–area relationship of the Caribbean islands and Sri Lanka, but is lower than the islands of Papua New Guinea (476,000 km^2) and Borneo (ca 800,000 km^2), which are home to more than 90 different species (Bonaccorso, 1998; Struebig et al., 2010). The species richness of terrestrial mammals is also around what would be expected for an island the size of Madagascar (Goodman et al., 2008), while amphibians and reptiles are relatively species rich and the number of bird species is surprisingly low (Langrand, 1990).

The number of species described for Madagascar has soared over the past two decades (Figure 2.2). This is largely due to the efforts of field biologists and taxonomists and the very intensive inventories they have carried out across the whole island (Goodman and Benstead, 2003). New genetic techniques and standards in species definition have also played a part, leading to the subdivision of forms into distinct species that had previously been

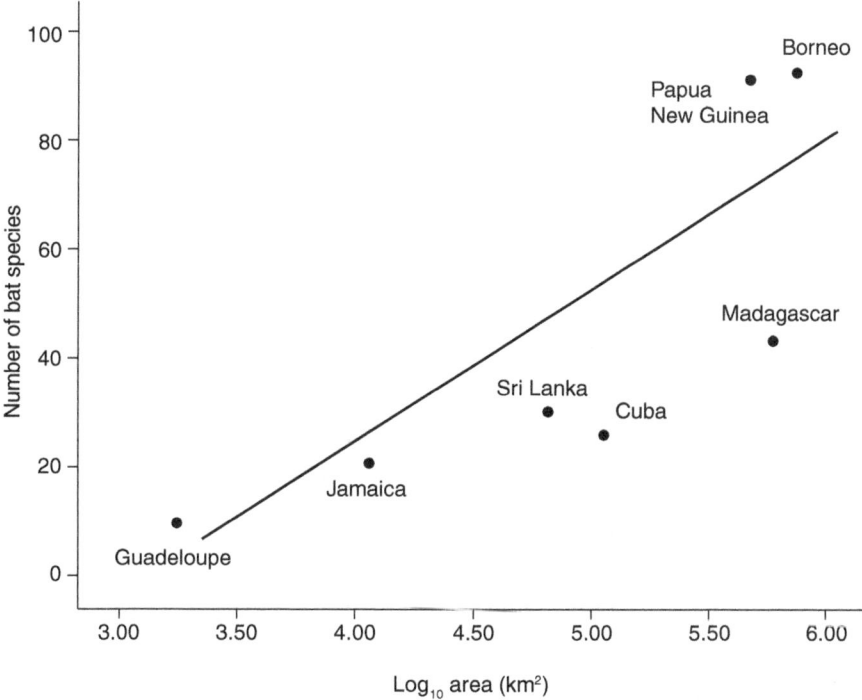

Figure 2.1 Species–area relationship for bats (including flying foxes) on islands of different size

Source: data from Morgan and Woods, 1986; Bonaccorso, 1998; Struebig et al., 2010; Goodman, 2011.

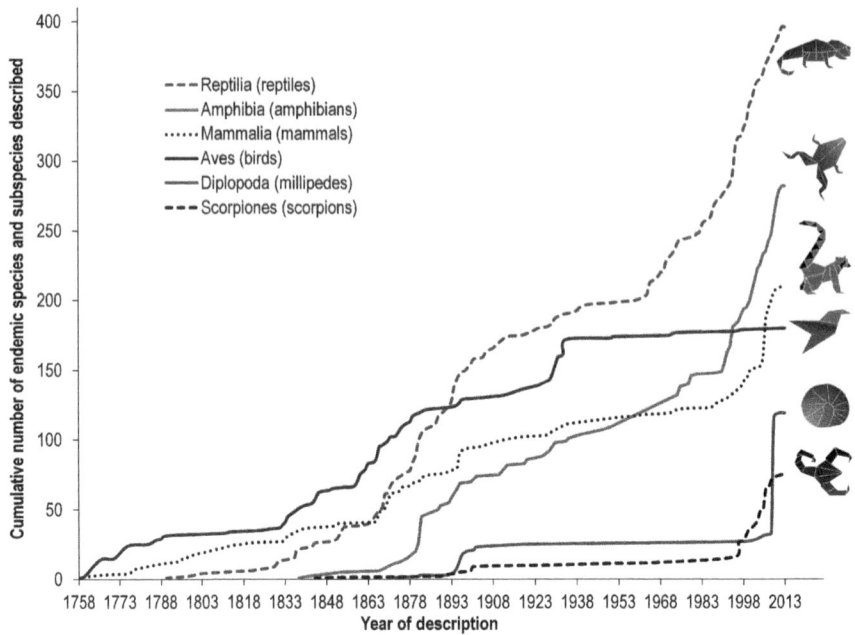

Figure 2.2 Number of species described for Madagascar (since Linnaeus in 1758) in a range of vertebrate and invertebrate taxonomic groups

Source: Wilmé (2012).

considered single species (Tattersall, 2007; Markolf *et al.*, 2011). Irrespective of the number of species that will eventually be distinguished, the number of plant and vertebrate species is lower than in other biodiversity hotspots (Table 2.1). Madagascar's uniqueness is thus based on the phylogenetic distinctiveness of higher level taxa (families and genera) rather than on the total number of species, reflecting old lineages that diverged from their relatives a long time ago, with radiation within these taxa leading to the high degree of species endemism (Ceballos and Brown, 1995; Barthlott *et al.*, 1996; Kreft and Jetz, 2007).

In general, island biodiversity hotspots are characterized by higher degrees of endemism than mainland hotspots due to the geographic isolation of islands. This facilitates allopatric speciation, a process whereby populations of the same species become separated from each other (often because of physical barriers) and diverge genetically over time in different geographical areas so that the different forms no longer mix. Similar-sized biodiversity hotspots that are part of continents have lower percentages of endemism because many of these species are also likely to occur outside the geographical boundary of the hotspot. Understanding Madagascar's isolation is therefore the key to understanding how its flora and fauna came to be so distinct.

Table 2.1 Endemic plant and vertebrate families, genera and species in some global biodiversity hotspots in the tropics and subtropics. On the species level, the numbers indicate the total number of species (upper row) and percentage of endemic species (lower row)

Region	Endemic plant and vertebrate families and genera		Total number of species and percentage of endemism						
	Endemic families	Endemic genera	Plants	Mammals	Birds	Reptiles	Amphibians	Freshwater fish
Island situation								
Madagascar and Indian Ocean islands	8 + 16	310 + 168	13,000 89%	155 93%	313 58%	381 96%	228 99%	164 59%
Carribean islands	1 + 4	205 + 64	13,000 50%	89 46%	607 28%	499 94%	165 99%	161 40%
Sundaland	3 + 1	117 + 82	25,000 60%	381 45%	771 19%	449 54%	242 71%	950 37%
Continental situation								
Cape Floristic region	5 + 0	160 + 2	9,000 69%	90 4%	324 2%	100 22%	51 31%	34 41%
Mesoamerica	3 + 2	65 + 73	17,000 17%	440 15%	1,124 19%	686 35%	575 61%	509 67%
Southwest Australia	4 + 3	87 + 13	5,571 53%	57 21%	285 4%	177 15%	33 19%	20 50%

Source: Mittermeier et al. (2004, 2010).

Geological history and explanations for Madagascar's macro-endemism

According to traditional biogeographical classification, Madagascar is part of the Paleotropis, one of the six biogeographic units of the world based on climatic zonation, phylogenetic relationships and species similarities (Lomolino et al., 2010). These units correspond to the continents as we know them today, with subdivisions between the tropics and the temperate zone. The Paleotropis covers most of the African continent along with India, South Asia and Indonesia.

This classification of large biogeographic units has been revised recently by putting emphasis more explicitly on phylogenetic relationships, the age of separation of the various lineages of plants and animals and their evolutionary uniqueness (Holt et al., 2013). This new classification distinguishes Madagascar as one of 20 distinct zoogeographic regions of the world with an evolutionary uniqueness that is only topped by the whole of Australia, emphasizing its prominent position as one of the prime global biodiversity hotspots.

The evolutionary uniqueness of Madagascar's biota is rooted in the plate tectonics of the Mesozoic (the geological era from 251 to 65 million years ago that includes the Triassic, Jurassic and Cretaceous). At the beginning of the Mesozoic the landmasses formed a single 'continent' called Pangaea that, by the end of the Cretaceous, had broken up into the continents we know today. Some 166 million years ago, Madagascar was part of the southern continent of Gondwana, together with South America, Africa, Antarctica, Australia and India (Figure 2.3). At this time, Gondwana started to split up into the various sub-units. Some 88 million years ago, Madagascar and India started to separate, leaving Madagascar as an isolated landmass at about its present position, while India continued to drift towards the north. Madagascar has thus been isolated from any other landmass for at least 88 million years.

At the time of separation from Gondwanaland, the major terrestrial taxa now found on Madagascar had either not yet evolved or their evolution was in its infancy. Some 65 million years ago, a huge asteroid hit the Earth. This marks the time of mass extinction, the demise of dinosaurs and the transition from the Cretaceous to the Tertiary, named the K-T event, after the German term *Kreide* for Cretaceous (O'Leary et al., 2013; Yoder, 2013a).[3] The K-T event seems to have eliminated most faunal elements present on Madagascar at that time, leaving the island impoverished (de Wit, 2003; Ali and Krause, 2011). The majority of animal species present on Madagascar thus evolved from colonization events after the K-T event rather than from any stock present on the landmass at the time of isolation.

Our understanding of geological palaeohistory relies on the discovery of fossil material, mainly excavated from stratified rocks within sedimentary basins. Madagascar's three main sedimentary basins lie on the western

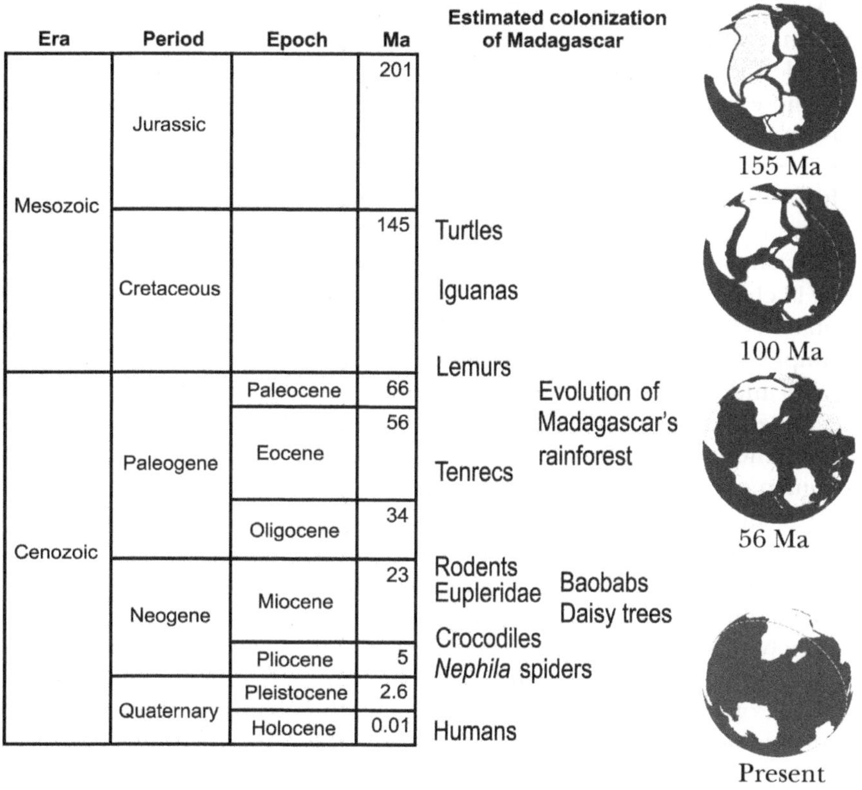

Era	Period	Epoch	Ma	Estimated colonization of Madagascar
Mesozoic	Jurassic		201	
	Cretaceous		145	Turtles
				Iguanas
Cenozoic	Paleogene	Paleocene	66	Lemurs
		Eocene	56	Evolution of Madagascar's rainforest
				Tenrecs
		Oligocene	34	
	Neogene	Miocene	23	Rodents Eupleridae Baobabs Daisy trees Crocodiles Nephila spiders
		Pliocene	5	
	Quaternary	Pleistocene	2.6	
		Holocene	0.01	Humans

155 Ma

100 Ma

56 Ma

Present

Figure 2.3 Geological time scales, phases of continental drift and colonization of Madagascar by selected vertebrate groups

Notes: iguanas and freshwater turtles have closest relatives with South America; all terrestrial mammalian groups derive from single colonization events from Africa during times when ocean currents were favourable for rafting from Africa to Madagascar; and crocodiles as an example of one of the few taxa that colonized Madagascar from Africa despite unfavourable ocean currents. Most amphibians and reptiles colonized Madagascar from Africa during the Paleogene. Species with airborne dispersal states, such as spiders of the genus *Nephila* or Daisy trees (*Psiadia* spp.) could arrive without the help of ocean currents. Numbers are in million years before present (Ma).

Sources: modified from Krause (2003); Lomolino *et al.* (2010); Kuntner and Agnarssen (2011); Crottini *et al.* (2012); Samonds *et al.* (2012); Strijk *et al.* (2012).

slope running from the far north to the southern tip of the island. The main Paleogene and Neogene sedimentary deposits for Madagascar are almost entirely of marine origin. In concert with the intense weathering of tropical soils, this greatly reduces the chance of recovering any terrestrial fossils in the upper layers (Besairie and Collignon, 1972).

According to Krause et al. (1997, 1998, 1999), none of the major endemic vertebrate lineages currently found on the island have been recorded in the Cretaceous. Therefore, these groups must have evolved elsewhere before colonizing Madagascar. However, given the lack of fossils for most of the subsequent Paleogene, the reconstruction of the colonization events that contributed to Madagascar's current flora and fauna must rely on the reconstruction of the geographic constellation of land masses and other geographical features, processes and events. These geological reconstructions can be supplemented with data from molecular clocks to date likely divergences. Molecular clocks are based on the assumption that mutations occur in DNA at a steady and predictable rate over time. For example, by looking at the genetic differences between current species of lemurs and using a standard rate of mutation over time, the time elapsed since their common ancestor can be estimated. This allows us to align the arrangements of landmasses and key geological events with the temporal reconstruction of the evolutionary splits of various species (Vences et al., 2001; Yoder et al., 2005; Crottini et al., 2012; Yoder, 2013b).

While molecular techniques provide exciting insights into evolutionary history and phylogeny, it is important to remember that the rate of change in nucleotides that genetic clocks are based on differs between organisms and between coding and non-coding DNA sequences, and must be calibrated using other known and dateable standards (mostly fossils). Given the lack of fossils of the present lineages on Madagascar, the exact dates of divergence from their mainland relatives and the subsequent radiation remain uncertain.

Having established that Madagascar's high levels of endemism are principally related to its isolation and that the mass extinction 65 million years ago removed a large part of the island's flora and fauna, how did the island come to be colonized from other land masses? Two main scenarios have been formulated for the colonization of the island by terrestrial vertebrates. One hypothesis has suggested 'island hopping' either from Africa across a narrower Mozambique Channel, or between the Antarctic remains of Gondwanaland and Madagascar during the Late Cretaceous, when the land masses had not drifted apart as far as they have today (Ali and Huber, 2010). Looking first at possible island hopping routes across the Mozambique Channel since the Cretaceous, the distance between eastern Mozambique and western Madagascar has been at least 230km, even when sea levels were at their lowest during the Pleistocene glacio-eustatic oscillations associated with continental glaciation (Ali and Huber, 2010). It has thus been a substantial barrier for a significant period of time.

With regards to possible island hopping from the Antarctic remains of Gondwanaland, modern techniques based on the orientation of the Earth's magnetic field allow us to reconstruct the timing of the spread of the Earth's crust (Ali and Krause, 2011; Figure 2.4). This reconstruction indicates that by the end of the Cretaceous, Madagascar was separated from Antarctica

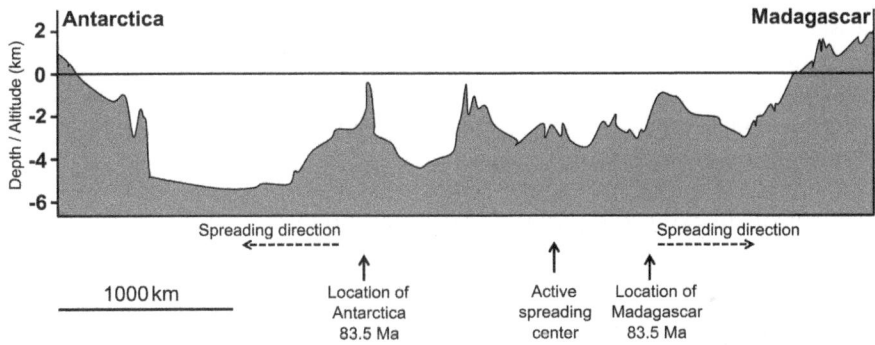

Figure 2.4 Bathymetric profile (underwater topography) between Antarctica and Madagascar. The age of the seafloor and seafloor spread can be dated based on changes in the earth's geomagnetic field

Source: modified from Ali and Krause (2011).

by about 2,000 km and deep water. In addition, sea levels were about 100 m higher than they are today. There is therefore little evidence to support the idea that Madagascar was connected with Africa or Antarctica by a landbridge or a series of islands that could serve as stepping stones for terrestrial colonizers. This leaves the 'rafting' of organisms across the Mozambique Channel as the only remaining option.

Rafting is assumed to be possible when large portions of entangled vegetation are flushed into the ocean, for example after exceptional weather events, tsunamis or, more commonly, when parts of gallery forests break off along river beds and are transported into the sea. These floating tangles of vegetation can achieve substantial dimensions. While the formation of 'rafts' large enough to transport animals alive between Africa and Madagascar is certainly not a frequent phenomenon on a human time scale, it is sufficiently frequent over thousands or millions of years (de Queiroz, 2005). Following the K-T event, ocean currents were favourable for the overwater dispersal of terrestrial taxa from Africa to Madagascar for a period of about 30–40 million years. The rafting scenario is thus consistent with the timing of when the Madagascar lineages of vertebrates split off from their continental relatives (Crottini *et al.*, 2012). The passage would not have taken more than 30 days. The window of maximum opportunity for rafting closed about 20 million years ago when the oceanic surface currents changed from a south-western direction (i.e. from Tanzania towards Madagascar) to the north–south direction in the Mozambique Channel that we see today. This would have made the colonization of Madagascar difficult (Ali and Huber, 2010; Krause, 2010; Samonds *et al.*, 2012).

There is evidence to support rafting as a mechanism for the colonization of islands. For example, the dispersal of several green iguanas on mats of logs and uprooted trees has been witnessed on the island of Anguilla in the Caribbean (Censky et al., 1998). Since several of the modern Malagasy vertebrate taxa, such as lemurs and tenrecs, are known to enter torpor or extended forms of reduced metabolic rates (Dausmann et al., 2009), these groups would have had the prerequisite characteristics to increase their chances of surviving the passage (Kappeler, 2000; for a diverging view see Masters et al., 2006).

Larger mammals, such as ungulates, did not make the passage, except for semi-aquatic species such as hippopotamuses (MacPhee, 1994). In general, rafting seems to have favoured taxa that could cling on to vegetation, as all four mammalian lineages that made it to Madagascar (primates, tenrecs, rodents and the carnivorous Eupleridae) represent animals that are partially or exclusively arboreal. This also applies to reptiles and amphibians, with terrestrial reptiles often found swimming or on logs in the middle of the sea (Censky et al., 1998; Gerlach et al., 2006; Rocha et al., 2006). For amphibians, a passage in salt water ought to be problematic as they are unlikely to have been able to tolerate the osmotic stress imposed by the salt on their skin. Despite the problems of overseas dispersal in saltwater, rafting is still considered more likely for amphibian dispersal than as yet unproven mechanisms, such as the transportation of amphibian eggs attached to the legs or feathers of waterbirds (Vences et al., 2003, 2004; Measey et al., 2007).

While the majority of vertebrate taxa on Madagascar are estimated to be between 19–79 million years old, crocodiles reached Madagascar only five million years ago. These animals are able to tolerate and swim across salt water. Some groups of frogs and geckos are also rather young and must have reached Madagascar despite unfavourable ocean currents. Chameleons have taken the reverse route, originating in Madagascar and colonizing Africa and Asia (Raxworthy et al., 2002; Crottini et al., 2012).

There are other key differences in colonization by different taxonomic groups. While amphibians and reptiles have colonized Madagascar several times independently (Vences et al., 2003; Crottini et al., 2012), higher mammalian orders have colonized Madagascar only once (Yoder et al., 2003; Krause, 2010). Thus, four successful colonization events from Africa explain the entire origin of the island's endemic land mammals (including lemurs, tenrecs, carnivorans and rodents). Flying vertebrates, such as birds, flying foxes and bats, also have Asian origins and do not show the same level of endemism as other vertebrate taxa of Madagascar (e.g. Warren et al., 2005; O'Brien et al., 2009; Goodman, 2011; Raherilalao and Goodman, 2011).

Although rafting from mainland Africa accounts for the majority of Madagascar's vertebrates, two groups – freshwater turtles and a relative of iguanas – seem to be 'leftovers' from the time when dispersal was still possible between South America and Madagascar. The island's freshwater fish also have Gondwanan ancestors rather than being derived from African colonizations (Sparks and Smith, 2005). A spectacular example is provided by the blind cave fish of south-western Madagascar, whose closest relatives are found in caves in Australia (Chakrabarty *et al.*, 2012).

Trans-oceanic dispersal not only accounts for the majority of Madagascar's fauna but is also a major component of the colonization of remote islands by plants (e.g. Nathan *et al.*, 2008; Michalak *et al.*, 2010). Madagascar's famous baobabs (*Adansonia* spp.) are thought to have diverged from their African relatives about 7–17 million years ago (Baum *et al.*, 1998; Pock Tsy *et al.*, 2009) and thus should have missed the favourable currents from Africa to Madagascar. Other plants, such as the ancestors of Daisy trees (*Psiadia* spp., Asteraceae) made the passage from Africa to Madagascar about 10 million years ago, with extensive and rapid radiation within Madagascar as well as secondary dispersal and radiation to peripheral islands around Madagascar (Strijk *et al.*, 2012). Seeds of other plant species could have been transported by the wind or carried by birds and flying foxes, either by becoming attached to them or by being carried in their digestive tracts. Passive aerial dispersal from Africa has also occurred recently with small airborne invertebrates such as the spider *Nephila* spp., which arrived on Madagascar from Africa about 2.5 million years ago (Kuntner and Agnarsson, 2011).

Contemporary biogeography, species radiation and micro-endemism

Unfortunately, our understanding of how Madagascar's flora and fauna diversified following colonization is hampered by the fact that there is a huge gap in the fossil record between the island's isolation during the Mesozoic; the asteroid impact 65 million years ago; changing ocean currents; and the forces contributing to the contemporary biogeography of Madagascar's biota. Continuous records based on pollen or carbon-dated subfossils (see Chapter 3 by Dewar) are available only for the last 40,000 years (Gasse and Van Campo, 2001; Burney *et al.*, 2004). Therefore, interpretations of the drivers that led to the contemporary patterns of distribution are largely based on mechanisms known from recent geological times.

One of the main geographical features of the island is its topographic asymmetry, with an eastern slope and a western slope extending from a mountain range that runs north–south along the eastern part of the island. The two slopes represent 27 per cent and 73 per cent of the total area of the island respectively. The eastern slope rises steeply from sea level to over 1,500 m within a distance of only 100 km. From the crest of the mountain chain (between 1,200 m and 1,600 m above sea level), the land forms a central highlands that decline more gently towards the west coast. Madagascar's highest peak is less than 3,000 m but the whole island is mountainous, with over half of Madagascar's land area more than 500 m above sea level.

Highlands at altitudes between 1,000 m and 1,500 m are widely spread in the central part, but also in the north and south.

Madagascar's topography plays a key role in shaping its distinct climatic zones. The eastern mountains capture the humid trade winds blowing from the east. Once the winds hit the island, they rise, cool off and discharge their humidity along the eastern flanks of the mountain chain (this is called orographic rainfall). The east thus receives rainfall all year round. During the southern hemisphere winter, winds do not contain enough moisture to deliver rain towards the drier western part of the island (i.e. there is a rain 'shadow', or 'foehn', effect). However, during the southern hemisphere summer, winds are sufficiently hot and humid so that rain is able to reach the western parts of the island, resulting in a seasonal monsoon climate (Jury, 2003). These abiotic conditions have shaped three distinct forest formations, namely eastern humid forests, western dry-deciduous forests, and the spiny forests and thickets of the south and southwest, all separated by a mosaic of woodlands and grasslands (Lowry et al., 1997; Moat and Smith, 2007; Bond et al., 2008; Figure 2.5).

Madagascar's different forest types are primarily the result of total annual rainfall and its seasonal distribution. In turn, animals have adapted to the abiotic conditions as well as the resources provided by plants during the different seasons (Hemingway and Bynum, 2005). In concert, these adaptations of plants and animals gave rise to the distinct ecosystems depicted in Figure 2.5. Since the mosaic of woodlands, grasslands and wetlands of the central part of the island has been reduced substantially (see Chapter 3 by Dewar and Chapter 4 by McConnell and Kull for more details on land cover change), the extant forest formations appear geographically more distinct and show less prominent transitional zones than there would have been historically.

Madagascar's distinct climatic and vegetation zones have played a significant part in driving speciation and shaping biological diversity. The evolution of species in situ seems to be favoured in heterogeneous environments where subpopulations encounter different abiotic and biotic conditions. Eventually they adapt to local conditions and evolve into distinct species. However, heterogeneity is a question of spatial scale and perspective. A tree may represent a specific habitat for a sedentary lizard but not for a carnivore with a large home range. Therefore, speciation is inversely related to the dispersal abilities of organisms (Kisel and Barraclough, 2010) and groups with poor dispersal ability radiate into distinct species on much smaller spatial (and probably also temporal) scales than comparable taxa with large home ranges and good dispersal abilities. In global analyses, reptiles and amphibians have radiated into distinct species on islands as small as Jamaica (12,000 km^2). Distinctive radiation of small mammals occurred on islands such as Cuba (115,000 km^2) and large mammals, such as carnivores, require even larger areas to produce distinctive species, such as those of the islands of New Guinea (786,000 km^2) and Madagascar (587,041 km^2) (Terborgh, 1992).

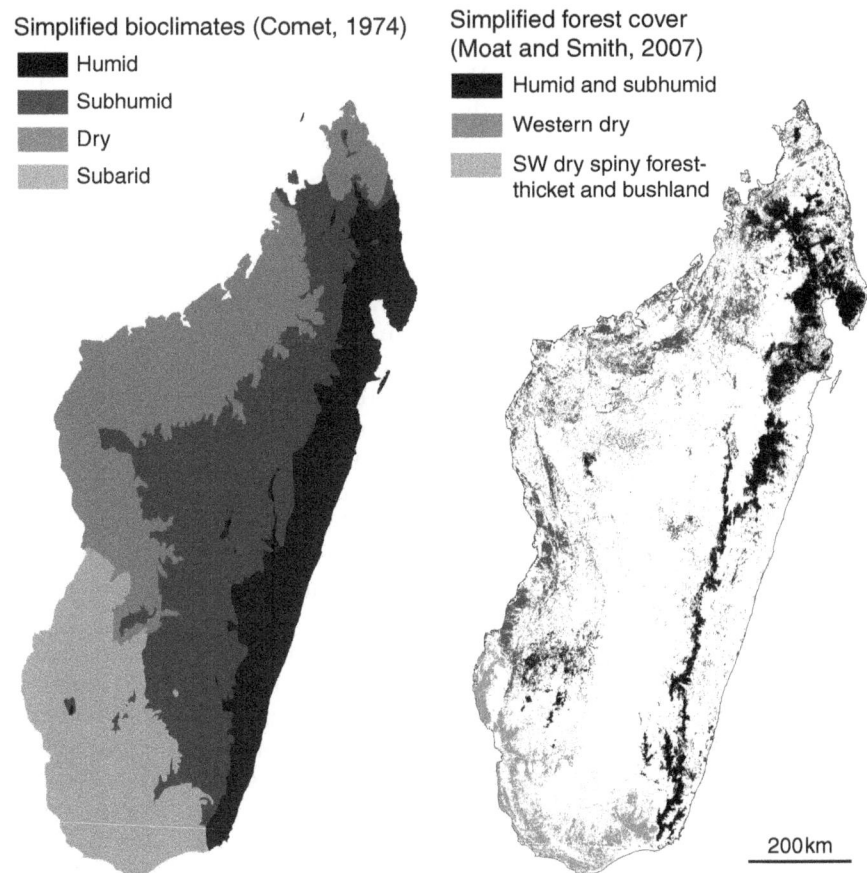

Figure 2.5 Vegetation formations and climatic conditions forming distinct ecotypes of Madagascar

Sources: Cornet (1974); Moat and Smith (2007).

The simplified categorization of Madagascar's phytogeography as 'grassland', 'humid', 'dry-deciduous' and 'spiny' forest, with associated animal species, has been refined over the past few years as they described the contemporary situation rather than the conditions under which ecosystems evolved. For example, many animal communities, ranging from butterflies to lemurs, show distinct differences in species richness and species composition between the northern and southern parts of the continuous eastern rain forest (Lees *et al.*, 1999). Similar subdivisions have also been defined for western Madagascar (Martin, 1995; Ganzhorn *et al.*, 2006; Thalmann, 2007). New inventories and analyses have revealed that the western part of Madagascar contains a mixture of 'eastern' and 'western' species and that some eastern species of the humid forest have closer genetic affinities to western species of the dry forest than to their congeneric species (i.e. species belonging to the same genus) within the same broad type of vegetation (Raxworthy and Nussbaum, 1997; Yoder et al., 2000). In addition, many species have extremely small known distributions within any given vegetation type that cannot be explained simply by adaptation to a broad vegetation formation (Wollenberg et al., 2008; Vieites et al., 2009).

Several hypotheses have been put forward to explain the processes that led to the present pattern of species distribution and the evolution of micro-endemics, i.e. endemic species with very small areas of occurrence (reviewed by Pearson and Raxworthy, 2009; Vences et al., 2009). The 'mid-domain-effect' hypothesis has its roots in the species–area relationship. It argues that species richness should be highest at mid-altitudes because these regions have the largest area and therefore the highest potential for the speciation and persistence of a diverse array of species. In addition, they can be colonized by both lowland and highland species (Lees et al., 1999). For Madagascar, this pattern has been illustrated with a variety of taxa, from butterflies to lemurs (Lees et al., 1999; Goodman and Ganzhorn, 2004a; Soarimalala and Goodman, 2011). However, critical tests of the processes and mechanisms that might have led to these patterns are still lacking (Kerr et al., 2006; Vences et al., 2009).

The 'river hypothesis' postulates that rivers have formed geographical barriers that have prohibited dispersal and led to allopatric speciation (Petter et al., 1977; Martin, 1995; Goodman and Ganzhorn, 2004b). While rivers can be used to delimit some geographical subunits, the assumption that rivers in Madagascar have been stable elements that have persisted long enough to lead to speciation does not seem to hold. Based on the current hydrological system and the fragmented state of forest cover, it is always possible to identify a river that delineates the known distribution of a forest species. This seems most obvious on the western slope and along the eastern coastal portion of the island (Moat and Smith, 2007). However, as species distributions become better known, most of the rivers proposed as barriers to dispersal could have not have played this role (Jenkins et al., 2003; Craul et al., 2008; Mittermeier et al., 2010; Knopp et al., 2011).

The 'climate hypothesis' assumes adaptations to local climatic and associated vegetation traits as the driving force for species radiation (Dewar and Richard, 2007). This hypothesis goes beyond the classical idea of adaptation to a broad vegetation formation, adding altitudinal and latitudinal climatic gradients and assuming substantial heterogeneity within these vegetation formations, driven by unpredictable climatic variation. Madagascar's climate has been described as hypervariable, with rainfall in particular showing considerable intra- and interannual variation compared with other regions with comparable rainfall (Dewar and Richard, 2007). This might have driven speciation in a number of ways. It has been suggested, for example, that the island's fauna has been heavily influenced by unpredictable patterns of fruiting and flowering, which in turn are driven by unpredictable climate (Dewar and Richard, 2007). This has been proposed as a possible explanation for the fact that Madagascar has comparatively few species of frugivorous birds and mammals.

Climate, and more specifically Pleistocene climate change, is likely to have played a role in other ways, with changes in temperature and humidity being particularly important. During colder phases, more of the Earth's water is stored as ice (especially at higher latitudes) and sea levels drop. At lower latitudes, the climate becomes drier and forests recede. During warmer phases, ice melts, rainfall increases and sea levels rise, with forests expanding. Remaining forest areas thus become important refugia as climate changes. Higher latitude refugia are classically the areas not covered by ice, while at lower latitudes refugia are areas where humid conditions have been maintained during the dry phases (Stewart and Lister, 2001; Bennett and Provan, 2008).

The 'watershed chorological hypothesis'[4] postulates that such climatic changes did not impact every habitat in a uniform way but that some pockets of habitats were able to maintain constant conditions and retain humidity during dry phases of climate oscillations (Wilmé et al., 2006, 2012; Wilmé, 2012). In this scenario, major river valleys at higher altitudes, which receive orographic rainfall, acted as refugia. By contrast, areas at lower elevation were isolated for longer periods during the dry episodes of palaeoclimatic oscillations. When the climate became wet again, the hydrological system reconnected in the larger watersheds with sources at high elevation, while the smaller watersheds with sources at low elevation remained isolated. Animals or plants surviving in pockets in dry areas were thus isolated for longer periods of time and required the evolution of new adaptations, leading to speciation and thus creating centres of endemism, while gallery and high mountain forests provided routes for dispersal and retreat (Wilmé et al., 2006, 2012; Wilmé and Callmander, 2006; Wilmé, 2012). In refugia, species could either retain their traits or evolve new adaptations. In concert, this resulted in a mosaic of distinct animal and plant communities within but also between vegetation formations (Figure 2.6).

As well as climatic variability, disturbances due to stochastic events, such as cyclones or fires, increase the patchiness of a region and might allow the coexistence of more species than either a highly disturbed area or a vegetation formation that has been stable for a long time. The former would be inhabited by colonizing species with a high potential for dispersal but a low potential for competition, while the latter would be dominated by species with high competitive potential. The highest number of species would be expected at intermediate levels of disturbance where colonizing and competitive species coexist in non-equilibrium. This 'intermediate disturbance hypothesis' (Connell, 1978; Fox, 2013) is hard to evaluate on a general community level, as it is difficult to quantify disturbances and to define what 'low' and 'high' levels of disturbance are for a community as a whole. But weather phenomena seem to be less predictable in Madagascar than

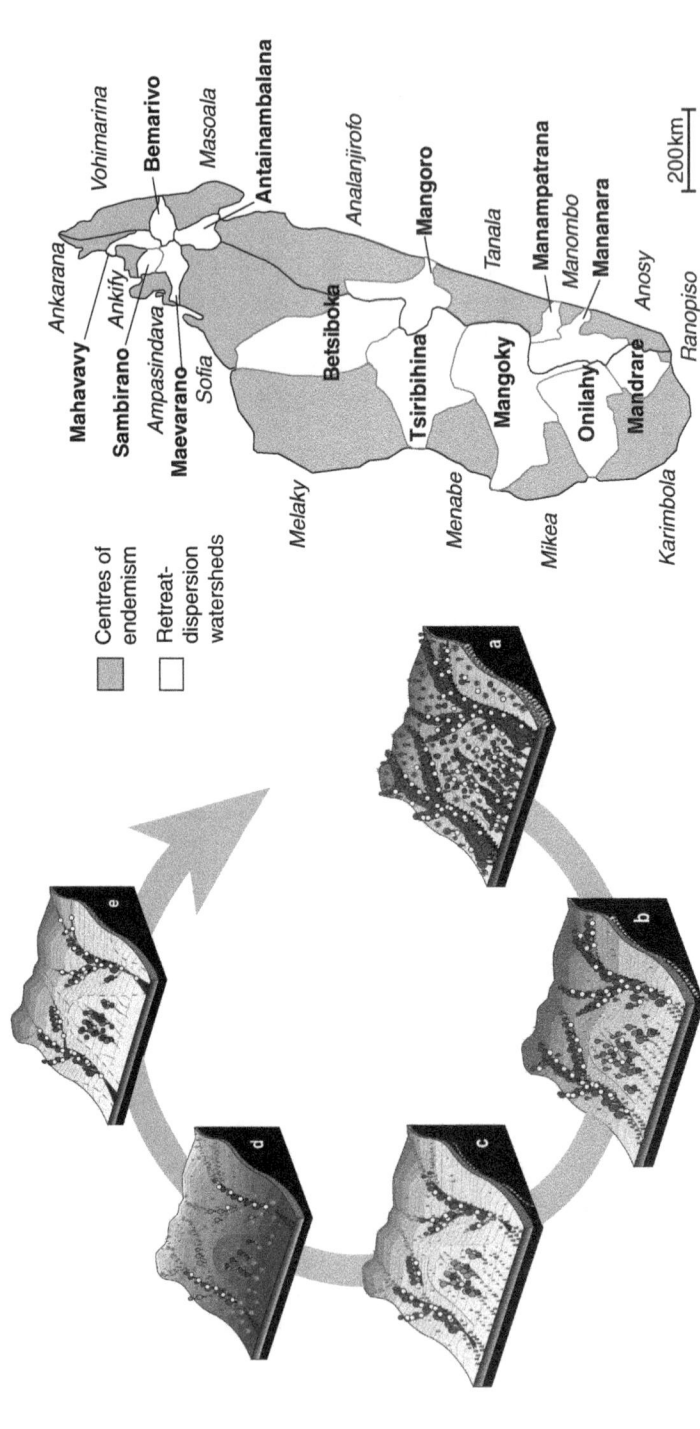

Figure 2.6 Left: schematic sequence of the retreat of forests during cold dry periods and forest expansion during warm, humid episodes. White circles represent species adapted to humid environments; dark circles are species adapted to dry conditions; a–e different states of a cycle from humid (warm) to dry (cold) conditions. Right: resulting centres of endemism and retreat-dispersion watershed

Source: left: Wilmé (2012); right: proposed by Wilmé *et al.* (2006, 2012).

in comparable regions of the world (Wright, 1999; Jolly et al., 2002; Dewar and Richard, 2007), and since stochastic cyclones seem an important components of Madagascar's environment (Ganzhorn, 1995; Birkinshaw and Randrianjanahary, 2007), disturbances can play an important role for the evolution of micro-endemics or special adaptations (Wright, 1999; Wright et al., 2012) and thus the persistence of diverse communities.

In conclusion, the evolution and distribution of species in Madagascar following isolation and colonization is best explained by a combination of: i) adaptation to the north–south bands of vegetation types that have a similar physiognomy (such as humid and dry forests); ii) dispersal and retreat during climatic and vegetation changes; and iii) the separation of populations by various geographical barriers within the island. These processes can be reconstructed in part for the last few thousand years (Burney et al., 2004) but, if the molecular clocks applied today are correct, speciation is much older than this. For example, the radiations within the genera of lemurs all predate the last glacial conditions some 20,000–40,000 years ago by two orders of magnitude (Pastorini et al., 2003; Yoder et al., 2005; Johnson et al., 2008; Ramaromilanto et al., 2009; Weisrock et al., 2012).

Madagascar as a model for biotic evolution

The research and material discussed so far show that Madagascar can be used as a model for testing various evolutionary processes and mechanisms. This is most obvious in the case of convergent evolution, where unrelated lineages that occupy similar niches evolve similar biological traits. For example, lemurs have evolved the same range of sizes as their continental primate relatives, ranging from the smallest primate in the world (Madame Berthe's mouse lemur [*Microcebus berthae*], which weighs about 30 g) to the gorilla-sized *Archaeolemur* that went extinct about 500 years ago. Several species of lemur also find ecological counterparts among marsupials that have evolved similar adaptations in Australia (Smith and Ganzhorn, 1996). Most obviously, the fork-marked lemurs (*Phaner* spp.) of Madagascar and the Australian sugar glider (*Petaurus breviceps*) not only look alike externally but both genera feed preferentially on tree exudates and have similar specializations of their digestive tract to cope with this rather special diet. The Malagasy aye-aye (*Daubentonia madagascariensis*) has a specialized third finger to extract grubs living in trees. It shares this feature with members of the genus *Dactylopsila*, marsupial mammals from New Guinea and northern Australia (but in this case it is the fourth finger that is specialized). Both groups are thought to fill the niche of woodpeckers, a family of birds that is absent from Madagascar as well as Australia and New Guinea.

Evidence of convergent evolution can be found in many other taxa. The spiny tenrecs, for example, have evolved spines and look like small versions of the European hedgehog (*Erinaceus europaeus*), even though they are derived from different ancestors. Among birds, the flightless and by-now extinct elephant birds (*Aepyornis* spp., *Mullerornis* spp.) have their ecological counterparts on all continents, including the extinct moas of New Zealand. A spectacular example for convergent evolution is provided by the Malagasy *Mantella* frogs. Not only have they evolved similar warning colouration and poisons as South American poison-arrow frogs but they also achieve their poisonous defence through processing compounds they obtain by consuming ants, although they are a different species to those of their American counterparts (Clark *et al.*, 2005).

The most famous example that evolution is acting in similar ways throughout the world has been provided by Darwin himself. Darwins's orchid (*Angraecum sesquipedale*), a species endemic to Madagascar, has a 40 cm nectar spur. When Darwin first saw the species, its pollinator was not known and pollination only seemed possible by a moth with a tongue of similar size. Darwin postulated the existence of such a moth, which was discovered and described some 40 years later as the hawkmoth *Xanthopan morganii praedicta* (Arditti *et al.*, 2012).

Implications of Madagascar's biodiversity and evolutionary history for conservation

Global conservation priorities are defined by threats, combined with high species richness and weighed by the uniqueness of species. The best-known approaches are the World Wide Fund for Nature's (WWF) 'Global 200 Ecoregions' (Olson and Dinerstein, 1998) and the identification of 'Biodiversity Hotspots' by Myers *et al.* (2000; updated by Mittermeier *et al.*, 2004; Zachos and Habel, 2011). The 'Global 200' identified areas that would protect global biodiversity most efficiently, while biodiversity hotspots are regions with a high degree of endemism that have suffered over 70 per cent habitat loss. Madagascar matches both of these definitions, although the exact extent of forest loss is contested (see Chapter 3 by Dewar and Chapter 4 by McConnell and Kull).

Large proportions of Madagascar's endemic flora and fauna are found exclusively in forests. Most of the island's remaining forests are highly fragmented with more than 45 per cent of forests existing in patches of less than 500 km^2 and more than 80 per cent of the forest area being less than 1 km from an edge (Harper *et al.*, 2007). In many areas of the world, forest 'edges' represent a different ecosystem than forest interiors and do not allow the persistence of true forest species (Laurance, 2000). The fragmentation of the island's forests is thus a matter of great concern for the persistence of Madagascar's forest species.

In Madagascar, as in most parts of the world, conservation priorities have been defined by subjective decisions. When the island was still a colony, the French colonial government used the educated guesses of their fine naturalists to establish a number of national parks and protected areas that covered most ecosystems of the island (see Chapter 5 by Scales for more on the colonial history of Madagascar's protected areas). Over the course of the past few decades, new protected areas have been installed, mainly based on individual initiatives or the discovery of charismatic species, such as bamboo lemurs (Goodman and Benstead, 2003) rather than on a general and science-based conservation strategy. Today, conservation priorities are defined primarily by biotic richness (Hannah et al., 1998, 2008; Andreone et al., 2008; Kremen et al., 2008), which in many parts coincides with the 'educated guesses' of the past and the postulated centres of endemism (Wilmé et al., 2006, 2012). New conservation approaches integrate the concepts of species richness, metapopulations, networks and corridors to allow plants and animals to disperse from their isolated protected area to other sites. Given the effect climate change has had in the past on Madagascar's flora and fauna, and the importance of forest refugia and corridors in allowing species to persist and adapt, the goal is to achieve a network of protected areas that are linked by suitable habitats and allow the exchange of individuals. This is supposed to buffer species against extinction due to small population size and genetic erosion and should allow species to retreat into suitable habitats or maintain viable populations (Hannah et al., 2008; Kremen et al., 2008; Irwin et al., 2010).

While it would be desirable to define conservation priorities on scientific grounds, reality has outrun any academic exercise. In 2003, the former president, Marc Ravalomanana, announced in the 'Durban Vision' that Madagascar's protected area would be tripled. After the initial excitement of the conservation community, it turned out that there was hardly enough land available to achieve this goal. Thus, government organizations and non-governmental organizations (NGOs) are still struggling to identify suitable areas for conservation, based more on feasibility and socio-economic aspects than on biodiversity arguments (for more on the Durban Vision, see Chapter 9 by Corson and Chapter 10 by Virah-Sawmy).

Conclusion

The diversity of Madagascar's biota is largely the result of geological isolation, colonization events and subsequent radiation into niches provided by abiotic and biotic heterogeneity. The principle reason for Madagascar's endemism is its geological history, with continental drift leading to isolation prior to the radiation of most major biota, especially higher plants and vertebrates. After its isolation, Madagascar's flora and fauna evolved without the influence of ruminants, large carnivores or humans, i.e. they have evolved not only under isolation but under very different conditions to mainland Africa, thus creating a huge potential for radiation that resulted in the globally unique ecosystems seen today.

Acknowledgements

We thank Jason Ali for providing figures from his previous publications. Ivan Scales provided excellent comments on the manuscript. Monika Hänel helped with figures.

Notes

1 Endemic species can come about either by the extinction of a species in all but a restricted area, or by the local evolution of new species with restricted distributions.
2 Phylogeny refers to the evolutionary history of a species or higher taxonomic grouping of organisms while phylogenetics is the study of the evolutionary relationship between groups of organisms.
3 According to the new nomenclature, the lower Tertiary is now named 'Paleogene' and the corresponding time of asteroid impact is called the K-Pg-event.
4 Chorology is the study of the spatial distribution of species and its causes.

References

Ali, J. R. and Huber, M. (2010) 'Mammalian biodiversity on Madagascar controlled by ocean currents', *Nature*, vol 463, pp653–656.

Ali, J. R. and Krause, D. W. (2011) 'Late Cretaceous bioconnections between Indo-Madagascar and Antarctica: refutation of the Gunnerus Ridge causeway hypothesis', *Journal of Biogeography*, vol 38, pp1855–1872.

Andreone, F., Carpenter, A. I., Cox, N., du Preez, L., Freeman, K., Furrer, S., Garcia, G., Glaw, F., Glos, J., Knox, D., Kohler, J., Mendelson, J. R., Mercurio, V., Mittermeier, R. A., Moore, R. D., Rabibisoa, N. H. C., Randriamahazo, H., Randrianasolo, H., Raminosoa, N. R., Ramilijaona, O. R., Raxworthy, C. J., Vallan, D., Vences, M., Vieites, D. R. and Weldon, C. (2008) 'The challenge of conserving amphibian megadiversity in Madagascar', *Plos Biology*, vol 6, pp943–946.

Arditti, J., Elliottt, J., Kitching, I. J. and Wasserthal, L. T. (2012) 'Good Heavens what insect can suck it – Charles Darwin, *Angraecum sesquipedale* and *Xanthopan morganii praedicta*', *Botanical Journal of the Linnean Society*, vol 169, pp403–432.

Barthlott, W., Lauer, W. and Placke, A. (1996) 'Global distribution of species diversity in vascular plants: towards a world map of phytodiversity', *Erdkunde*, vol 50, pp317–327.

Baum, D. A., Small, R. I. and Wendel, J. F. (1998) 'Biogeography and floral evolution of baobabs (*Adansonia*, Bombacaceae) as inferred from multiple data sets', *Systematic Biology*, vol 47, pp181–207.

Bennett, K. D. and Provan, J. (2008) 'What do we mean by 'refugia?'', *Quaternary Science Reviews*, vol 27, pp2449–2455.

Besairie, H. and Collignon, M. (1972) 'Géologie de Madagascar. I. Les terrains sédimentaires. Nouvelle carte géologique au 1/500 000e', *Annales Géologiques de Madagascar, Tananarive*, vol 35, pp1–463.

Birkinshaw, C. and Randrianjanahary, M. (2007) 'The effects of cyclone Hudah on the forest of Masoala Peninsula, Madagascar', *Madagascar Conservation and Development*, vol 2, pp17–20.

Bonaccorso, F. J. (1998) *Bats of Papua New Guinea*, Conservation International, Washington, DC.
Bond, W. J., Silander Jr, J. A., Ranaivonasy, J. and Ratsirarson, J. (2008) 'The antiquity of Madagascar's grasslands and the rise of C4 grassy biomes', *Journal of Biogeography*, vol 35, pp1743–1758.
Burney, D. A., Pigott Burney, L., Godfrey, L. R., Jungers, W. L., Goodman, S. M., Wright, H. T. and Timothy Jull, A. J. (2004) 'A chronology for late prehistoric Madagascar', *Journal of Human Evolution*, vol 47, pp25–63.
Ceballos, G. and Brown, J. H. (1995) 'Global patterns of mammalian diversity, endemism, and endangerment', *Conservation Biology*, vol 9, pp559–568.
Censky, E. J., Hodge, K. and Dudley, J. (1998) 'Over-water dispersal of lizards due to hurricanes', *Nature*, vol 395, pp556.
Chakrabarty, P., Davis, M. P. and Sparks, J. S. (2012) 'The first record of a transoceanic sister-group relationship between obligate vertebrate troglobites', *PLoS ONE*, vol 7, e44083.
Clark, V. C., Raxworthy, C. J., Rakotomalala, V., Sierwald, P. and Fisher, B. L. (2005) 'Convergent evolution of chemical defense in poison frogs and arthropod prey between Madagascar and the Neotropics', *Proceedings of the National Academy of Sciences USA*, vol 102, pp11617–11622.
Connell, J. H. (1978) 'Diversity in tropical rain forests and coral reefs', *Science*, vol 199, pp1302–1309.
Cornet, A. (1974) 'Essai de cartographie bioclimatique à Madagascar', *Notice explicative no 55*, ORSTOM, Paris.
Craul, M., Radespiel, U., Rasolofoson, D. W., Rakotondratsimba, G., Rakotonirainy, O., Rasoloharijaona, S., Randrianambinina, B., Ratsimbazafy, J., Ratelolahy, F., Randrianamboavaonjy, T. and Rakotozafy, L. (2008) 'Large rivers do not always act as species barriers for *Lepilemur* sp', *Primates*, vol 49, pp211–218.
Crottini, A., Madsen, O., Poux, C., Strauß, A., Vieites, D. R. and Vences, M. (2012) 'Vertebrate time-tree elucidates the biogeographic pattern of a major biotic change around the K-T boundary in Madagascar', *Proceedings of the National Academy of Sciences USA*, vol 109, pp5358–5363.
Dausmann, K. H., Glos, J. and Heldmaier, G. (2009) 'Energetics of tropical hibernation', *Journal of Comparative Physiology B-Biochemical Systemic and Environmental Physiology*, vol 179, pp345–357.
de Queiroz, A. (2005) 'The resurrection of oceanic dispersal in historical biogeography', *Trends in Ecology and Evolution*, vol 20, pp68–73.
de Wit, M. J. (2003) 'Madagascar: heads it's a continent, tails it's an island', *Annual Review of Earth and Planetary Sciences*, vol 31, pp213–248.
Dewar, R. E. and Richard, A. F. (2007) 'Evolution in the hypervariable environment of Madagascar', *Proceedings of the National Academy of Sciences USA*, vol 104, pp13723–13727.
Fox, J. W. (2013) 'The intermediate disturbance hypothesis should be abandoned', *Trends in Ecology and Evolution*, vol 28, pp86–92.
Ganzhorn, J. U. (1995) 'Cyclones over Madagascar': fate or fortune?', *Ambio*, vol 24, pp124–125.
Ganzhorn, J. U., Goodman, S. M., Nash, S. and Thalmann, U. (2006) 'Lemur biogeography', in S. Lehman, and J. G. Fleagle (eds) *Primate Biogeography*, Plenum/Kluwer Press, New York.

Gasse, F. and Van Campo, E. (2001) 'Late quaternary environmental changes from a pollen and diatom record in the southern tropics (Lake Tritrivakely, Madagascar)', *Palaeogeography, Palaeoclimatology, Palaeoecology*, vol 167, pp287–308.

Gerlach, J., Muir, C. and Richmond, M. D. (2006) 'The first substantiated case of trans-oceanic tortoise dispersal', *Journal of Natural History*, vol 40, pp2403–2408.

Goodman, S. M. (2011) *Les chauves-souris de Madagascar*, Association Vahatra, Antananarivo.

Goodman, S. M. and Benstead, J. P. (2003) *The Natural History of Madagascar*, University of Chicago Press, Chicago, IL.

Goodman, S. M. and Benstead, J. P. (2005) 'Updated estimates of biotic diversity and endemism for Madagascar', *Oryx*, vol 39, pp73–77.

Goodman, S. M. and Ganzhorn, J. U. (2004a) 'Elevational ranges of lemurs in the humid forests of Madagascar', *International Journal of Primatology*, vol 25, pp331–350.

Goodman, S. M. and Ganzhorn, J. U. (2004b) 'Biogeography of lemurs in the humid forests of Madagascar: the role of elevational distribution and rivers', *Journal of Biogeography*, vol 31, pp47–55.

Goodman, S. M., Ganzhorn, J. U. and Rakotondravony, D. (2008) 'Les mammifères', in S. M. Goodman (ed.) *Paysages Naturels et Biodiversité de Madagascar*, Muséum national d'histoire naturelle, Paris.

Hannah, L., Rakotosamimanana, B., Ganzhorn, J., Mittermeier, R. A., Olivier, S., Iyer, L., Rajaobelina, S., Hough, J., Andriamialisoa, F., Bowles, I. and Tilkin, G. (1998) 'Participatory planning, scientific priorities, and landscape conservation in Madagascar', *Environmental Conservation*, vol 25, pp30–36.

Hannah, L,. Dave, R., Lowry II, P. P., Andelman, S., Andrianarisata, M., Andriamaro, L., Cameron, A., Hijmans, R., Kremen, C., MacKinnon, J., Randrianasolo, H. H., Andriambololonera, S., Razafimpahanana, A., Randriamahazo, H., Randrianarisoa, J., Razafinjatovo, P., Raxworthy, C., Schatz, G. E., Tadross, M. and Wilmé, L. (2008) 'Climate change adaptation for conservation in Madagascar', *Biology Letters*, vol 4, pp590–594.

Harper, G. J., Steininger, M. K., Tucker, C. J., Juhn, D. and Hawkins, F. (2007) 'Fifty years of deforestation and forest fragmentation in Madagascar', *Environmental Conservation*, vol 34, pp1–9.

Hemingway, C. and Bynum, N. (2005) 'The influence of seasonality on primate diet and ranging', in C. van Schaik and D. Brockman (eds) *Seasonality in Primates: Studies of Living and Extinct Human and Non-human Primates*, Cambridge University Press, Cambridge.

Holt, B. G., Lessard, J.-P., Borregaard, M. K., Fritz, S. A., Araújo, M. B., Dimitrov, D., Fabre, P.-H., Graham, C. H., Graves, G. R., Jønsson, K. A., Nogués-Bravo, D., Wang, Z., Whittaker, R. J., Fjeldså, J. and Rahbek, C. (2013) 'An update of Wallace's zoogeographic regions of the world', *Science*, vol 338, pp74–78.

Irwin, M. T., Wright, P. C., Birkinshaw, C., Fisher, B. L., Gardner, C. J., Glos, J., Goodman, S. M., Loiselle, P., Rabeson, P., Raharison, J.-L., Raherilalao, M.-J., Rakotondravony, D., Raselimanana, A., Ratsimbazafy, J., Sparks, J. S., Wilmé, L. and Ganzhorn, J. U. (2010) 'Patterns of species change in anthropogenically disturbed forests of Madagascar', *Biological Conservation*, vol 143, pp2351–2362.

Jenkins, R. K. B., Brady, L. D., Bisoa, M., Rabearivony, J. and Griffiths, R. A. (2003) 'Forest disturbance and river proximity influence chameleon abundance in Madagascar', *Biological Conservation*, vol 109, pp407–415.

Johnson, S. E., Lei, R., Martin, S. K., Irwin, M. T. and Louis, E. E. (2008) 'Does Eulemur cinereiceps exist? Preliminary evidence from genetics and ground surveys in southeastern Madagascar', *American Journal of Primatology*, vol 70, pp372–385.

Jolly, A., Dodson, A., Rasamimanana, H. M., Walker, J., O'Connor, S., Solberg, M. and Prel, V. (2002) 'Demography of *Lemur catta* at Berenty Reserve, Madagascar: effects of troop size, habitat and rainfall', *International Journal of Primatology*, vol 23, pp327–353.

Jury, M. R. (2003) 'The climate of Madagascar', in S. M. Goodman and J. P. Benstead (eds) *The Natural History of Madagascar*, University of Chicago Press, Chicago, IL.

Kappeler, P. M. (2000) 'Lemur origins: rafting by groups of hibernators?', *Folia Primatologica*, vol 71, pp422–425.

Kerr, J. T., Perring, M. and Currie, D. J. (2006) 'The missing Madagascan mid-domain effect', *Ecology Letters*, vol 9, pp149–159.

Kisel, Y. and Barraclough, T. G. (2010) 'Speciation has a spatial scale that depends on levels of gene flow', *American Naturalist*, vol 175, pp316–334.

Knopp, T., Rahagalala, P., Miinala, M. and Hanski, I. (2011) 'Current geographical ranges of Malagasy dung beetles are not delimited by large rivers', *Journal of Biogeography*, vol 38, pp1098–1108.

Krause, D. W. (2003) 'Late Cretaceous vertebrates of Madagascar: a window into Gondwanan biogeography at the end of the age of dinosaurs', in S. M. Goodman and J. Benstead (eds) *The Natural History of Madagascar*, University of Chicago Press, Chicago, IL.

Krause, D. W. (2010) 'Washed up in Madagascar', *Nature*, vol 463, pp613–614.

Krause, D. W., Asher, R. J., Buckley, G., Gottfried, M. and Laduke, T. C. (1998) 'Biogeographic origins of the non-dinosaurian vertebrate fauna of Madagascar: new evidence from the Late Cretaceous', *Journal of Vertebrate Paleontology*, vol 18, p57A.

Krause, D. W., Prasad, G. V. R., von Koenigswald, W., Sahni, A. and Grine, F. E. (1997) 'Cosmopolitanism among Gondwanan Late Cretaceous mammals', *Nature*, vol 390, pp504–507.

Krause, D. W., Rogers, R. R., Forster, C. A., Hartman, J. H., Buckley, G. A. and Sampson, S. D. (1999) 'The Late Cretaceous vertebrate fauna of Madagascar: implications for Gondwanan paleobiogeography', *Geological Society of America Today*, vol 9, pp1–7.

Kreft, H. and Jetz, W. (2007) 'Global patterns and determinants of vascular plant diversity', *Proceedings of the National Academy of Science USA*, vol 104, pp5925–5930.

Kremen, C., Cameron, A., Moilanen, A., Phillips, S. J., Thomas, C. D., Beentje, H., Dransfield, J., Fisher, B. L., Glaw, F., Good, T. C., Harper, G. J., Hijmans, R. J., Lees, D. C., Louis Jr. E., Nussbaum, R. A., Raxworthy, C. J., Razafimpahanana, A., Schatz, G. E., Vences, M., Vieites, D. R., Wright, P. C. and Zjhra, M. L. (2008) 'Aligning conservation priorities across taxa in Madagascar with high-resolution planning tools', *Science*, vol 320, pp222–226.

Kuntner, M. and Agnarsson, I. (2011) 'Phylogeography of a successful aerial disperser: the golden orb spider *Nephila* on Indian Ocean islands', *BMC Evolutionary Biology*, vol 11, pp119.

Langrand, O. (1990) *Guide to the Birds of Madagascar*, Yale University Press, New Haven, CT.

Laurance, W. F. (2000) 'Do edge effects occur over large spatial scales?', *Trends in Ecology and Evolution*, vol 15, pp134–135.

Lees, D. C., Kremen, C. and Andriamampianina, L. (1999) 'A null model for species richness gradients: bounded range overlap of butterflies and other rainforest endemics in Madagascar', *Biological Journal of the Linnean Society*, vol 67, pp529–584.

Lomolino, M. V. B. R., Riddle, R. J., Whittaker, R. and J. H. Brown. (2010) *Biogeography*, Sinauer Associates, Sunderland, MA.

Lowry II, P. P., Schatz, G. E. and Phillipson, P. B. (1997) 'The classification of natural and anthropogenic vegetation in Madagascar', in S. M. Goodman and B. D. Patterson (eds) *Natural Change and Human Impact in Madagascar*, Smithsonian Institution Press, Washington, DC.

MacArthur, R. H. and Wilson, E. O. (1967) *The Theory of Island Biogeography*, Princeton University Press, Princeton, NJ.

MacPhee, R. D. E. (1994) 'Morphology, adaptations, and relationships of Plesiorycteropus, and a diagnosis of a new order of eutherian mammals', *Bulletin of the American Museum of Natural History*, vol 220, pp1–214.

Magurran, A. E. and McGill, B. J. (2010) *Biological Diversity: Frontiers in Measurement and Assessment*, Oxford University Press, Oxford.

Markolf, M., Brameier, M. and Kappeler, P. M. (2011) 'On species delimitation: yet another lemur species or just genetic variation?', *BMC Evolutionary Biology*, vol 11, p216.

Martin, R. D. (1995) 'Prosimians: from obscurity to extinction?', in L. Alterman, G. A. Doyle and M. K. Izard (eds) *Creatures of the Dark: The Nocturnal Prosimians*, Plenium Press, New York.

Masters, J. C., de Wit, M. J. and Asher, R. J. (2006) 'Reconciling the origins of Africa, India and Madagascar with vertebrate dispersal scenarios', *Folia Primatologica*, vol 77, pp399–418.

Measey, G. J., Vences, M., Drewes, R. C., Chiari, Y., Melo, M. and Bourles, B. (2007) 'Freshwater paths across the ocean: molecular phylogeny of the frog *Ptychadena newtoni* gives insights into amphibian colonization of oceanic islands', *Journal of Biogeography*, vol 34, pp7–20.

Michalak, I. L., Zhang, B. and Renner, S. S. (2010) 'Trans-Atlantic, trans-Pacific and trans-Indian Ocean dispersal in the small Gondwanan Laurales family Hernandiaceae', *Journal of Biogeography*, vol 37, pp1214–1226.

Mittermeier, R., Gil, P., Hoffmann, M., Pilgrim, J., Brooks, T., Goetsch Mittermeier, C., Lamoreux, J. and da Fonseca, G. (2004) *Hotspots Revisited*, CEMEX, Mexico City.

Mittermeier, R. A., Louis Jr. E. E., Richardson, M., Schwitzer, C., Langrand, O., Rylands, B., Hawkins, F., Rajaobelina, S., Ratsimbazafy, J., Rasoloarison, R., Roos, C., Kappeler, P. M. and Mackinnon, J. (2010) *Lemurs of Madagascar*, Conservation International, Arlington, VA.

Moat, J. and Smith, P. (2007) *Atlas of the Vegetation of Madagascar: Atlas de la végétation de Madagascar*, Royal Botanic Gardens, Kew.

Morgan, G. S. and Woods, C. A. (1986) 'Extinction and the zoogeography of West-Indian land mammals', *Biological Journal of the Linnean Society*, vol 28, pp167–203.

Myers, N., Mittermeier, R. A., Mittermeier, C. G., da Fonseca, A. B. and Kent, J. (2000) 'Biodiversity hotspots for conservation priorities', *Nature*, vol 403, pp853–858.

Naeem, S., Bunker, D. E., Hector, A., Loreau, M. and Perrings, C. (2009) *Biodiversity, Ecosystem Functioning, and Human Wellbeing*, Oxford University Press, Oxford.

Nathan, R., Schurr, F. M., Spiegel, O., Steinitz, O., Trakhtenbrot, A. and Tsoar, A. (2008) 'Mechanisms of long-distance seed dispersal', *Trends in Ecology and Evolution*, vol 23, pp638–647.
O'Brien, J., Mariani, C., Olson, L., Russell, A. L., Say, L., Yoder, A. D. and Hayden, T. J. (2009) 'Multiple colonisations of the western Indian Ocean by *Pteropus* fruit bats (Megachiroptera: Pteropodidae): the furthest islands were colonised first', *Molecular Phylogenetics and Evolution*, vol 51, pp294–303.
O'Leary, M. A., Bloch, J. I., Flynn, J. J., Gaudin, T. J., Giallombardo, A., Giannini, N. P., Goldberg, S. L., Kraatz, B. P., Luo, Z.-X., Meng, J., Ni, X., Novacek, M. J., Perini, F. A., Randall, Z. S., Rougier, G. W., Sargis, E. J., Mary, T., Silcox, M. T., Simmons, N. B., Spaulding, M., Velazco, P. M., Weksler, M., Wible, J. R. and Cirranello, A. L. (2013) 'The placental mammal ancestor and the Post-K-Pg radiation of placentals', *Science*, vol 339, pp662–667.
Olson, D. M. and Dinerstein, E. (1998) 'The global 200: a representation approach to conserving the earth's most biologically valuable ecoregions', *Conservation Biology*, vol 12, pp502–515.
Pastorini, J., Thalmann, U. and Martin, R. D. (2003) 'A molecular approach to comparative phylogeography of extant Malagasy lemurs', *Proceedings of the National Academy of Sciences USA*, vol 100, pp5879–5884.
Pearson, R. G. and Raxworthy, C. J. (2009) 'The evolution of local endemism in Madagascar: watershed vs. climatic gradient hypotheses evaluated by null biogeographic models', *Evolution*, vol 63, pp959–967.
Petter, J.-J., Albignac, R. and Rumpler, Y. (1977) *Faune de Madagascar: Mammifères Lémuriens*, ORSTOM, Paris.
Phillipson, P. B., Schatz, G. E., Lowry II, P. P. and Labat, J.-N. (2006) 'A catalogue of the vascular plants of Madagascar', in S. A. Ghazanfar and H. J. Beentje (eds) *Taxonomy and Ecology of African Plants: Their Conservation and Sustainable Use*, Proceedings XVIIth AETFAT Congress, Royal Botanic Gardens, Kew.
Pock Tsy, J.-M. L,. Lumaret, R., Mayne, D., Vall, A. O. M., Abutaba, Y. I. M., Sagna, M., Rakotondralambo Raoseta, S. O. and Danthu, P. (2009) 'Chloroplast DNA phylogeography suggests a West African centre of origin for the baobab, *Adansonia digitata* L. (Bombacoideae, Malvaceae)', *Molecular Ecology*, vol 18, pp1707–1715.
Raherilalao, M. J. and Goodman, S. M. (2011) *Histoire naturelle des familles et sous-familles endémiques d'oiseaux de Madagascar*, Association Vahatra, Antananarivo.
Ramaromilanto, B., Lei, L., Engberg, S. E., Johnson, S. E., Sitzmann, B. D. and Louis Jr., E. E. (2009) 'Sportive lemur diversity at Mananara-Nord Biosphere Reserve, Madagascar', *Occasional Papers, Museum of Texas Tech University*, vol 286, pp1–22.
Raxworthy, C. J. and Nussbaum, R. A. (1997) 'Biogeographic patterns of reptiles in Eastern Madagascar', in S. M. Goodman and B. D. Patterson (eds) *Natural Change and Human Impact in Madagascar*, Smithsonian Institution Press, Washington, DC.
Raxworthy, C. J., Forstner, M. R. J. and Nussbaum, R. A. (2002) 'Chameleon radiation by oceanic dispersal', *Nature*, vol 415, pp784–787.
Rocha, S., Carretero, M. A., Vences, M., Glaw, F. and Harris, D. J. (2006) 'Deciphering patterns of transoceanic dispersal: the evolutionary origin and biogeography of coastal lizards (*Cryptoblepharus*) in the Western Indian Ocean region', *Journal of Biogeography*, vol 33, pp13–22.

Rosenzweig M. L. (1995) *Species Diversity in Space and Time*, Cambridge University Press, Cambridge.

Samonds, K. E., Godfrey, L. R., Ali, J. R., Goodmand, S. M., Vences, M., Sutherland, M. R., Irwing, M. T. and Krause, D. W. (2012) 'Spatial and temporal arrival patterns of Madagascar's vertebrate fauna explained by distance, ocean currents, and ancestor type', *Proceedings of the National Academy of Sciences USA*, vol 109, pp5352–5357.

Smith, A. P. and Ganzhorn, J. U. (1996) 'Convergence in community structure and dietary adaptation in Australian possums and gliders and Malagasy lemurs', *Austral Journal of Ecology*, vol 21, pp31–46.

Soarimalala, V. and Goodman, S. M. (2011) *Les petits mammifères de Madagascar*, Association Vahatra, Antananarivo.

Sparks, J. S. and Smith, W. L. (2005) 'Freshwater fishes, dispersal ability, and nonevidence: "Gondwana life rafts" to the rescue', *Systematic Biology*, vol 54, pp158–165.

Stewart, J. R. and Lister, A. M. (2001) 'Cryptic northern refugia and the origins of modern biota', *Trends in Ecology and Evolution*, vol 16, pp608–613.

Strijk, J. S., Noyes, R. D., Strasberg, D., Cruaud, C., Gavory, F., Chase, M. W., Abbott, R. J. and Thébaud, C. (2012) 'In and out of Madagascar: dispersal to peripheral islands, insular speciation and diversification of Indian Ocean Daisy trees (*Psiadia*, Asteraceae)', *PLoS ONE*, vol 7, e42932.

Struebig, M. J., Christy, L., Pio, D. and Meijaard, E. (2010) 'Bats of Borneo: diversity, distributions and representation in protected areas', *Biodiversity and Conservation*, vol 19, pp449–469.

Tattersall, I. (2007) 'Madagascar's lemurs: cryptic diversity or taxonomic inflation?', *Evolutionary Anthropology*, vol 16, pp12–23.

Terborgh, J. (1992) *Diversity and the Tropical Rain Forest*, Scientific American Library, W. H. Freeman, New York.

Thalmann, U. (2007) 'Biodiversity, phylogeography, biogeography, and conservation: lemurs as an example', *Folia Primatologica*, vol 78, pp420–443.

Vences, M., Freyhof, J., Sonnenberg, R., Kosuch, J. and Veith, M. (2001) 'Reconciling fossils and molecules: cenozoic divergence of cichlid fishes and the biogeography of Madagascar', *Journal of Biogeography*, vol 28, pp1091–1099.

Vences, M., Kosuch, J., Rodel, M. O., Lotters, S., Channing, A., Glaw, F. and Bohme, W. (2004) 'Phylogeography of *Ptychadena mascareniensis* suggests transoceanic dispersal in a widespread African-Malagasy frog lineage', *Journal of Biogeography*, vol 31, pp593–601.

Vences, M., Vieites, D. R., Glaw, F., Brinkmann, H., Kosuch, J., Veith, M. and Meyer, A. (2003) 'Multiple overseas dispersal in amphibians', *Proceedings of the Royal Society B*, vol 270, pp 2435–2442.

Vences, M., Wollenberg, K. C., Vieites, D. R. and Lees, D. C. (2009) 'Madagascar as a model region of species diversification', *Trends in Ecology and Evolution*, vol 24, pp456–465.

Vieites, D. R., Wollenberg, K. C., Andreone, F., Köhler, J., Glaw, F. and Vences, M. (2009) 'Vast underestimation of Madagascar's biodiversity evidenced by an integrative amphibian inventory', *Proceedings of the National Academy of Sciences USA*, vol 106, pp8267–8272.

Warren, B. H., Bermingham, E., Prys-Jones, R. P. and Thebaud, C. (2005) 'Tracking island colonization history and phenotypic shifts in Indian Ocean bulbuls

(*Hypsipetes*: Pycnonotidae)', *Biological Journal of the Linnean Society*, vol 85, pp271–287.

Weisrock, D. W., Smith, S. D., Chan, L. M., Biebouw, K., Kappler, P. M. and Yoder, A. D. (2012) 'Concatenation and concordance in the reconstruction of Mouse lemur phylogeny: an empirical demonstration of the effect of allele sampling in phylogenetics', *Molecular Biology and Evolution*, vol 29, pp1615–1630.

Wilmé, L. (2012) *Biogeographic Evolution of Madagascar's Microendemic Biota: Analyse et Déconstruction*, Thèse de doctorat, Université de Strasbourg.

Wilmé, L. and Callmander, M. W. (2006) 'Les populations reliques de primates: les Propithèques', *Lemur News*, vol 11, pp24–31.

Wilmé, L., Goodman, S. M. and Ganzhorn, J. U. (2006) 'Biogeographic evolution of Madagascar's microendemic biota', *Science*, vol 312, pp1063–1065.

Wilmé, L., Ravokatra, M., Dolch, R., Schuurman, D., Mathieu, E., Schuetz, H. and Waeber, P. O. (2012) 'Toponyms for centers of endemism in Madagascar', *Madagascar Conservation and Development*, vol 7, pp30–40.

Wilson, E. O. and Peters, F. M. (eds) (1988) *Biodiversity*, National Academy Press, Washington, DC.

Wollenberg, K. C., Vieites, D. R., van der Meijden, A., Glaw, F., Cannatella, D. C. and Vences, M. (2008) 'Patterns of endemism and species richness in Malagasy cophyline frogs support a key role of mountainous areas for speciation', *Evolution*, vol 62, pp1890–1907.

Wright, P. C. (1999) 'Lemur traits and Madagascar ecology: coping with an island environment', *American Journal of Physical Anthropology*, vol 110(S29), pp31–72.

Wright, P. C., Erhart, E. M., Tecot, S., Baden, A. L., Arrigo-Nelson, S. J., Herrera, J., Morelli, T. L., Blanco, M. B., Deppe, A., Atsalis, S., Johnson, S., Ratelolahy, F., Tan, C. and Zohdy, S. (2012) 'Long-term lemur research at Centre Valbio, Ranomafana National Park, Madagascar', in P. M. Kappeler and D. P. Watts (eds) *Long-Term Field Studies of Primates*, Springer, Heidelberg.

Yoder, A. D. (2013a) 'Fossils versus clocks', *Science*, vol 339, pp656–658.

Yoder, A. D. (2013b) 'The lemur revolution starts now: the genomic coming of age for a non-model organism', *Molecular Phylogenetics and Evolution*, vol 66, pp442–452.

Yoder, A. D., Burns, M. M., Zehr, S., Delefosse, T., Veron, G., Goodman, S. M. and Flynn, J. J. (2003) 'Single origin of Malagasy Carnivora from an African ancestor', *Nature*, vol 421, pp734–737.

Yoder, A. D., Rasoloarison, R. M., Goodman, S. M., Irwin, J. A., Atsalis, S., Ravosa, M. J. and Ganzhorn, J. U. (2000) 'Remarkable species diversity in Malagasy mouse lemurs (primates, *Microcebus*)', *Proceedings of the National Academy of Sciences USA*, vol 97, pp11325–11330.

Yoder, A. D., Olson, L. E., Hanley, C., Heckman, K. L., Rasoloarison, R., Russell, A. M., Ranivo, J., Soarimalala, V., Karanth, K. P., Raselimanana, A. P. and Goodman, S. M. (2005) 'A multidimensional approach for detecting species patterns in Malagasy vertebrates', *Proceedings of the National Academy of Sciences USA*, vol 102, suppl. 1, pp6587–6594.

Zachos, F. E. and Habel, J. C. (2011) *Biodiversity Hotspots*, Springer, Heidelberg.

3 Early human settlers and their impact on Madagascar's landscapes

Robert E. Dewar

Introduction

The deep history of environmental variability and change in Madagascar provides a benchmark against which to assess conditions today and design management interventions for the future (Willis et al., 2007). As knowledge of the island's palaeoecology and the role played by people in changes over the past few thousand years has increased, it has forced abandonment of the widespread but simplistic story of a lost Garden of Eden, destroyed by the Malagasy people. Recent work is steadily replacing that narrative with an account of the dynamic and complex interplay of climate, the arrival and activities of people, and the nature of the landscapes they encountered and shaped. This chapter gives a glimpse of the research that supports these new interpretations, which offer valuable insights for environmental policy in the years to come. Recounting some of my own work with colleagues in Madagascar, I also try to evoke the interest, excitement, and unpredictability of field archaeology.

The last Ice Age and its aftermath as prologue

Throughout the Pleistocene, or Ice Age (about 2.5 million to 12,000 years ago), global temperatures rose and fell, in both the temperate zones and the tropics. Changing climates led to alterations in the distribution of plant and animal species and biological communities. Whereas in temperate zones their distribution often moved towards the poles in warmer periods and away from them during cooler times, in the tropics the principal shifts of species were to higher elevations in warm periods and to lower elevations in cooler ones. Madagascar was no exception to this general pattern. Burney (1997) estimates that at the last glacial maximum (18,000 years ago), all of Madagascar above 1,000m in elevation had a vegetation of ericaceous heathland, similar to the modern vegetation above 2,000m on Madagascar's highest mountains. Madagascar's forests, of whatever types, apparently retreated towards the warmest zones near sea level. It is interesting to note that many of the best-known and preserved modern tropical forests of Madagascar today are at or above 1,000m in elevation: Ranomafana, Andasibe/Mantadia, and Zahamena; these places would have been unrecognizably different only 15,000 years ago and almost certainly without any of their modern fauna.

The most complete record of changing plant communities at the end of the Ice Age comes from palaeoecological studies of sediment columns from lakes and bogs. In Madagascar, the longest record is from Lake Tritrivakely, 15km northwest of modern Antsirabe (Gasse and Van Campo, 2001; see Figure 3.1 for the location of this and other palaeontological and archaeological sites). The 40m of sediment extracted by coring the lake span about 150,000 years. Over this time, there were six cycles of warming and cooling, with a progressive shift from heathland to grassland to grassland–forest mosaic and forest as temperatures climbed, followed by a return to the ericaceous heath in the coldest periods. The climax of the most recent cold period (ca. 18,500 years ago) was followed by a steady warming about 15,000 years ago, and by the spread of grasses and then trees characteristic of mid-elevation forests today.

The Holocene (ca. 12,000 years ago to the present) saw further fluctuations in the island's natural communities, some in response to global temperature changes and others to changes in rainfall patterns, and, along the coasts, changes in relative sea level. For the central highlands, Burney (1987a) interpreted another pollen column from Tritrivakely as showing an unstable mixture of heathland species and grasses until about 5,000 years ago, when a savannah of grasses and some dry-adapted trees was established, with an increasing presence of trees through to about the time of the appearance of pollen from species introduced by people. For the southeastern coastal plains, Virah-Sawmy et al. (2009, 2010) describe the past 6,000 years of vegetational history on the basis of four sedimentary sequences. They report complex, shifting patterns of coastal forest, woodland and grassland throughout their sequences, and identify episodes of aridity and a sea level surge as the most obvious causes of vegetational change. In this coastal zone, the effects of temperature changes are not evident. In northern Madagascar, Holocene droughts have been identified as the likely cause of a major contraction in the golden-crowned sifaka (*Propithecus tattersalli*) population well before humans arrived on the island (Quéméré et al., 2012). The challenge of distinguishing between natural and anthropogenic drivers of change will be taken up later in this chapter.

Modern Madagascar is strikingly varied in climate and natural communities, ranging from the arid bush of the southwest, with usually less than half a meter of rain in a year, to the humid east coast where rain forests may receive more than 4m of rainfall each year. The island's climates are also all especially variable on a year-to-year basis, when compared to areas of similar rainfall in Africa (Dewar and Richard, 2007). This variability takes two general forms: in the drier areas of the south and west, where rainfall occurs only in a short rainy season, there are huge differences in annual

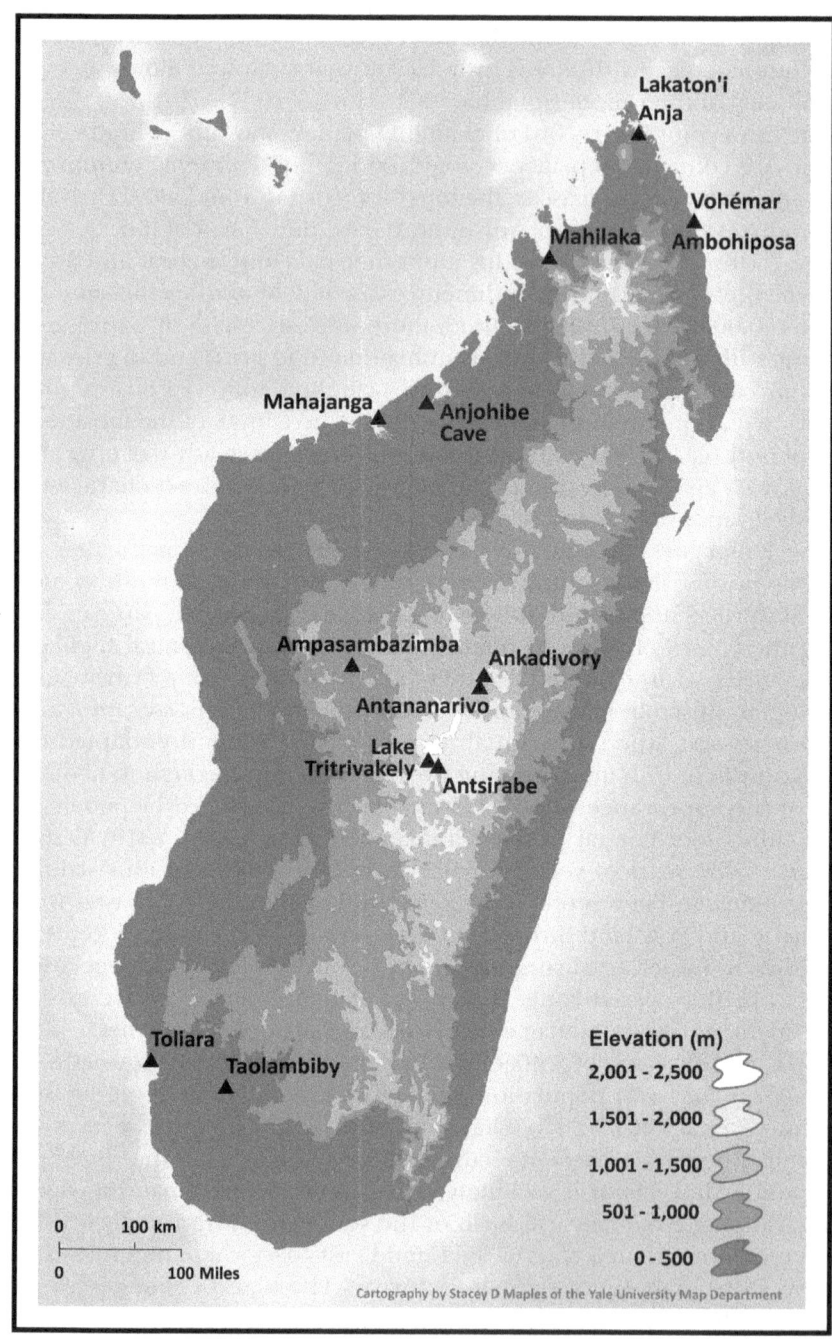

Figure 3.1 Location of key palaeontological and archaeological sites

rain fall – some years very dry, others much wetter than normal. In the moist regions of the east, yearly precipitation varies less, but the timing, or seasonality, of the rain is unpredictable. The causes of this variability are complex, but clearly relate to large-scale geophysical processes in the Indian and Pacific Oceans established several million years ago.

A second contributor to climatic variability is that Madagascar is almost perfectly centered on the pathway followed by violent Indian Ocean cyclones (Mavume et al., 2009), which bring very high winds and torrential downpours. Cyclones hit some places on Madagascar about twice a year, triggering massive floods and landslides, stripping trees of leaves and fruit, and destroying crops in the fields.

In short, leaving aside the arrival of people, the modern plants and animals of Madagascar have had to adapt to the dramatic climatic changes of the Ice Age and to warmer but still oscillating conditions thereafter, as well as deal with the unusual degree of unpredictability that characterizes the island. Their primary response to climatic oscillations seems to have involved retreating to low-elevation refuges near the coasts during the coldest periods, with expansions to higher elevations in the interior during the warmer periods (Wilmé et al., 2006). This pattern of expansion 'inward and upward' when climate conditions permitted is the best explanation for lemur subfossils found at the famous site of Ampasambazimba, near Lac Itasy in the central highlands. At 1,100m elevation, it is unlikely any lemurs were present 15,000 years ago, and none are found today, but during the Holocene the bones of at least 15 genera of lemurs accumulated there. Pollen from the site suggests they lived in a mosaic of bushland, woodlands, and savannah (MacPhee et al., 1985).

The extremely unpredictable nature of Madagascar's environments presents a different kind of evolutionary challenge to plants and animals. Biologists studying Madagascar's endemic mammals have described unusual patterns of reproduction and longevity, and these life-history traits may be a response to this challenge (Dewar and Richard, 2007). For example, some lemurs seem to cope with year-to-year variability by limiting their investment in reproduction each year, and compensating for that by extended life spans. This pattern, known as 'bet-hedging', may also have characterized some of the now-extinct giant lemurs (Catlett et al., 2010).

Human settlement in Madagascar

The warming of the Holocene expanded opportunities for many of the native species of Madagascar. The Holocene also saw the arrival of people of diverse backgrounds, and a suite of animals that people brought with them, by intention or not. All these new arrivals had to adapt to the environments they found, and their activities affected, and continue to affect, Madagascar's varied ecosystems and native species. Just as climates and landscapes vary dramatically across Madagascar, so also the timing of human occupation, local population histories, and the nature of economic systems varied. In consequence, it is likely that the role of people in shaping environmental histories varied substantially across the island too. The past 30 years have seen a major increase in archaeological and palaeoecological research aimed at elucidating these issues.

The first question commonly asked about the Malagasy people is 'Where did they come from?' and the second, 'When did they come?' The brief answer to these questions is that people have been coming to Madagascar for at least a few thousand years, and that they have come from several different places and at different times. Clues to the overseas homelands of arriving Malagasy can be found in their language, genetics, tools, crops and domestic animals, and oral traditions, which point to various places in the Indian Ocean world. Yet the synthesis of all these connections, of the varied cultures, traditions, and economies, almost certainly happened in Madagascar, and it was there that the authentic roots of the Malagasy people were established. Rakotoarisoa (1986, p89) noted that Malagasy people present a certain paradox, whose undoubted unity seems to reside in their undoubted diversity, and Ottino (1974, p12) explained Malagasy culture as a historical amalgam whereby the cultures of immigrants of different origins, at diverse times, were modified both by their interactions and by necessary adaptation to new ecological circumstances.

Malagasy origins: biological and linguistic evidence

The modern scholarly consensus holds that the Malagasy must have had both Asian and African ancestors at least (Allibert, 2008), and indigenous histories from many parts of the island emphasize origins from overseas, often specifically pointing to the Middle East (Beaujard, 1991–1992; Rakotoarisoa, 1998). All Malagasy groups sampled in recent genetic studies have Asian and African ancestors, and the genetic data suggest that Asian and African immigrants included roughly equal numbers of men and women (Hurles et al., 2005; Forster et al., 2008; Ricaut et al., 2009; Tofanelli et al., 2009; Razafindrazaka et al., 2010). Modern Malagasy are quite variable in appearance, with some individuals and some groups bearing a strong resemblance to people from Southeast Asia, while others look more like East Africans. This contrast is reflected in genetic studies, with groups differing in the relative proportions of Asian and African ancestors.

Today, the people of Madagascar overwhelmingly speak Malagasy, though with significant dialectical differences. Malagasy is an Austronesian language, its closest relationships with Barito Valley languages of Borneo (Adelaar, 2009). In the early seventeenth century, however, people along Madagascar's west coast spoke an African language while an Austronesian language was spoken in the rest of the island (Grandidier et al., 1903–1920, pp21–22). The former would have been a Bantu group language related to Swahili; the latter an early form of Malagasy. Blench (2007) has called attention to early twentieth-century reports of languages spoken in the southwest, which were apparently neither Austronesian nor Bantu in origin, and which he believes were likely related to non-Bantu African languages. Today, languages of African origin have essentially disappeared, although there are small west coast communities where Makoa, a language of Mozambique, is still spoken and a village on Nosy Be where, at least until the 1980s, Swahili was a second language (Gueunier, nd, p5).

Lexical borrowings into Malagasy come from many sources. The most obvious are recent, from French and English. Older borrowings are from Arabic, Swahili, Malay, Javanese, and languages of South Sulawesi (Adelaar, 1995; Beaujard, 2003). Phonological, morphological, and syntactic changes in Malagasy indicate strong interactions with a Swahili language (Dahl, 1988; Adelaar, 2009). These kinds of changes are evidence of a long period of close contact between speakers of early Malagasy and those who spoke a Swahili language. While some scholars have proposed that this period of contact may have occurred in East Africa, others prefer to argue that it occurred on Madagascar. Malagasy has been written in Arabic script in the southeast of the island from at least the sixteenth century (Beaujard, 1998), and this is almost certainly related to the very long history of Arab merchants visiting the island for trade.

Archaeological evidence of the earliest people on the island

Until recently, syntheses of the prehistory of Madagascar, including my own, were very clear that Madagascar was settled 1) no more than about 2,000 years ago, perhaps less, and 2) by people already regularly employing iron tools, since no stone tool industry had ever been found. In collaboration with Chantal Radimilahy, Henry Wright, and other colleagues, our recent discoveries have shown that both of these 'facts' of Malagasy prehistory are wrong (Dewar et al., 2013). My purpose in this section is not only to present a brief summary of these discoveries, but also to illustrate the role played by chance and serendipity in exploring the past.

The first site with stone tools we found was a very small rock shelter just east of Vohémar, on a hill known as Ambohiposa – 'fossa [i.e., *Cryptoprocta ferox*, the largest native carnivoran] hill' (Dewar et al., 2013). It was a moment of luck and good archaeological instincts combined, as Henry Wright climbed the hill looking for caves, while the rest of the team waited for the return of a man we sought to interview. With only a few minutes to work, he collected a small soil sample. Back at the Museum in Antananarivo, Chantal Radimilahy set in motion a meticulous process of screening and examination, which showed that the sample contained small flakes of stone. Subsequent excavation revealed a shallow deposit that yielded a small number of stone tools and flakes, of a variety of types of stone that must have been carried to the cave by people, because they do not occur there naturally. The stone items are tiny, and were only recovered thanks to the painstakingly intensive screening technique: all the excavated sediment was screened through window screen, and the residue was carefully washed and then sorted under magnification. The two radiocarbon dates for the tool-bearing deposit, when calibrated, fall in the tenth and thirteenth centuries AD. These are contemporary with the two oldest phases of village sites in the Vohémar region. We searched for additional rock shelters and caves in the Vohémar region that might yield more stone tools, but were unsuccessful.

In 2011, we revisited the site of Lakaton'i Anja in the Montagne des Français, southeast of the Bay of Antsiranana (formerly Diego-Suarez). Lakaton'i Anja was first excavated in 1986 (Dewar and Rakotovololona, 1992). From that work, it was clear that there were at least two periods of occupation: an upper layer with abundant imported ceramics from the Near East and East Asia, and concordant radiocarbon dates that firmly established it as roughly twelfth–thirteenth century AD. Below this was a layer that yielded some bone, and very few ceramics of any sort, for which there were radiocarbon dates that extended back to roughly the fourth–sixth centuries AD. This was and remains the oldest known archaeological site on Madagascar. Lakaton'i Anja is a commodious rockshelter with a level, sandy floor, quite close to a permanent stream, and with very deep deposits. We returned in 2011 because we suspected that the recovery techniques employed in the 1980s would not have found stone tools such as those we found at Ambohiposa.

The 2011 excavations, employing intensive techniques, yielded stone tools from both the upper layers and from deeper deposits (Dewar et al., 2013). The tools are similar to those found at Ambohiposa, though the types of rock employed are different, probably reflecting differences in local availability. The excavated layers were dated by optically stimulated luminescence, which dates the last exposure of hundreds of individual sand grains to sunlight. Tools were found as deep as layers dated to about 2000 BC. In 2012, we returned to the site for further excavations, but the analyses are not yet complete.

The stone tool assemblages from Ambohiposa and Lakaton'i Anja are technologically very similar. The tools were made either by retouching microblades previously detached from a prepared core, or by a very expedient 'smash and grab' bashing of cores to yield irregular flakes with sharp edges for immediate use and discard. These assemblages bear a reasonable resemblance to some East African Late Stone Age (early to middle Holocene) assemblages; they are completely unlike any known contemporary stone tool assemblages from Southeast Asia.

At present, we can say that foragers were visiting rock shelters in northern Madagascar from at least 2000 BC. Most likely these foragers had their origin in East Africa. But this is on the basis of just two sites at which stone tools have been found; many more sites will have to be discovered and excavated in order to establish when people began coming to the island, how long they occupied particular sites, and how widespread these early forays were. It would be a mistake to conclude that northern Madagascar was the earliest area settled until sites in other areas have been carefully examined. Further data are also needed to tell us more about the activities of those who arrived. Hopefully, more sites will soon be discovered now that archaeologists are aware that they may find stone tools.

Other clues to early arrivals

Three other types of evidence may bear upon the timing of the earliest arrival of people in Madagascar: paleontology, oral tradition, and linguistics. Perez et al. (2003) reported a radiocarbon date that calibrates to 402–204 BC on a 'cut-marked' bone of an extinct lemur from the paleontological site of Taolambiby, located near the Onilahy River of the arid southwest. Gommery et al. (2011) described possibly cut-marked hippo bones associated with a calibrated date of 2288–2035 BC from Anjohibe Cave along the northwest coast north of Mahajanga. The Taolambiby bone was recovered in the early twentieth century and has no stratigraphic context; the damage has not been described and an unknown preservative contaminated the bone (Perez et al., 2005). The Anjohibe bone is from an assemblage previously described as the result of a catastrophic death of a herd of hippos (Burney et al., 1997). In neither case have butchery tools or other evidence of human presence been found, and the interpretation of these paleontological finds remains uncertain and inconclusive.

Equally tantalizing are the oral histories from many areas in Madagascar that refer to a group of original inhabitants, often called the *vazimba*. Although a century of scholarship surrounds the *vazimba*, it remains impossible to draw definitive conclusions from the diverse, and sometimes clearly mythic, traditions reported and discussed. Language, the potential third thread tugging us back into the distant past, yields only a hint: Blench (2007) suggests that the non-Bantu African language he detects in records from the southwest may be a vestige of early arrivals in that region.

Early agriculture, pastoralism, and trade in prehistory

Archaeology provides a well-documented record of villages and hamlets spreading and diversifying across the island from about AD 700, with the first European accounts of the island beginning about AD 1500. The period between AD 700 and 1500 saw the arrival of humans who brought with them an array of domesticated plants and animals, and commensals such as rats and mice; the widespread use and manufacture of iron tools; the establishment of ports in Madagascar as part of the Indian Ocean trade network; the building of the first city; and the establishment of human settlements across the island. From the perspective of an ecological historian, virtually the full spectrum of human-linked agents of environmental change was introduced during these centuries: cattle, goats, rats, swidden agriculture, and dense human population concentrations, to list the most obvious. For a cultural historian, these centuries saw the widespread establishment of the Malagasy language, implying one or more migrations from Southeast Asia, as well as a more geographically limited Bantu-speaking population with roots across the Canal of Mozambique. These movements also signaled the development of trade with the Swahili Coast of East Africa with merchants from the medieval Middle East, and possibly with Southeast Asia. From an economic perspective, this was the period when Madagascar developed distinctive regional economies.

These changes happened not as the result of some slow process of in situ cultural evolution or development, as for example in the Near East or China; rather, they were imports by peoples already accustomed to fishing, farming, herding, forging iron, and the life of cities. Current archaeological data cannot precisely date all of these arrivals and adaptations to local circumstances, but at least the broad outlines are clear (Dewar and Wright, 1993).

The oldest villages currently identified date to about AD 700 on the Northeast coast near Mananara and on Nosy Be (Wright and Fanony, 1992). Their occupants were likely swidden farmers who would also have exploited native plants and animals. Over the next 500 years, villages and hamlets are recorded around the coasts and hinterlands of Madagascar in every area that has been archaeologically surveyed: in the extreme north, along the east coast, in the southeast near Fort Dauphin, in the arid spiny bush of the deep south, and along the west coast. The economic base of many of these villages was farming. The crops grown are largely unknown, but must have been drawn from the array of Old World tropical cultigens: rice, millets, taro, bananas, coconuts, and the greater yam (Beaujard, 2010). In the south, there were coastal settlements of maritime fishermen who also hunted native species, possibly including the now-extinct elephant birds. More striking than these coastal sites, in the southern interior were the large (20–30 ha), walled proto-urban *manda* villages of the tenth–thirteenth centuries whose occupants had herds of cattle and sheep/goats, and hunted tenrecs (Parker Pearson, 2010). Meanwhile, the 60ha city of Mahilaka, likewise enclosed by a wall, was flourishing in the eleventh and twelfth centuries on the northwest coast (Radimilahy, 1998).

Clear signs indicate that at the same time people were establishing villages with economies suited to local conditions, Madagascar maintained close links with the world of the Indian Ocean. Imported ceramics of the eleventh to thirteenth centuries from the Near East and China have been found in large sites all the way around the coast. They are most common in the north, and especially at Mahilaka, which was certainly a major trading port. In fact, no contemporary sites on the Swahili Coast of East Africa are as large as Mahilaka. The first appearance of the black rat (*Rattus rattus*) and house mouse (*Mus musculus*) provides evidence of a different kind of link with the rest of the world: remains of these ubiquitous commensals have been found at Mahilaka, signaling their arrival in Madagascar between the eleventh and fourteenth centuries, and possibly earlier (Aplin et al., 2011). Malagasy house mice are closely related to the mice of Oman: inadvertent transport by mariners traveling between Madagascar and the Near East seems very likely (Duplantier et al., 2002).

Near the end of this period, in the thirteenth century, villages were established in the central highlands of Madagascar. The oldest known, Ankadivory, northeast of Antananarivo, is a small community of farmers who kept cattle, and probably grew rice in the nearby marshes (Rakotovololona, 1993).

After AD *1500*

In terms of ecological transformation, the period after AD 1500 is important for two developments. First, in the central highlands there was substantial and fairly continuous population growth, best demonstrated in the area just northeast of Antananarivo (Wright, 2007). The establishment of large areas of very productive irrigated rice fields supported this growth. It is the most densely populated region of Madagascar today, as it seems to have been for the past few hundred years. There is little evidence of sustained population growth in coastal areas after AD 1500. Particularly along the east coast, many settlements were abandoned as people retreated inland, often to villages apparently sited with an eye to defense. Beginning in the sixteenth century, Madagascar was a source of slaves for the Mascareignes and the Cape Colony in South Africa, and it is likely that slave-raiding was common. One consequence of this cataclysmic development was the abandonment by farmers of areas they had already brought into cultivation and the establishment of new fields in safer areas inland and upland, often on steeper slopes. The human ecological footprint likely spread into the interior as a result of this population shift more quickly than would otherwise have happened.

European records of visits to Madagascar's coasts begin just after AD 1500, and are often focused on the ports of the northwest coast. These accounts make clear that rice, cattle, and slaves were among the most important exports, as they would continue to be well into the nineteenth century. Individually, the northwest coast ports had relatively limited life spans: different ports dominated during each century and important ports of one century were often abandoned in the next (Vérin, 1986). Competition between ports was undoubtedly intense, for the revenues derived from controlling trade were the foundation of all the important polities of the time, but environmental degradation in the immediate hinterlands of these port cities may have played a role too. Mahilaka was largely abandoned after the fourteenth century and local environmental degradation has been suggested as a cause (Wright et al., 2005).

Trading ports were important along the east coast as well. Vohémar, the first archeological site reported in Madagascar, was reputed for its fifteenth- and sixteenth-century tombs containing many imported goods, and especially East Asian ceramics (Vérin, 1986). Writing of his experiences in Madagascar around AD 1650, Flacourt mentioned Vohémar as a center for trade in gold (1661 [1995, p134]), but he did not visit it. In the 1660s, the French established a trading center at Fenérive seeking, with some success, cattle and rice to supply their embattled colony at Fort Dauphin (Martin, 1990). Overseas traders sought cattle, beef, hides, rice, and slaves from Tamatave, Foulepoint, and Fenérive along the east coast well into the nineteenth century. Efforts to control the trade were among the most important drivers of the political and military alliances and conflicts of this period.

Along with the traders came a suite of New World crops: corn [maize], sweet potatoes, manioc, and tobacco, to name only the most important. In return, some evidence indicates that North Carolina rice farming began with the import of Malagasy seed.

The distribution of people across the island was markedly uneven during this momentous and in some ways calamitous period of growth. Although few archaeological sites are as big as 5 ha in size before AD 1400, even in settled areas, the exceptions are extreme. The arid south had 20–30 ha walled villages from the tenth to thirteenth centuries, and the port city of Mahilaka, 60 ha in extent, had an estimated population of 5,000–10,000. This unequal distribution continues today, with many areas of the island still sparsely populated.

Human impact on the environment: the data

The reality of anthropogenic change in Malagasy landscapes is evident to the eye: clearings where forest has been turned into fields; stacks of hardwood logs awaiting export; bags of charcoal for sale along roadsides; fire licking across grasslands; *lavaka* (erosion gullies) lacerating hillsides in the highlands; fields of imported crops, and woodlands of Mexican pines and Australian eucalyptus; and cattle and goats seemingly everywhere. Much of the change in vegetative cover is rapid and dramatic, easily observed over the past 40 years through satellite imagery (Harper et al., 2007).

Modern observations and processes are not uniformly good guides to past anthropogenic effects, however, and retrospective inference from the present is a weak foundation for understanding the past. Three readily observable phenomena illustrate this point. First, fires set by people are a scourge of many modern landscapes in Madagascar, yet fire has a deep history in some parts of the island, long pre-dating human presence (Gasse and Van Campo, 2001). Second, all grasslands in Madagascar have been viewed as 'unnatural' landscapes, created by people burning and clearing forests for pasture and fields as happens today. Yet close analysis of grassland plants and animals suggests that the blanket characterization of these communities as biologically impoverished and recent artifacts is overdrawn. Pollen records indeed show that grasslands have a long history in some parts of the island, and likely invaded Madagascar millions of years ago as part of a worldwide expansion of grassy biomes (Bond et al., 2008; Willis et al., 2008). Finally, although some *lavaka* can be directly linked to recent human activity, *lavaka* were widespread in the central highlands at least a thousand years ago (Cox et al., 2009) and some are many thousands of years old, well pre-dating the arrival of people (Wells and Andriamihaja, 1993).

Flawed understandings and reconstructions of the past can be as misleading as inference from the present. Many popular discussions contrast modern landscapes with the 'original vegetation' of Madagascar as part of a narrative in which people arrived on the island, destroyed the forest, and thereby unleashed a wave of extinctions. This is a problematic way to frame the discussion, in light of mounting palaeoecological evidence of landscapes in continuous, if usually slow, change. The phrase 'original vegetation' is commonly used to refer to the vegetative cover at the very beginning of human activity in Madagascar, but this implies a more confident understanding of the early period of Malagasy prehistory than we currently possess. A related usage links the term to the vegetation present at the time when, according to palaeoecological evidence, important anthropogenic changes started to occur. This requires clear criteria for distinguishing anthropogenic effects from those caused by climatic change, and palaeoecologists differ among themselves on these (contrast Burney, 1999 and Virah-Sawmy et al., 2010). I conclude that 'original vegetation' is not a useful concept (see also Chapter 4 by McConnell and Kull for a discussion of debates over recent forest cover change and the problem with the concept of 'original forest cover' in estimates of forest loss).

Over the past few decades, palaeoecological and archaeological research has provided new insights and shaped debates about the impact of people on Madagascar's landscapes. This research has focused on changes in vegetative cover, the role played by people, and how habitat destruction and fragmentation along with hunting precipitated the extinction of so many species of animals. A growing consensus holds that these effects interacted in complex ways, locally and regionally shaped by the island's diverse environments, and that the nature of their impact on animal species would also have varied (Dewar, 1997; Burney, 1999; Crowley, 2010; Virah-Sawmy et al., 2010). I draw upon a cross-section of studies to illustrate this point in relation to the earlier phases of human settlement outlined in previous sections, considering first vegetation change and then the extinctions.

Vegetation change

The most commonly proposed anthropogenic causes of changes in vegetative cover in the past are: 1) changes in fire frequency and pattern either because of clearance for fields, management of pastures, or as an indirect result of changes in herbivore communities; 2) the transformation of natural communities into agricultural fields; 3) changes in Madagascar's herbivore community by the elimination of native species (ratites, hippos, giant tortoises, terrestrial lemurs) and their replacement by cattle, sheep, and goats; and 4) the introduction of exotic plant species. Here, I focus on the first two, about which we have the most information.

Until recently, it was supposed that the first people on Madagascar imported fire, and the result was a gigantic conflagration utterly destructive to a forested but fragile landscape (e.g. Morat, 1973, p192). That view now appears wrong for at least two reasons. First, the palaeoecological research of the last quarter century makes clear that periodic fires have been an important element of many Malagasy ecosystems for tens of thousands of years. Burney (1987b) showed that fire was frequent throughout the past 10,000 years at Tritrivakely, and that it was common on Montagne d'Ambre in the north about 35,000 years ago, and highly variable there between 18,000 and 20,000 years ago. Interestingly, at Tritrivakely, the highest concentrations of charcoal are from the Holocene, 4,000–10,000 years ago, followed by a decline for the next 3,000 years, and then by a resurgence for the last 1,000 years. Change in the rate of charcoal deposition in pollen cores is the most commonly employed index of the onset of human activity (Burney, 1999; Burney et al., 2003, 2004), and this resurgence has been attributed to human activity. The linkage is quite plausible in my view, although it is rarely possible to make tight chronological correlations between cores or with the archaeological record, and so it is difficult to determine clearly the roles played by climatic and anthropogenic drivers. Note too that the concentrations of charcoal in recent centuries never reached those that rather uniformly characterized the early Holocene.

A second reason for rejecting the 'gigantic conflagration' view is that Malagasy plant formations are not uniformly vulnerable to fire. At Virah-Sawmy et al.'s (2009) sites, all located within about 50 km, one site had regular intervals of fire throughout the past 6,000 years; another showed little evidence of fire at any point; and a third showed a substantial episode of fire only for a few centuries centered around the fifteenth century, which was almost certainly the result of human activity.

Replacing 'gigantic conflagration' is evidence that specific vegetational changes over the past 2,000 years have many causes, some related to pastoralism, some to the introduction of crops and fields, some to forestry and logging, and some to substantial environmental degradation in the vicinity of high populations, such as Mahilaka. Much research is still needed to piece together accurate, place-specific accounts, and major puzzles remain. For example, the French slaver and trader Nicolas Mayeur described life in the interior of Madagascar more than a century before the colonial period (1785 [1913]). The landscape he described in the highlands in the 1780s and 1790s was not very different to what was observed around 1900. If that landscape is largely anthropogenic, the anthropogenic cause is quite deeply buried in the past, with little archaeological evidence as yet of human population levels or activities commensurate with such a total transformation of a huge area.

Animal extinctions

During the Holocene, more than 40 species of mammals, reptiles, and birds became extinct in Madagascar (Goodman and Benstead, 2003). All of these species are known from palaeontological sites and they are usually referred to as subfossils, since their bones are so young that they have not mineralized. Most famous are the 16 species of extinct subfossil lemurs, all as large, and often much larger than any surviving lemurs; the three species of pygmy hippopotamus; a half dozen or so species of elephant birds assigned to *Aepyornis* and *Mullerornis*; and two giant tortoises (*Dipsochelys*). Together, these have been described, with justice, as Madagascar's extinct megafauna. Many of these species were active by day, more terrestrial than surviving native species, and they were overwhelmingly grazers or browsers. The guild of native terrestrial herbivores of Madagascar was essentially wiped out with these extinctions. The subfossil lemurs were more frequently arboreal than terrestrial, and thus the guild of arboreal herbivores was significantly reduced as well.

Bones of most of the extinct species have been dated by radiocarbon (Crowley, 2010), and many of these survived after AD 0, and in some cases until about the fourteenth or fifteenth century. Flacourt, writing of his experiences in Fort Dauphin in the mid-seventeenth century, presented a list of native animals, some recognizable modern species, some clearly fantastical, and a few that may be now only known palaeontologically: the *tretretre* may have been a giant lemur and the *vouronpatra* an ostrich-like bird (Flacourt, 1661 [1995, pp219–229]). Twentieth-century descriptions of large animals by rural populations may also refer to extinct species, including a pygmy hippopotamus (Godfrey, 1986; Burney and Ramilisonina, 1998). These records should not be quickly dismissed, but they are difficult to interpret. In fact, establishing a date for final extinction of any species is difficult, since one must be sure that all viable habitats are examined. Using radiometric dates on bones does not eliminate the difficulty since relatively few individuals of any species are buried in places where their bones are preserved, far fewer are actually found, fewer still dated, and it is unlikely that these will include bones of the last surviving individuals.

The sites where subfossils have been found are not well distributed among the different regions of the island. Apart from one coastal site with hippos, there are no palaeontological or archaeological sites with subfossil remains in any of the humid forests of the east coast (see Figure 3.1). As a result, for significant regions of the island, we have little or no information about the species composition of their natural communities, let alone evidence as to the timing and causes of extinctions. The best-sampled areas are the central highlands (particularly near Antsirabe and Lac Itasy), the arid south and southwest (particularly along the coast from Morondava to Fort Dauphin), and the extreme north and northwest (Anjohibe, the Ankarana, and Montagne des Français). All discussions of the timing and ecological impact of the extinctions of the subfossils has to be limited to these areas, and the biological history of unsampled areas may be different. Equally, it cannot be assumed a priori that all extinct species disappeared simultaneously (as a 'wave'), or for the same reasons. Table 3.1 presents the latest calibrated radiocarbon date on bone available for each subfossil mammal, reptile, and bird species, except the ratites, which are discussed below.

The ratites, the giant ostrich-like elephant birds of Madagascar, are assigned to a disputed number of species in two genera that differ in size: *Aepyornis* (which may have weighed 400–500 kg) and the smaller (rhea or emu sized) *Mullerornis*. Their bones are quite common in some subfossil sites, and particularly in ancient marshes of the central highlands. Fragments of the eggs (and rarely whole ones) are extremely common on southwestern coastal beaches. There are only three radiocarbon dates on ratite bones, the youngest being AD 661–961 from a site near Belo-sur-Mer (Crowley, 2010). By contrast, there are more than two dozen dates on ratite eggshell (Crowley, 2010; Parker Pearson, 2010), including some found in human occupation sites in the south. Interpreting ratite eggshell dates is difficult for two reasons. First, eggshell fragments are extremely resistant to decay or dissolution and may persist apparently unaltered for tens of thousands of years as a distinctive (and individually datable) part of natural sediment. Second, it seems likely that ostrich eggshell 'dates' in South Africa are about 180 ± 120 years older than apparently contemporary charcoal dates (Vogel et al., 2001) and *Aepyornis* eggshell dates in southern Madagascar show an even stronger tendency to be too early: 740 ± 125 years (Parker Pearson, 2010, p88). Parker Pearson proposes that ratite eggs were collected by humans in the south from about the tenth to fourteenth centuries AD and that after AD 1400 there were likely few surviving ratites (2010, p98).

On balance, taking into account the uncertainties and the difficulties of using radiocarbon dates to estimate extinction dates, it seems likely that the period of AD 500–1500 saw a substantial reduction in the number of native animal species, and their average size, in the regions for which we have data: the south, southwest, northwest, north, and central highlands.

Conclusion

For some, destruction of the simple story told for so long about Madagascar's past may be a disappointment, but for many – and certainly for me – the research of recent decades is opening up much more interesting and exciting vistas. I would not have become an archaeologist if I had not believed that studying the past is of value in its own right. In addition, insights from

Table 3.1 Youngest radiocarbon determination for extinct species in Madagascar

Species				Last calibrated date	Number of specimens dated
Mammals	Lemurs		*Archaeoindris fontoynanti*	402–167 BC	2
			Archaeolemur edwardsi	**AD 988–1177**	12
			Archaeolemur majori	**AD 711–892**	24
			Babakotia radofilai	3327–2874 BC	1
			Hadropithecus stenognathus	**AD 544–874**	8
			Megaladapis edwardsi	**AD 1296–1487**	12
			Megaladapis grandidieri	AD 980–1177	1
			Megaladapis madagascariensis	**AD 423–561**	14
			Mesopropithecus globiceps	AD 437–643	4
			Mesopropithecus pithecoides	AD 607–771	1
			Pachylemur insignis	**AD 715–988**	17
			Pachylemur jullyi	**AD 682–874**	8
			Palaeopropithecus ingens	**AD 1316–1628**	31
			Palaeopropithecus maximus	2462–2201 BC	2
	Carnivorans		*Cryptoprocta spelea*	**AD 552–647**	9
	Artiodactyls		*Hippopotamus lemerlei*	**AD 778–969**	22
	Rodents		*Hypogeomys australis*	AD 442–651	3
	Bibymalagasy		*Plesiorychteropus* sp.	350 BC–AD 4	1
Birds			*Alpochen sirabensis*	AD 550–895	4
			Centrornis majori	15603–15243 BC	1
			Coua primavea	52 BC–AD 240	1
Reptiles			*Dipsochelys abrupta*	AD 654–1952	1
			Dipsochelys grandidieri	AD 688–970	1
			Dipsochelys sp.	**AD 435–652**	13

Notes: all dates are calibrated at 2 σ with Oxcal 4.1 (Ramsey et al., 2010) and southern hemisphere atmospheric curve (McCormac et al., 2004). Dates are from Crowley (2010). Taxa listed in bold have at least eight available dates.

Madagascar's long and dynamic environmental history, and the complicated history of how people settled and interacted with the island's landscape, have a real bearing on its future (Dewar and Richard, 2012). Efforts to trace these complex dynamics will surely continue, as they must. Knowledge of the massive environmental disruptions over the past millennium must be accompanied by acknowledgment of the continuing reverberations of societal traumas experienced during the past 150 years (Graeber, 2007; Randriamamonjy, 2009). Different in their causes and consequences, the historic perturbations of Malagasy nature and society have helped shape the present challenges and affect the capacity of society to respond. As I have tried to emphasize in this chapter, it is important to undertake these studies with proper humility, recognizing the complexity of social and environmental systems, the incompleteness of our evidence of both the present and the past, the limitations of the tools we possess for exploring them, and the continuing role that happenstance and luck play in all of this. With diligent and rigorous research, Madagascar's past has much to teach us.

References

Adelaar, A. (1995) 'Malay and Javanese loanwords in Malagasy, Tagalog and Siraya (Formosa)', *Bijdragen tot de Taal-, Land en Volkenkunde*, vol 150, pp50–65.

Adelaar, A. (2009) 'Towards an integrated theory about the Indonesian migrations to Madagascar', in P. Peregrine, I. Peiros and M. Feldman (eds) *Ancient Human Migrations*, University of Utah Press, Salt Lake City.

Allibert, C. (2008) 'Austronesian migration and the establishment of the Malagasy civilization: contrasted readings in linguistics, archaeology, genetics and cultural anthropology', *Diogenes*, vol 218, pp7–16.

Aplin, K. P., Suzuki, H., Chinen, A. A., Chesser, R. T., ten Have, J., Donnellan, S. C., Austin, J., Frost, A., Gonzalez, J. P., Herbreteau, V., Catzeflis, F., Soubrier, J., Fang, Y. P., Robins, J., Matisoo-Smith, E., Bastos, A. D. S., Maryanto, I., Sinaga, M. H., Denys, C., Van Den Bussche, R. A., Conroy, C., Rowe, K. and Cooper, A. (2011) 'Multiple geographic origins of commensalism and complex dispersal history of black rats', *PLoS One*, vol 6, no 11, pp26257–26357.

Beaujard, P. (1991–1992) 'Islamés et systèmes royaux dans le sud-est de Madagascar : les exemples Antemoro et Tanala', *Omaly sy Anio*, vols 33–36, pp235–286.

Beaujard, P. (1998) *Le Parler Secret Arabico-Malgache du Sud-est de Madagascar: Recherches Etymologique*, L'Harmattan, Paris.

Beaujard, P. (2003) 'Les arrivées austronésiennes à Madagascar: vagues ou continuum? (partie 1)', *Etudes Océan Indien*, vols 35–36, pp59–128.

Beaujard, P. (2010) 'Océan Indien, le grand carrefour', *L'Histoire*, vol 355, pp30–35.

Blench, R. M. (2007) 'New palaeozoogeographical evidence for the settlement of Madagascar', *Azania*, vol 42, pp69–82.

Bond, W. J., Silander, J. A. Jr., Ranaivonasy, J. and Ratsirarson, J. (2008) 'The antiquity of Madagascar's grasslands and the rise of C4 grassy biomes', *Journal of Biogeography*, vol 35, pp1743–1758.

Burney, D. A. (1987a) 'Late Holocene vegetational change in central Madagascar', *Quaternary Research*, vol 28, pp130–143.

Burney, D. A. (1987b) 'Late Quaternary stratigraphic charcoal records from Madagascar', *Quaternary Research*, vol 28, pp274–280.
Burney, D. A. (1997) 'Theories and facts regarding Holocene environmental change before and after human colonization', in S. M. Goodman and D. B. Patterson (eds) *Natural Change and Human Impact in Madagascar*, Smithsonian Press, Washington DC.
Burney, D. A. (1999) 'Rates, patterns, and processes of landscape transformation and extinction in Madagascar', in R. D. E. MacPhee (ed.) *Extinction in Near Time*, Kluwer/Plenum, New York.
Burney, D. A., Burney, L. P., Godfrey, L. R., Jungers, W. L., Goodman, S. M., Wright, H. T. and Jull. A. J. T. (2004) 'A chronology for late prehistoric Madagascar', *Journal of Human Evolution*, vol 47, pp25–63.
Burney, D. A., James, H., Grady, F., Rafamantanantsoa, J.-G., Ramilisonina, Wright, H. T. and Cowart, J. B. (1997) 'Environmental change, extinction and human activity: evidence from caves in NW Madagascar', *Journal of Biogeography*, vol 24, pp755–767.
Burney, D. A. and Ramilisonina (1998) 'The Kilopilopitsofy, Kidoky, and Bokyboky: accounts of strange animals from Belo-Sur-Mer, Madagascar, and the megafaunal "extinction window"', *American Anthropologist*, vol 100, no 4, pp957–966.
Burney, D. A., Robinson, G. S. and Burney, L. P. (2003) 'Sporormiella and the late Holocene extinctions in Madagascar', *Proceedings of the National Academy of Sciences*, vol 100, no 19, pp10800–10805.
Catlett, K. K., Schwartz, G. T., Godfrey, L. R. and Jungers, W. L. (2010) '"Life history space": a multivariate analysis of life history variation in extant and extinct Malagasy lemurs', *Amererican Journal of Physical Anthropology*, vol 142, pp391–404.
Cox, R., Bierman, P., Jungers, M. C. and Rakotondrazafy, A. F. M. (2009) 'Erosion rates and sediment sources in Madagascar inferred from ^{10}Be analysis of lavaka, slope, and river sediment', *Journal of Geology*, vol 117, no 4, pp363–376.
Crowley, B. E. (2010) 'A refined chronology of prehistoric Madagascar and the demise of the megafauna', *Quaternary Science Reviews*, vol 29, pp2591–2603.
Dahl, O. C. (1988) 'Bantu substratum in Malagasy', *Etudes Océan Indien*, vol 9, pp91–132.
Dewar, R. E. (1997) 'Were people responsible for the extinction of Madagascar's subfossils, and how will we ever know?', in S. M. Goodman and B. D. Patterson (eds) *Natural Change and Human Impact in Madagascar*, Smithsonian Institution Press, Washington, DC.
Dewar, R. E. and Rakotovololona, H. F. S. (1992) 'La chasse aux subfossiles: les preuves du XIeme siècle au XIIIeme siècle', *Taloha*, vol 11, pp4–15.
Dewar, R. E. and Richard, A. F. (2007) 'Evolution in the hypervariable environment of Madagascar', *Proceedings of the National Academy of Sciences*, vol 104, no 34, pp13723–13727.
Dewar, R. E. and Richard, A. F. (2012) 'Madagascar: a history of arrivals, what happened, and will happen next', *Annual Review of Anthropology*, vol 41, pp495–517.
Dewar, R. E. and Wright, H. T. (1993) 'The culture history of Madagascar', *Journal of World Prehistory*, vol 7, pp417–466.
Dewar, R. E., Radimilahy, C., Wright, H. T., Jacobs, Z., Kelly, G. O. and Berna, F. (2013) 'Stone tools and foraging in northern Madagascar challenge Holocene extinction models', *Proceedings of the National Academy of Sciences*, vol 110, 12583–12588.

Duplantier, J. M., Orth, A., Catalan, J. and Bonhomme, F. (2002) 'Evidence for a mitochondrial lineage originating from the Arabian peninsula in the Madagascar house mouse (*Mus musculus*)', *Heredity*, vol 89, pp154–158.

Flacourt, É. de (1661 [1995]) *Histoire de la Grande Ile de Madagascar*, C. Allibert (ed.), INALCO/Karthala, Paris.

Forster, P., Matsumara, S., Vizuette-Forster, M., Blumbach, P. B. and Dewar, R. E. (2008) 'The genetic prehistory of Madagascar's female Asian lineages', in P. Forster, S. Matsumara and C. Renfrew (eds) *Simulations, Genetics and Human Prehistory*, MacDonald Institute, Cambridge.

Gasse, F. and Van Campo, E. (2001) 'Late Quaternary environmental changes from a pollen and diatom record in the southern tropics (Lake Tritrivakely, Madagascar)', *Palaeogeography, Palaeoclimatology, Palaeoecology*, vol 167, pp287–308.

Godfrey, L. R. (1986) 'What were the subfossil indriids of Madagascar up to', *American Journal of Physical Anthropology*, vol 69, no 2, pp205–206.

Gommery, D., Ramanivosoa, B., Faure, M., Guérin, C., Kerloch, P., Sénégas, F. and Randrianantenaina, H. (2011) 'Oldest evidence of human activities in Madagascar on subfossil hippopotamus bones from Anjohibe (Mahajunga Province)', *Comptes Rendus Palevol*, no 10, pp271–278.

Goodman, S. M. and Benstead, J. P. (2003) *The Natural History of Madagascar*, University of Chicago Press, Chicago, IL.

Graeber, D. (2007) *The Lost People: Magic and the Legacy of Slavery in Madagascar*, Indiana University Press, Bloomington, IN.

Grandidier, A., Charles-Roux, J., Delhorbe, C., Froidevaux, H. and Grandidier, G. (1903–1920) *Collection des Ouvrages Anciens Concernant Madagascar*, vols i–ix, Comité de Madagascar, Paris.

Gueunier, N. J. (nd) *Contes de la Côte Ouest de Madagascar*, Karthala, Ambozontany/Antananarivo/Paris.

Harper, G. J., Steininger, M. K., Tucker, C. J., Juhn, D. and Hawkins, F. (2007) 'Fifty years of deforestation and forest fragmentation in Madagascar', *Environmental Conservation*, vol 34, pp325–333.

Hurles, M. E., Sykes, B. C., Jobling, M. A. and Forster, P. F. (2005) 'The dual origin of the Malagasy in Island Southeast Asia and East Africa: evidence from maternal and paternal lineages', *American Journal of Human Genetics*, vol 76, no 5, pp894–901.

MacPhee, R. D. E., Burney, D. A. and Wells, N. A. (1985) 'Early Holocene chronology and environment of Ampasambazimba, a Malagasy subfossil lemur site', *International Journal of Primatology*, vol 6, no 5, pp463–489.

Martin, J. (1990) *L'Empire Triomphant: L'Aventure Coloniale de la France. Maghreb, Indochine, Madagascar, Iles et Comptoirs*, Denoël, Paris.

Mavume, A. F., Rydberg, L., Rouault, M. and Lutjeharms, J. R. E. (2009) 'Climatology and landfall of tropical cyclones in the South-West Indian Ocean', *Western Indian Ocean Journal of Marine Science*, vol 8, no 1, pp15–36.

Mayeur, N. (1785 [1913]) 'Voyage au pays d'Ancove', in Dumaine (ed.) *Bulletin de l'Académie Malgache*, vol xii, 2ᵉ partie, pp13–42.

McCormac, F. G., Hogg, A.G., Blackwell, P.G., Buck, C.E., Higham, T. F. G. and Reimer, P. J. (2004) 'SHCal04 southern hemisphere calibration 0-11.0 cal BP', *Radiocarbon*, vol 46, pp1087–1092.

Morat, P. (1973) *Les Savanes du Sud-ouest de Madagascar*, Office de la Recherche Scientifique et Technique Outre-Mer, Paris.

Ottino, P. (1974) *Madagascar: Les Comores et le Sud-Ouest de l'Océan Indien*, Université de Madagascar, Antananarivo, Madagascar.
Parker Pearson, M. (2010) *Pastoralists, Warriors and Colonists: The Archaeology of Southern Madagascar*, Archaeopress, Oxford.
Perez, V. R., Burney, D. A., Godfrey, L. R. and Nowak-Kemp, M. (2003) 'Box 4: butchered sloth lemurs', *Evolutionary Anthropology*, vol 12, p260.
Perez, V. R., Godfrey, L. R., Nowak-Kemp, M., Burney, D. A., Ratsimbazafy, J. and Vassey, N. (2005) 'Evidence of early butchery of giant lemurs in Madagascar', *Journal of Human Evolution*, vol 49, no 6, pp722–742.
Quéméré, E., Amelot, X., Pierson, J., Crouau-Roy, B. and Chikhi, L. (2012) 'Genetic data suggest a natural prehuman origin of open habitats in northern Madagascar and question the deforestation narrative in this region', *Proceedings of the National Academy of Sciences*, vol 109, no 32, pp13028–13033.
Radimilahy, C. (1998) *Mahilaka: An Archaeological Investigation of an Early Town in Northwestern Madagascar*, Department of Archaeology and Ancient History, Uppsala.
Rakotoarisoa, J. A. (1986) 'Principaux aspects des formes d'adaptation de la société traditionelle Malgache', in C. P. Kottak, J. A. Rakotoarisoa, A. Southall and P. Vérin (eds) *Madagascar: Society and History*, Carolina Academic Press, Durham, NC.
Rakotoarisoa, J. A. (1998) *Mille Ans d'Occupation Humaine dans le Sud-Est de Madagascar*, L'Harmattan, Paris.
Rakotovololona, H. F. S. (1993) 'Ankadivory et la période Fiekena: début d'urbanisation à Madagascar', *Données archéologiques sur l'origine des villes à Madagascar-Mombassa*, Musée d'Art et d'Archéologie, Antananarivo, Madagascar.
Ramsey, C. B., Dee, M., Lee, S., Nakagawa, T. and Staff, R. A. (2010) 'Developments in the calibration and modeling of radiocarbon dates', *Radiocarbon*, vol 52, no 3, pp953–961.
Randriamamonjy, F. (2009) *Histoire de Madagascar, 1895 – 2002*, Trano Printy Fiangonana Loterana Malagasy, Antananarivo, Madagascar.
Razafindrazaka, H., Ricaut, F. X., Cox, M. P., Mormina, M., Dugoujon, J. M., Randriamarolaza, P., Guitard, E., Tonasso, L., Ludes, B. and Crubézy, E. (2010) 'Complete mitochondrial DNA sequences provide new insights into the Polynesian motif and the peopling of Madagascar', *European Journal of Human Genetics*, vol 18, no 5, pp575–581.
Ricaut, F. X., Razafindrazaka, H., Cox, M. P., Dugoujon, J. M., Guitard, E., Sambo, C., Mormina, M., Mirazon-Lahr, M., Ludes, B. and Crubézy, E. (2009) 'A new deep branch of eurasian mtDNA macrohaplogroup M reveals additional complexity regarding the settlement of Madagascar', *BMC Genomics*, vol 10, p605.
Tofanelli, S., Bertoncini, S., Castri, L., Luiselli, D., Calafell, F., Donati, G. and Paoli, G. (2009) 'On the origins and admixture of Malagasy: new evidence from high-resolution analyses of paternal and maternal lineages', *Molecular Biology and Evolution*, vol 26, no 9, pp2109–2124.
Vérin, P. (1986) *The History of Civilization in North Madagascar*, A. A. Balkema, Rotterdam/Boston, MA.
Virah-Sawmy, M., Gillson, L. and Willis, K. J. (2009) 'How does spatial heterogeneity influence resilience to climatic changes? Ecological dynamics in southeast Madagascar', *Ecological Monographs*, vol 79, no 4, pp557–574.
Virah-Sawmy, M., Willis, K. J. and Gillson, L. (2010) 'Evidence for drought and forest declines during the recent megafaunal extinctions in Madagascar', *Biogeography*, vol 37, no 3, pp506–519.

Vogel, J. C., Visser, E. and Fuls, A. (2001) 'Suitability of ostrich eggshell for radiocarbon dating', *Radiocarbon*, vol 43, no 1, pp133–137.
Wells, N. A. and Andriamihaja, B. (1993) 'The initiation and growth of gullies in Madagascar: are humans to blame?', *Geomorphology*, vol 8, no 1, pp1–46.
Willis, K. J., Gillson, L. and Virah-Sawmy, V. (2008) 'Nature or nurture: the ambiguity of C4 grasslands in Madagascar', *Journal of Biogeography*, vol 35, pp1741–1742.
Willis, K. J., Araújo, M. B., Bennett, K. D., Figueroa-Rangel, B., Froyd, C. A. and Myers, N. (2007) 'How can a knowledge of the past help to conserve the future? Biodiversity conservation and the relevance of long-term ecological studies', *Philosophical Transactions of the Royal Society B: Biological Sciences*, vol 362, no 1478, pp175–187.
Wilmé, L., Goodman, S. M. and Ganzhorn, J. U. (2006) 'Biogeographic evolution of Madagascar's microendemic biota', *Science*, vol 312, no 5776, pp1063–1065.
Wright, H. T. (ed.) (2007) *Early State Formation in Central Madagascar: An Archaeological Survey of Western Avaradrano*, no 43, Museum of Anthropology, University of Michigan, Ann Arbor, MI.
Wright, H. T. and Fanony, F., trans. Alibert, C. (1992) 'L'évolution des systèmes d'occupation des sols dans la vallée de la rivière Mananara au nord-est de Madagascar', *Taloha*, vol 11, pp16–64.
Wright, H. T., Radimilahy, C. and Allibert, C. (2005) 'L'évolution des systèmes d'installation dans la baie d'Ampasindava et à Nosy-Be', *Taloha*, vols 14–15, p29.

Part 2

Paradise lost?

The myths, narratives and received wisdoms at the heart of conservation research and policy

4 Deforestation in Madagascar

Debates over the island's forest cover and challenges of measuring forest change

William J. McConnell and Christian A. Kull

Introduction

Forests – and their absence – are linchpins between Madagascar's biotic richness, its degradation by humans, and conservation action. Much of the justification for conservation action depends on descriptions of previously extensive forests being cut and burned, on documentation of the threats to the patches that remain, and on success in slowing or stopping deforestation. Forest cover change is a highly iconic outcome of the different kinds of human interactions with the environment since the island was settled (Figures 4.1 and 4.2; see also Chapter 3 by Dewar and Chapter 5 by Scales). This chapter investigates measurements of forest cover and its change on the island, and focuses attention on the technical challenges and social context of that scientific effort.

The evidence for the deforestation of indigenous woody vegetation in certain places and times is strong: early settlers' fires reduced the woody cover of the island's interior over a millennium ago (Burney et al., 2004), a deforestation frontier has progressively moved inland from the east coast over recent centuries (Brand, 1998), and dramatic clearance for maize cultivation has affected the southwest in recent decades (Harper et al., 2007). Yet the science of measuring forest area in different time periods, and of assessing change to that forest cover, using historical data, air photos, and satellite images is surprisingly messy and difficult. There are difficulties with categories, with scale, and with other aspects of change assessment. Furthermore, the science of measuring and quantifying deforestation takes place in a social context, one where certain discourses and metaphors condition the kinds of questions that are asked and the kinds of results that are highlighted (Larson, 2011). In the case of Madagascar, there is a dominant normative understanding of environmental change in which an idyllic and nature-rich island is rapidly despoiled by a burgeoning human population, trapped in a spiral of poverty and degradation (Kull, 2000). This has encouraged, we argue, the curious persistence in the literature of certain dubiously sourced and now outdated statistics claiming with relative certainty that 80 or 90 percent of the island's original forest is gone. The

Figure 4.1 Forest patches in the highlands landscape. Are these images of deforestation or reforestation? Top left: NE of Antananarivo, trees on the ridge are clearly eucalypts and other species planted in grassland by people. Yet in the hillside hollow, which is also cultivated, are these remnant 'native' trees or planted fruit trees or both? Top right: eastwards view in Ialatsara forest station (southern highlands): large area of native forest (cut by rice fields) with large pine plantations in the distance. Bottom left: north of Anjozorobe, this is a landscape in process of colonization and afforestation (mostly with eucalypts). Bottom right: ecotone of native grassland and forest at the eastern end of Alaotra basin.

Source: C Kull (1996–2010).

loss of forest in some portions of the island is, as we have reviewed earlier, dramatic enough that such exaggerations are unnecessary. These exaggerations are even potentially harmful in that they can undermine scientific authority, put blinders on the types of questions that are asked, and push to the sidelines important debates about the impacts of strong conservation policies on rural people.

An assessment of deforestation in Madagascar requires an analysis of forest cover at different time points. We review the data and debates over historical and contemporary forest cover. This involves reviewing the theories and evidence for pre-settlement forest cover, and the published estimates of more recent conditions, beginning with the work of colonial botanists.

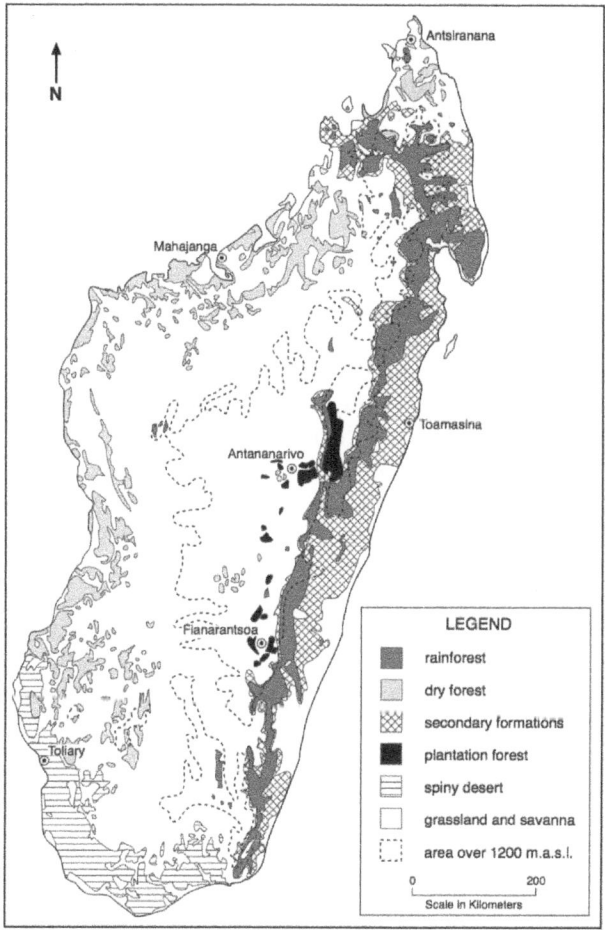

Figure 4.2 Current vegetation zones of Madagascar

Source: Kull (2000). Map is generalized from AGM (1969), Green and Sussman (1990), CI et al. (1995), and, in the case of the plantation forests, personal observations (meant to be representative, not necessarily precise).

Then we turn to subsequent analyses employing increasingly sophisticated technology for observing the land surface, first from the air and later from space. We explore the degree of divergence between the resulting estimates and then focus in on the challenge of estimating change. All along the way, our story becomes intertwined – of necessity – with a history and sociology of science, as the ideas that various authors argued for or against, and the approaches taken, reflect the worldviews and political contexts in which those authors were operating. This leads us to investigate, in the penultimate section, the social context in which a particular idea (the 80 or 90 percent deforestation hypothesis) has persisted. We conclude by presenting a number of recommendations for improved understanding in the spirit of evidence-based policy making.

How much forest was there 'originally'?

The question of what the landscapes of Madagascar looked like 'originally' is far from settled (see Chapter 3 by Dewar). The answer is likely to be quite different depending on the specific region concerned, and whether 'originally' is taken to mean before colonial records, before widespread evidence for settlement, before first human visits, at the end of the last ice age, or at some other point in time. Over the past 150 years, published claims about the island's 'original' vegetation have varied greatly, with some commentators presuming the existence of a dense, island-wide forest, and others defending the pre-human presence of grassy biomes, particularly in the highlands. The former claims were often rooted in ideological perspectives ranging from climax-based ecological theory, to temperate–climate conceptions of 'fire as disturbance', or colonial attitudes towards indigenous people's resource management. Evidence for these differing claims ranged from casual empirical observations of forest islands in grassland zones, to place names referring to no-longer extant forests, to oral histories of a 'great fire', or to rigorous analyses of biogeographical distributions, floral functional types, pedology, fossil beds, lake or marsh sediment pollens and charcoal.[1]

Particularly relevant to this chapter are the stances of colonial naturalists Perrier de la Bâthie (1921, 1936) and Humbert (1927, 1949, 1955), who argued that forest vegetation covered nearly all of Madagascar before human settlement.[2] Their conclusion has largely been taken at face value and repeated in numerous publications and reports, shaping, as we show later, the reporting of forest change statistics. Meanwhile, scientific debate about pre-human vegetation continues, with new forms of evidence emerging (Burney et al., 2003, 2004; Wilmé et al., 2006; Bond et al., 2008; Virah-Sawmy, 2009; Quémeré et al., 2012). Extrapolating from such research, it appears possible to make several conclusions. First, the strong version of the Perrier–Humbert island-wide forest hypothesis has been falsified, in that heathlands, grasslands, and a variety of non-forest vegetation covers predate humans on the island, maintained by lightning, moments of drier climate, and now-extinct mega-herbivores. Second, humans have certainly altered the vegetation cover of the island since their arrival, likely increasing grassland cover at the expense of woody vegetation. Third, the vegetation cover was quite dynamic before human arrival in response to climate shifts, with forest types expanding and contracting. Fourth, the story is quite different in different geographic regions. Finally, any estimate of prehistoric forest cover must be treated more as conjecture than as fact.

How much forest was there historically?

While it is not surprising that the extent and nature of the 'original' forest cover before written records remains contested, it is also the case that forest cover during the late pre-colonial and colonial periods (late 1800s–1960) is still not fully settled. The estimates of missionaries, colonial scientists, and foresters – based at first on field assessments and, after 1949, on measurements from air photos – resulted at times in wildly different numbers, reflecting differences in approach and materials, and these estimates have often been reproduced uncritically in more recent assessments.

Early mapping efforts can be attributed to missionary scientists such as James Sibree, whose 1879 map shows the distribution of dense forests (Figure 4.3). Like such maps, the quality of the first published *quantitative* estimates of Madagascar's vegetative mantle was limited by the size of the island and the inaccessibility of many areas. Estimates varied widely and exhibited no clear trend (Table 4.1).

Guichon (1960) made the first attempt to use remote sensing data to comprehensively estimate the island's forest cover. Advances in aerial photography during World War II had enabled the French to acquire imagery between 1949 and 1957 that served as the basis for the country's 1:100,000 topographic map series and a more systematic quantification of Madagascar's land cover. The cartography was not yet complete when Guichon's paper was published, but he drew reassurance that his 'first approximation' of 12,472,923 ha of lightly and non-degraded forest corresponded with the estimate of Girod-Genet and the later estimate of Lavauden, and that his estimate of 16,731,722 ha of total forest (including degraded stands) closely matched the more recent estimate of Perrier de la Bathie (1936). When including *savane arborée*, or wooded savanna, Guichon's estimate rose to 19,380,722 ha.[3]

The same air photos were later used by H. Humbert and colleagues to create a landmark map of vegetation zones. In the report accompanying the map, they estimate 19,819,000 ha of 'forest formations' (Humbert and Cours Darne, 1965, after p82), but do not specify what vegetation categories are included or excluded. Subsequent work based on the same aerial photography data led to somewhat different totals, illustrating the confusion that is quite common surrounding the use and re-use of land cover data by different researchers. This can be illustrated by the authoritative 1996 report on the national forest inventory (*Inventaire Ecologique Forestier National*, or IEFN). In its 'Table 5.02', the IEFN report provided four different forest cover estimates for 1949–1957, presumably all based on the same aerial photographs (DEF et al., 1996). What is striking is the inconsistency with which these estimates were handled. As Table 4.2 shows, different categories of forest are included (or not) in the 'total forest cover' figures, inconsistent figures (due to methodological differences or rounding, we presume) are not explained, and the source references are confused.

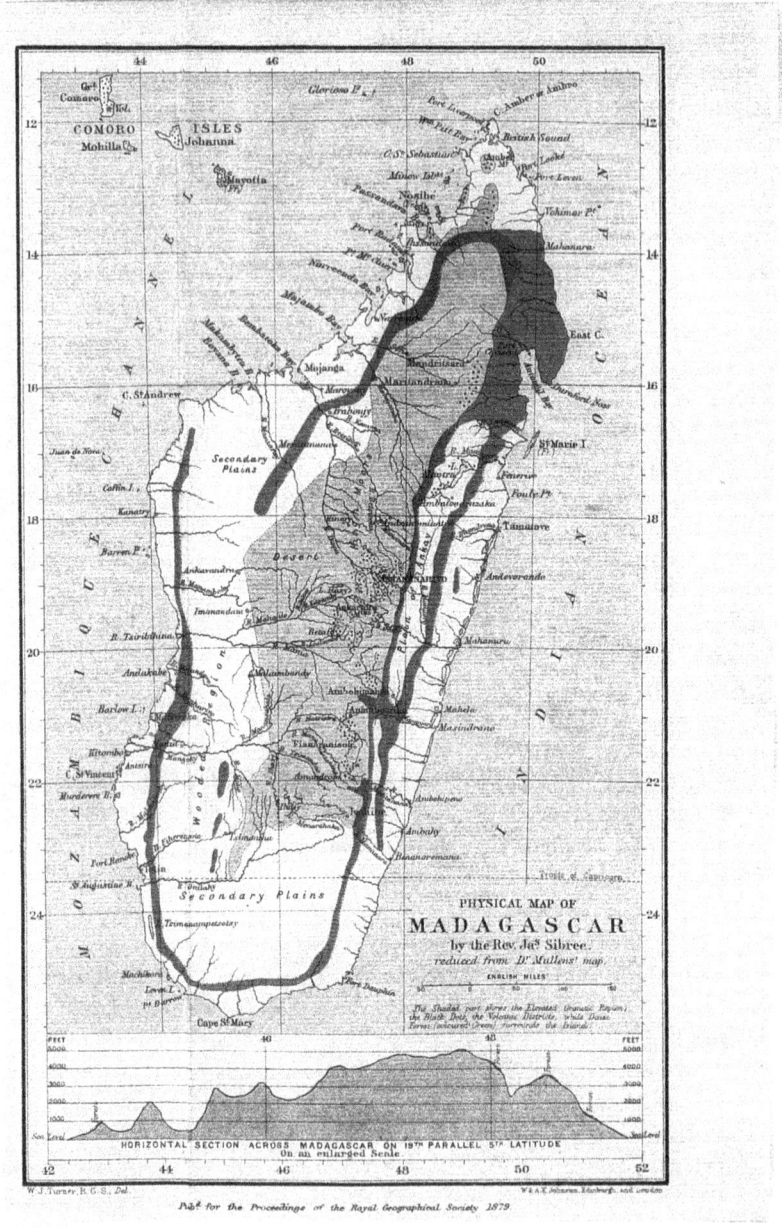

Figure 4.3 James Sibree's (1879) Physical Map of Madagascar emphasizes the extent of barren hills encountered by early visitors ('desert' and 'high moors') and gives a potentially reasonable, albeit rough, representation of the distribution of 'dense forests' at the time in dark shading. Light shading denotes the 'elevated granitic regions'.

Table 4.1 Colonial period estimates of Madagascar's forest cover

Auteurs	Date	Superficie en millions d'hectares	
LAVAUDEN	1895	20	(1)
GIROD-GENET	1899	12	
P. DE LA BATHIE	1921	7	
Prof. HUMBERT	1927	2 ou 3	(3)
LAVAUDEN	1931	10	
P. DE LA BATHIE	1936	17	
M. R. HEIM	1955	1,5	

Notes: reproduced with permission from Guichon (1960, p409). Estimates vary widely, and show no clear trend, particularly considering Guichon's notes in which he suggests that (1) Lavauden's 1895 estimate was probably exaggerated and (2) that Humbert's 1927 estimate excluded several important types of forest cover.

A second example of how old data are used and misused comes from Harper et al. (2007), which contains the most recent published estimate we are aware of based on the same air photographs. They digitized the Humbert and Cours Darne (1965) map and present an estimate of 1949–1957 forest cover different again from all previous interpretations. Echoing the inconsistencies in the IEFN report, Harper et al. leave a confused reference trail and fully explain neither their choice of forest types included in the figure nor how their numbers were derived (Table 4.2).

The examples above show how maddeningly frustrating efforts to quantify forest cover have been. Not only have scientists struggled with the difficult and vast terrain and the enormity of the task of nation-wide air photo analysis, but their tendency to discriminate, lump together, and emphasize different categories of forest cover have led to much confusion, which, it appears, is exacerbated by later inattention in re-using older work.

So, how much forest was there historically? Clearly, the eyeball estimates from the late 1800s onwards are quite divergent and not usable without diligent triangulation with other region- and date-specific sources. As far as the air photos acquired between 1949 and 1957, one may possibly conclude that the areal extent of the main 'forest' categories (excluding savannas) was on the order of 16 million hectares. The highly variable forest cover figures circulating in the literature result from differing definitions of forest, with changes in cartographic technology also contributing. These issues are developed more fully below, following a review of the work estimating forest cover from space.

How much forest was there recently?

A new round of nation-wide estimates of forest cover emerged in the 1980s, benefitting from the new, space age technologies of satellite-based remote sensing. As far as we can ascertain the first published map of Madagascar's

Table 4.2 Inconsistent re-interpretations of total forest cover according to the 1949–1957 air photos (and the topographic maps derived from them)

Forest cover 1949–1957	Source	Types of forest included	Other comments
As presented by IEFN (DEF et al., 1996) in 'Table 5.02' as total of evergreen and other forest formations			
16,695,000 ha	Guichon, 1960	No explanation provided in IEFN; our assessment is that the figure presented is for all forest *except* raphia and savanna.	Slight divergence from Guichon's own figure likely due to rounding.
19,148,000 ha	Humbert and Cours-Darne, 1965	No explanation provided in IEFN; based on Dufils (2003), we take this figure to include degraded and/or secondary forests, as well as mangrove, gallery, and plantation forests.	Divergence from Humbert and Cours-Darne's own (1965) estimate (19,819,000 ha) probably due to different methods (they likely used a manual planimeter, while the IEFN likely calculated from digitized versions of the maps).
12,378,000 ha	Eaux et Forêts, 1953–1974	No explanation provided in IEFN.	Confusion over what the actual source was: the IEFN bibliography only refers to 'Direction des Eaux et Forêts et de la Conservation des Sols (1971) Inventaire forestier à Madagascar, 70 pages'.

Forest cover 1949–1957	Source	Types of forest included	Other comments
10,300,000 ha	Lanley [sic], 1986 [sic]	According to the IEFN, only closed-canopy broadleaf formations, and based on incomplete 1:100,000 map series, perhaps excluding southern forest and bush/scrub.	Considerable confusion over source and interpretation of data: the IEFN cites Nelson and Horning (1993), who in turn report having obtained the Lanley [sic] (1981) estimate from Grainger (1984). The actual source appears to be Lanly (1981) which indicates that this forest cover figure is meant to be an estimate based on the 1949–1957 air photos and *extrapolated to 1980* based on deforestation trends.

As presented by Harper et al. (2007) in 'Table 2' as 'total forest area'

Forest cover 1949–1957	Source	Types of forest included	Other comments
15,995,900 ha (159,959 km²)	Blasco 1965; Humbert and Cours-Darne, 1965	According to Harper et al, this figure applies to 'forest and mangrove suffering little or no degradation'. Strangely, the terms 'little or no degradation' are not found in the Humbert and Cours Darne (1965) map, though they are reminiscent of Guichon's (1960) table. Harper et al. explicitly state that their study considers forest cover to consist of 'primary vegetation dominated by tree cover at least seven meters in height'.	Harper et al.'s bibliographic references are problematic. Blasco (1965) is actually a chapter in the Humbert and Cours-Darne (1965) volume bearing the title they provide (Blasco is also the first of four cartographers named on the Humbert and Cours-Darne (1965) maps, and perhaps this is the reason he is cited). The figure 15,995,900 ha does not appear anywhere in the document and must derive from Harper et al.'s digitization of the maps (and the difference with other figures above is due to not including savanna and to methodological divergences between digital areal calculations versus measurements made with a manual planimeter).

forest cover from satellite imagery employed visual interpretation of printed images from the Landsat Multispectral Scanner (MSS) sensor acquired in the 1970s (Faramalala, 1988a). It did not include a numeric estimate of forest cover, explaining that cloud cover made this impossible (Faramalala, 1988a, p147). However, the map was later digitized (Faramalala, 1995) and used by Du Puy and Moat (1999), who report that it revealed 10,784,000 ha of 'primary vegetation' including evergreen and deciduous forest, as well as mangrove and marshland. The IEFN report (DEF et al., 1996, p75) cites 'statistics published by Faramalala (1995)' as the source for an estimate of 15,812,000 ha of forest.[4] This includes 10,676,000 ha of evergreen forest, likely differing from the Du Puy and Moat (1999) estimate in the exclusion of mangrove and/or marshland. Later, Mayaux et al. (2000) reported that data from Faramala (1981)[sic][5] reveal 10,603,200 ha of dense humid and dense dry forest, as well as mangrove. When they add 'secondary complex' they find a total of 15,484,400 ha. It is difficult to understand the reasons for the discrepancies (albeit minor) between these figures and those reported by DEF et al. (1996), as they are presumably based on the same digital polygons.

More recently, Harper et al. (2007) exploited MSS data from the same period as Faramalala, reporting a total of 14,731,000 ha of humid, dry, and spiny forest (mangrove is listed as 'not available'). Their study's 'Table 3' contains an arithmetic error and the total of the three listed categories is actually 13,933,500 ha. The total forest estimate differs by as much as 12 percent from those reported by the IEFN (DEF et al., 1996) and Mayaux et al. (2000). It is likely that much of the difference is attributable to different methodologies: while Faramalala used visual interpretation of printed satellite images, Harper et al. re-classified images from the same time period digitally using a supervised classification technique (see Box 4.1).[6]

Subsequent island-wide studies have employed a variety of satellite sources, taking advantage of not only Landsat but also other platforms. For example, Nelson and Horning (1993) used three Local Area Coverage (LAC) scenes acquired in 1990 and 1991 by the Advanced Very High Resolution Radiometer (AVHRR) sensor to derive an estimate of 6,091,800 ha of forest in four bioclimatic vegetation zones (rainforest, hardwood, grass, and spiny). The study employed an automated classification procedure with Landsat MSS photoproducts serving to 'train' the classifier, and the Faramalala (1981[sic]) data as a reference map to evaluate the quality of the product.[7]

The IEFN report (DEF et al., 1996) contributed a new mapping of forest cover based on visual interpretation of Landsat TM (Thematic Mapper) data from 1990–1994. Importantly, this mapping was accompanied by extensive, detailed fieldwork in different vegetation types. This analysis resulted in an estimate of 13,260,000 ha of forest, including 6,062,000 ha of evergreen forest.

The next island-wide assessment (Mayaux et al., 2000) used data from the SPOT-4 VEGETATION instrument, which collects data at roughly the same spatial resolution as AVHRR-LAC. The study used 36 ten-day image composites from 1998 and 1999 (the ten-day composites minimize cloud contamination, while the time series takes advantage of vegetation phenology). As in the Nelson and Horning (1993) study, semi-automated classification was performed using Faramalala (1981[sic]) and other reference data. Mayaux et al. (2000) reported a total of 17,303,200 ha of forest, including 10,104,100 ha of 'dense humid', 'dense dry', and 'mangrove' along with 7,199,100 ha of 'secondary' forest.

The most recent assessment of Madagascar's land cover published in the peer-reviewed literature is presented in Harper et al. (2007). Forest cover in the 1990s was analyzed using Landsat TM data yielding a figure of 10,605,700 ha (combining four categories of 'primary' humid, dry, spiny, and mangrove forest); the situation around the year 2000, using Landsat ETM+ (Enhanced Thematic Mapper) yielded an estimate of just 8,982,100 ha. This is considerably less than the most comparable estimate (Mayaux et al., 2000).

A subsequent map published by a consortium of United States Agency for International Development (USAID) partners (CI et al., 2007) extends Harper et al.'s analysis to 2005 using much of the same data and techniques. Curiously, while the map's tabular estimate for c.1990 is quite similar to the figure published in Harper et al. (2007), the map's estimate for c.2000 (9,677,701 ha) is considerably higher than the earlier figure (8,982,100 ha) and closer to the Mayaux et al. (2000) estimate. They estimate forest cover c.2005 at 9,216,617 ha. No explanation is provided on the map of the reason for the revision. From these estimates, one may surmise that something like ten million hectares of 'primary' forest existed around the end of the twentieth century.

The divergent results of the above studies emphasize the ways in which different techniques – based on different definitions, assumptions, satellite data sources, and classification methodologies – influence measurements of current forest cover. In the next section, we see the implications of these issues for measuring forest cover change. First, however, we should mention that in addition to the national-scale remotely sensed analysis of forest cover reviewed above, numerous studies have undertaken original analyses of remote sensing data for sub-national land cover analysis in Madagascar. They are too numerous to review here (some important examples include Green and Sussman 1990, which we discuss below, as well as Laney, 2002; McConnell et al., 2004; Agarwal et al., 2005; Irwin et al., 2005; Vågen, 2006; Elmqvist et al., 2007; Scales, 2011; Quémére et al., 2012). With respect to these regional studies, what should always be kept in mind (particularly when dealing with *trends* in forest cover, the topic of the next section) is that choices over scale and boundaries can have a major impact on study findings; inclusion of large non-forest areas will lower forest cover rates substantially, as can coarsening the scale of analysis, by excluding small patches.

What are the *trends* in forest cover? The challenges of quantitative change assessment

The simplest way to represent change in forest cover is to compile estimates for particular time points and present them in tabular or graphic form. This is the approach employed by Guichon (1960) (Table 4.1) and subsequent authors, including Nelson and Horning (1993), the IEFN (DEF et al., 1996), McConnell (2002), and Dufils (2003). As explained above using the case of Guichon (1960), rather than contributing to an understanding of the evolution of forest cover on the island, this approach has instead served to illustrate the incommensurate nature of the individual estimates and to caution us against simple comparison. Guichon (1960, p408) suggested three key reasons for the significant disparity in the estimates: '(1) the definitions of different types of forest formations vary among the authors; (2) cartographic documentation was absent, insufficient or imprecise; and (3) sometimes, the figures were deformed, consciously or not, in order to prove a thesis or to justify a position' (authors' translation).

Below, we examine quantitative, remote-sensing-based estimates of forest cover change in Madagascar in light of Guichon's first two, rather technical, constraints. The following section then addresses the third, perhaps more contentious, issue. It is not just in Madagascar that widely varying definitions of forest have confounded those hoping to understand deforestation. Indeed, the FAO (the UN's Food and Agricultural Organization, the foremost authority on global forest cover) amended its canopy closure threshold in between two of its decadal global Forest Resource Assessments in order to be able to account for important changes in woody cover in semi-arid parts of Africa, much of which would have been missed under the prior definition. The effort required to revise prior estimates to match the new definition was substantial, and the procedure undermines confidence in the comparability of the data across assessments (Rudel et al., 2005).

This issue is accentuated in Madagascar by the wide range of environmental conditions in which trees grow on the island. While relatively undisturbed forest on the eastern escarpment would meet just about anyone's definition of forest, smaller patches in this landscape, as well as gallery forests and sparser woodlands of the highlands and west coast may not, particularly in analyses relying on satellite sensors that form images of the earth's surface in 30m, 80m, or even 1km pixels. The issue is even more pronounced in the 'spiny forest' of the arid south, while mangrove formations present their own unique set of challenges, due to the high degree of moisture below the canopy, complicating the spectral signature registered by the satellite. The problem of inconsistent characterization of Madagascar's land cover was perhaps most usefully addressed by Lowry et al. (1997), who review the history of vegetation classification in Madagascar and argue for an 'objective approach to chorological analysis and physiognomic classification in Madagascar' (p110). As described in the preceding section, mapping efforts in Madagascar have typically defined forest according to the capabilities of the sensors they employ and the analysts' particular goals. This has been a major constraint on efforts to precisely estimate change in the island's forest cover.

Perhaps the most widely cited study of deforestation in Madagascar using remotely sensed data was conducted by Green and Sussman (1990). The study compared 'original' forests (no source cited) and the Humbert and Cours Darne maps of forest cover (based on 1949–1957 aerial photography) with the authors' own interpretation of Landsat MSS imagery from 1985, concluding that the island's eastern rainforests diminished at a rate of 1.5 percent per year from the 1950s to the 1980s. We cannot know to what degree the forest cover types analyzed across periods differed, because of the second problem identified by Guichon: insufficiency of cartographic documentation. Unfortunately, in the Green and Sussman (1990) article the methods were described quite briefly, with no discussion of the comparability of the techniques employed to analyze the two very different data sources.

A subsequent, island-wide comparative study by Dufils (2003) also used data from the Humbert and Cours Darne (1965) maps, as well as those from the IEFN (DEF et al., 1996) and Mayaux et al. (2000).[8] Where Dufils differed from Green and Sussman (1990) was in taking into account the issue of forest cover types. Specifically, he attempted to isolate a comparable 'evergreen' class from the vegetation cover classes presented in these three sources, even as he acknowledged the difference in classification schemes across these studies. Unsurprisingly his harmonization of categories was imperfect and casts doubt on the trends reported.[9] A consequence is that, for example, the mismatch of forest types between the two source datasets would likely have led to the inflation of the deforestation rate of 1.6 percent per year which Dufils reports for the 1990s (as forest present in the later period was excluded). Unfortunately, the study is also afflicted with typographic errors,[10] and problematic assumptions about time points.[11] To the author's credit, the text acknowledges some of the key challenges in estimating rates of change, particularly the different estimates likely to come from different remote sensing systems. However, given the issues described here, the published conclusions must be used with caution.

In order to increase the comparability of single date estimates used in calculating change, a later study (Harper et al., 2007) undertook fresh interpretations of satellite imagery from the 1970s (MSS), c.1990 (TM) and c.2000 (ETM), in conjunction with the Humbert and Cours Darne (1965) maps. Laudably, the study set out to create comparable maps by applying a common set of criteria for defining forest (see Box 4.1).

Using this approach, they calculate annual deforestation rates in four geographic zones, over three periods. Results range from modestly negative (i.e. increase in forest cover) to strongly positive rates of nearly two percent per year in certain parts of the island. At the level of the entire country, their analysis suggests that deforestation was more rapid during the 1970s

Box 4.1 Using satellite imagery to classify land cover

Satellite images are made up of layers of data collected in different parts of the electromagnetic spectrum. These are referred to as spectral bands. Different surfaces absorb, radiate, and reflect electromagnetic radiation in different ways, yielding distinct 'spectral signatures'. Using these properties, each pixel in a satellite image can be classified into a land cover category (e.g. forest, grassland) based on the spectral signature of the ground cover (e.g. trees, grass, bare earth).

The science of deriving land cover maps from remotely sensed imagery (whether from an orbital or airborne platform) entails a number of key decisions that have major impacts on the outcome (Figure 4.4).

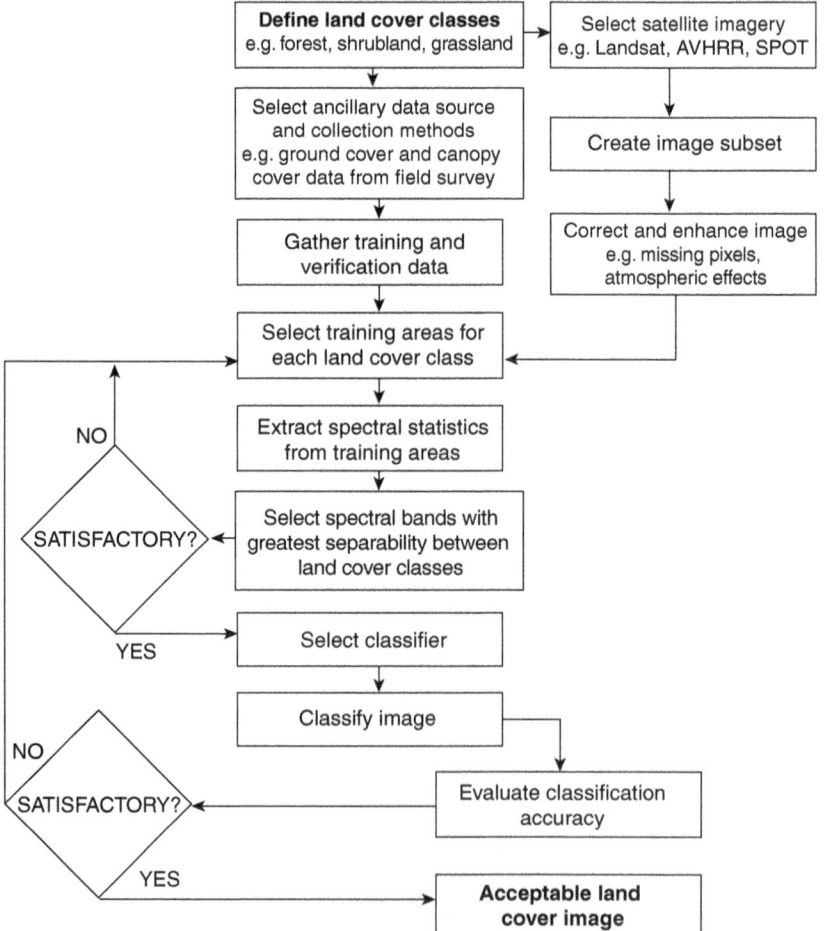

Figure 4.4 Key steps in the classification of satellite imagery
Source: adapted from Wilkie and Finn (1996).

1   Define land cover classes. While the appropriate set of land cover types for a given region may seem self-evident, they are anything but. Instead, the selection of categories is often a mix of the goals of the study and what is distinguishable in the imagery being used. Choices about which categories to analyze have profound, though often unrecognized, implications for policy (Kull, 2012). Globally, classification schemes vary enormously depending on bioclimatic zone. There exists no universal nomenclature and unambiguous definitions of a given land cover are rare. A laudable exception to this was a recent study whose authors specified that 'we defined "forest" as areas of primary vegetation dominated by tree cover at least seven meters in height, with neighbouring trees crowns touching or overlapping when in full leaf. In practice, this means that the canopy is at least 80% closed' (Harper et al., 2007, p2).[12]
2   Select imagery. Ideally this choice follows from the analytical needs, in an objective consideration of the spatial, spectral, and temporal characteristics of different sensors with respect to the particular study area and research questions. However, in practice other considerations often come into play, such as the analyst's familiarity with a particular sensor. On the other hand, sometimes it is the novelty of the sensor that motivates the land cover change study itself. This dynamic was in evidence in the testing of the new SPOT-4 VEGETATION instrument, with the results explicitly compared to previous exercises with the similar AVHRR instrument (Mayaux et al., 2000).
3   Create image subset. This chapter is limited to only island-wide studies, largely because the comparability of sub-national studies is hampered by incommensurate spatial bounds. When the spatial extent of a study is arbitrarily defined by the extent of map sheets, aerial photographs, or satellite images, this can have a major effect on the rate of change, since a relatively small change in study area extent – to include or exclude areas that are not representative of the overall change – can drastically alter the overall rate of change. There is no obvious solution to this, and all change rates calculated on such arbitrarily defined study areas must be treated with caution. It may be more appropriate in these cases to limit the reporting to the areal change in different periods (e.g. Vågen, 2006) or to graphical representation of change between time periods (e.g. Elmqvist et al., 2007).
4   Evaluate classification accuracy. In practice, the thresholds separating land cover categories (e.g. closed forest and open woodland) can be quite difficult to apply consistently and reliably without a

> great deal of very expensive field work. In addition, the arbitrary sampling of the landscape often results in pixels representing mixtures of land covers (especially when the heterogeneity of the landscape occurs at a finer scale than the sample rate (pixel size) of the sensor). As a result in remote sensing studies, 85 percent accuracy is considered pretty good, yet it is common to see studies that set out to explain single digit percentage changes in land cover from images classified with double digit errors; moreover the change is concentrated all along the edges of land cover patches where the bulk of the classification error occurs.

and 1980s – around 1.7 percent per year – than in the preceding and subsequent periods (0.3 percent in the 1950s and 1960s, and 0.9 percent per year in the 1990s). A map subsequently published using many of these same data and techniques (CI et al., 2007) tabulates an overall deforestation rate of 0.83 percent per year during the 1990s (why the rate for the 1990s differed between the two studies is unclear). It also presents new data indicating that this rate slowed to 0.53 percent per year during the period 2000–2005.[13]

This brief review suggests that our knowledge of forest cover at particular points in time is subject to high degrees of uncertainty, and that the hazards are compounded in the estimation of forest cover change. The studies reviewed above estimated rates of change ranging from less than half a percent per year to almost two percent per year. Unfortunately, no two studies cover the same forest types, in the same area, during the same period, and therefore none can serve to verify another. The studies also do not, typically, engage at all with confounding evidence of forest recovery or forest planting. We suspect that the unstated reason for this is that neither secondary forests nor plantations tend to be valued as highly for conservation purposes. In contrast, the IEFN (DEF et al., 1996) mapped a considerable extent of plantation forest from very similar data. Harper et al. (1997, p3) follow Green and Sussman (1990, p213) who admitted to being unable to distinguish large tracts of secondary forest and plantations from primary forest. Meanwhile, other studies (e.g. Rakoto Ramiarantsoa, 1995; Kull, 1998; McConnell and Sweeney, 2005; Elmqvist et al., 2007) acknowledge the local importance of reforestation and afforestation and set out to quantify and explain these phenomena.

In some cases, the individual studies provide information that can be used to judge the reliability of their estimates of forest cover at a given point in time, or of their estimates of change. We examine that briefly in the next section, before turning to Guichon's concern about the motivations of different analysts.

Methodological issues and accuracy assessments

Several issues repeatedly arise in the previous sections that complicate the measurement of forest area and changes in forest area. These range from data inconsistencies, to methodological assumptions, to definitional disagreements. Below we expand on one particular aspect: the importance of assessing the accuracy of the different forest mapping products. All empirical measurement approaches entail uncertainty, arising from error during the collection and analysis of data, and in the reporting of results. In this section, we briefly comment on some problematic cases concerning island-wide analyses of Madagascar's forest cover.

The analysis of the aerial photographic record, as embodied in the topographic map series and the Humbert and Cours Darne (1965) maps, generally relies on the analyst's expert opinion, often informed by direct, in situ, observations, as well as by knowledge of the ways in which different land covers reflect sunlight back to the camera, and experience with the typical appearance of certain features (Kull, 2012; Box 4.1). Given the time lapse since these analyses were conducted, it is virtually impossible to judge the accuracy of the product, since the landscape is bound to have changed and few reliable references (such as terrestrial photographs) exist.

In satellite-based remote sensing, one way to judge the quality of results is to compare classification results with reference data (e.g. maps or imagery from other sensors) used to 'train' the manual interpretation or automated classification algorithm. For example, Nelson and Horning (1993) used the Faramalala (1981[sic]) data as 'ground reference' to report 81 percent agreement between that reference and their AVHRR-based forest cover estimate. They go on to specify that while they achieve 94 percent correspondence between the non-forest pixels in the two datasets, the value for forest was just 62 percent, and describe the agreement in their hardwood stratum as 'abominable' (p1472). They ascribe much of the disagreement to differences in the ways in which the MSS and AVHRR sensors image the landscape, declining to speculate on what proportion of the difference might be attributable to actual change in the landscape between the 1970s and 1990/1991.

Mayaux et al. (2000) used the same MSS-based dataset to train their SPOT-4 based classification, but based their accuracy assessment on maps derived from contemporaneous TM data interpreted by local experts, as contained in a report commissioned by the authors' organization. This validation yielded a 'user's accuracy' (the probability that a pixel's land cover label in the map corresponds with the reference data) of 87.8 percent, and a 'producer's accuracy' (the probability that the land cover category of a certain location in the reference data was labeled as such in the map) of 85.6 percent. In other words, discrepancies exist between the map product and the reference data in approximately 12–15 percent of the cases. The authors note that part of the error is attributable to imperfect co-registration between their 1 km resolution SPOT-4 VEGETATION data and the 30 m reference data.

Similar procedures may be employed to assess accuracy in studies aiming to detect change in the landscape. For example, Harper et al. (2007, p329) used contemporary ground reference data to calibrate and to validate land cover maps derived from TM imagery, estimating '89.5% accuracy in identification of forest and non-forest'. If their c.1990 and c.2000 maps each contain errors of approximately 10 percent, this raises questions about their finding of 15 percent of forest loss between these two maps. While it is likely that many of the errors cancelled each other out (in the sense of the net change in forest cover), it must be admitted that the bulk of such classification errors occur at the forest edge, where so-called 'mixed' pixels (depicting a mixture of forest and non-forest cover) predominate, but also where most of change has occurred. It is important to keep these issues in mind when interpreting the results of forest cover change analyses; when our measurement error approaches or exceeds the observed dynamics, we should temper our confidence in the claims about the change said to have occurred.

The record of sub-national studies with respect to accuracy assessment is mixed, with some authors (e.g. Vågen, 2006; Scales, 2011) providing detailed information, while others (e.g. Green and Sussman, 1990; McConnell et al., 2004) provide less rigorous estimates or general caveats.

The curious persistence of the 90 percent claim

We have shown in the above sections that estimates of forest cover before the remote sensing record provide dubious benchmarks and, more than anything else, demonstrate fundamental problems that continued through much of the subsequent work. We have joined others (e.g. Nelson and Horning, 1993; Ingram and Dawson, 2005) in showing that technical challenges and inconsistent definitions, compounded by sometimes insufficient documentation of methods, leave considerable uncertainties around certain aspects of air photo and satellite-based analyses. Clearly, without denying that native forests – in general – are shrinking, more caution about estimates of forest area and loss is warranted than is generally accorded.

Here we turn to the question of how views of forest cover change have been shaped by the received wisdom of the 'island-wide forest'. We focus in particular on the persistent reproduction of the notion that 90 percent of the island's forests have been lost, and ask to what degree its persistence may be attributable to its usefulness in emphasizing dramatic deforestation. The 90 percent claim arises directly out of the island-wide forest hypothesis developed by Perrier de la Bâthie and Humbert (Burney, 1987; Kull, 2000). Specifically, Perrier de la Bâthie (1921) stated that the island was once completely covered by an 'arborescent' vegetation, and that nearly *nine-tenths* of this original vegetation had since been destroyed by shifting cultivation, grassland fires, and logging. This claim has been abundantly repeated and reworked. The statistic appears further confirmed by today's assessments that the island is about 10 percent forested, but only if one fallaciously assumes the rest was previously fully forested.

These ideas, used to contextualize the introduction of scientific articles (e.g. Hannah et al., 2008), to sell conservation activities in the promotional material of environmental agencies and simply to describe the island in media and tourism writing (e.g. Bradt et al., 1996; Wikipedia 'Madagascar' page, accessed 21 May 2012), is at best a problematic assertion. At times it is tempered or modified, with '90 percent' replaced by '80 percent', or 'forests' replaced by 'natural vegetation'. Indeed, words like 'virgin', 'primary', 'secondary', and 'natural' are frequently used without careful definition, tending to reaffirm the discourse initiated by Perrier de la Bâthie and Humbert.

Why this penchant to uncritically repeat a figure based on vague and contested ideas of the original vegetation? Why not use more recent figures? We showed above that the more reliable aerial photo analyses document that between 16 and 20 million hectares of the island were forested in the middle of the twentieth century, depending on which categories of woody vegetation cover are included. Island-wide studies using satellite images show that by the end of the century, somewhere between nine million and 17 million hectares of forest remained, again depending on the methods used and the categories included. From this, one might conclude that at most more than half of the forest cover may have been converted to other uses in the past 50 years, perhaps much less (cf. Table 4.3). The loss of even one-tenth of the island's forest in a lifetime is certainly alarming and arguably sufficient cause for action. Why, then, is the 90 percent figure repeated? Is a bigger number more attractive, does it better reinforce the claim that Madagascar deserves special 'hotspot' status by 'virtue' of having lost 90 percent or more of its natural vegetation? Below, we seek to answer these questions. First, we investigate the 'genealogy' of the 90 percent claim, picking apart its origins and its reproduction. Then, we comment on the discursive and political environment that facilitated this process.

Deconstructing the claim's origins and reproduction in the scientific literature

We focus our analysis on articles in scholarly publications that make a claim about 80 or 90 percent loss of forest or of natural vegetation more broadly. This claim is remarkably persistent in the conservation biology literature: we identified some 27 articles making such claims (Tables 4.3 and 4.4).

Four of the articles cite no source for the claim (Table 4.4). In some cases the claim is implicit, combining recent estimates of contemporary forest cover with conjecture about the conditions when humans first arrived, leaving the reader to surmise the gravity of the loss. For example, Durbin

Table 4.3 Analysis of articles in scholarly publications making the 90 percent claim

Article/journal	Claim	Their source or citation
Myers, 1988	'Nationwide, only 5 percent of the original vegetation remains' (p192).	Leroy, 1978; Guillaumet, 1984; Jolly et al., 1984; Lowry, 1986; Mittermeier, 1986; Jenkins, 1987.
Durbin and Ratrimoarisaona, 1996	'Most of [Madagascar's] biological diversity is concentrated in forests, which are believed to have covered around 90% of the island around 2000 years ago but … less than 11% of the island is now covered with forest' (p346).	Nelson and Horning, 1993.
Lowry et al., 1997	'Perhaps 10% or so of Madagascar might still be covered with native vegetation' (p117).	None.
DuPuy and Moat, 1998	'Over 80% of the island has already been stripped of its native vegetation cover … the majority of which is now very species-poor secondary grassland which is burnt annually and is subject to intense erosion' (p1).	Analysis of Faramalala (1988a, 1995) data.
Hannah et al., 1998	'Estimates of forest destruction indicate that 50–80% of Madagascar's original forest cover has disappeared in the 1500–2000 years since the arrival of humans' (p31).	Green and Sussman, 1990; Nelson and Horning, 1993; Faramalala, 1995.

Article/journal	Claim	Their source or citation
Myers et al., 2000	'Madagascar's remaining primary vegetation (% of original extent)'; shown as 9.9% in Table 1 (p854).	Calculated on the basis of 'Original extent of primary vegetation' of 594,150 km^2 (the entire island) and 'Remaining primary vegetation' of 59,038 km^2. While the sources of information on endemism are included in the Supplementary Materials (both experts and publications), no sources are cited for the estimates of 'original' or 'remaining' primary vegetation. The text stipulates that 'Additional details are available in ref. 16' which is Mittermeier et al. (1999), which contains no citation for its statement that 'Estimates vary, but it is thought that at least 85% and probably 90% or more of Madagascar's natural forest cover has already been lost …' (p198).
Ganzhorn et al., 2001	'Habitat loss [in Madagascar is] estimated at >90%' (p346).	Lowry et al., 1997.
Lehtinen et al., 2003	'At least 90% of the original forests have been lost' (p357).	Green and Sussman, 1990.
Goodman and Benstead, 2005	'This island nation … retains only an estimated 10% of the natural habitats that existed before human colonization' (p73).	None.
Lehman et al., 2005	'The loss of 80–90% of forest habitats in Madagascar' (p232).	Du Puy and Moat, 1998.
Bakoariniaina et al., 2006	'Over 90% of the Malagasy original forest is now gone' (p241).	None.

Continued

Table 4.3 Analysis of articles in scholarly publications making the 90 percent claim, *continued*

Article/journal	Claim	Their source or citation
Bollen and Donati, 2006	'Over 80% of the island has already been stripped of its native vegetation cover' (p57).	Du Puy and Moat, 1998.
Hume, 2006	'Less than 10 percent of primary growth vegetation remains in Madagascar' (p288).	Nelson and Horning, 1993; Du Puy and Moat, 1996; Myers et al., 2000.
Ingram and Dawson, 2006	'It has been estimated that the country retains only 9.9% of its primary vegetation' (p195).	Myers et al., 2000.
Lehman et al., 2006	'The loss of 80–90% of forest habitats in Madagascar' (p294).	Green and Sussman, 1990; Du Puy and Moat, 1998.
Norris, 2006	'Today Madagascar … has lost over 90 percent of its original forest cover' (p960).	None.
Sandy, 2006	'Only about 11 percent of the "original" forests remain' (p305).	WWF's *Living Planet Report* (Loh et al., 1999).
Hanski et al., 2007	'Currently, roughly 10% of the original forest cover remains' (p344).	10% claim not referenced (it is in abstract), elsewhere they cite Dufils, 2003.
Allnutt et al., 2008	'Recent analyses using remote sensing reveal that only 10–15% of original forest remains' (p174).	Harper et al., 2007.

Article/journal	Claim	Their source or citation
Andreone et al., 2008	'Ongoing habitat destruction has already led to destruction of 90% of the original vegetation' (p944).	Myers et al., 2000; Harper et al., 2007.
Hannah et al., 2008	'Deforestation has claimed approximately 90% of the island's natural forest' (p590).	Ingram and Dawson, 2005; Harper et al., 2007.
Craul et al., 2009	'Madagascar has already lost 90% of primary vegetation' (p2863).	None.
Whitehurst et al., 2009	'Madagascar's primary vegetation has decreased to 9.9% of its original extent' (p275).	Myers et al., 2000.
Barrett et al., 2010	'As much as 90% of the country's primary forest already lost' (p1109).	Myers (2000); Yoder and Nowak (2006); Harper et al. (2007).
Volampeno et al., 2010	'The loss of 80–90% of forest habitats on the island' (p306).	Du Puy and Moat, 1998; Mittermeier et al., 2010.
Durkin et al., 2011	'It has been estimated that continued habitat destruction has led to the disappearance of 90% of the original vegetation cover' (p114).	Harper et al., 2007.
Johnson et al., 2011	'Since the arrival of humans *ca* 2000 yr ago, Madagascar has lost 80–90 percent of its natural forest cover' (p371).	Harper et al., 2007.

Table 4.4 Description of the sources cited by articles in Table 4.3 to justify their 90 percent claims

Cited source	Relevance to 90 percent claim
Category 1: Primary sources of data (all discussed at length in chapter text)	
Humbert (1955); Humbert and Cours Darne (1965)	Vegetation cover analysis based on c.1950 air photos. No change analysis.
Faramalala (1981[sic], 1988a, 1995)	Visual interpretation of 1970's satellite images. Change analysis is implied in the mapping of secondary vegetation types, based on untested assumptions about the genesis of a broad range of vegetation types thought to be degraded.
Green and Sussman (1990)	Despite study's restricted focus on the humid forests of the east, it has often been cited in claims about overall deforestation. They calculate that 3.8 million hectares of eastern rainforest found in the 1980s represents 50 percent of that mapped in the 1950s, and 34 percent of the 'original' forests. No 90 percent claim is made.
Nelson and Horning (1993)	As a single date analysis, this may be cited to support '10% island-wide forest cover around 1990', but it does not imply or show the other 90 percent is former forest.
Harper et al. (2007)	State that 'forest covered 90% or more of the island' but go on to say that 'others argue that it was less'. In their discussion: 'By the 1950s, only 27% of Madagascar was forested and even a conservative estimate of pre-human forest cover suggests it had already lost more than half of its forest cover; the loss may have been as much as two-thirds, or more. Forest cover further declined to approximately 16% in c.2000, a loss of 40% in 50 years.'
Category 2: Secondary sources not already listed	
Leroy (1978)	A botanical review article.
Jolly et al. (1984); Guillaumet (1984)	General natural history reference book and chapter within it reviewing vegetation types.
Lowry (1986)	No relevance: dissertation on New Caledonia.
Mittermeier (1986)	A WWF action plan.
Jenkins (1987)	A natural history overview of Madagascar from IUCN, UNEP, and WWF.

Cited source	Relevance to 90 percent claim
Du Puy and Moat (1998)	Uses land cover data from Faramalala (1988a, 1995).
Loh et al. (1999)	WWF's *Living Planet Report*.
Ingram and Dawson (2005)	State that 'Considerable discrepancies exist between the estimated amounts and distribution of forest cover and loss in Madagascar' (p1449) and present results from analysis of 14 years of change using NOAA AVHRR.
Yoder and Nowak (2006)	A study on the evolution of Madagascar's fauna.
Mittermeier et al. (2010)	A 676pp lemur field guide.

and Ratrimoarisaona (1996) state that forests 'are believed to have covered around 90% of the island around 2000 years ago' (p346) without citing any source. They then cite Nelson and Horning's (1993) figure of 11 percent contemporary forest cover, leaving the reader to infer that around 88 percent of the island's forest disappeared 'since the arrival of man'.

In other cases, the claim is made explicitly, but without a source being cited. This is particularly problematic when such claims are subsequently used as sources for others. For example, Lowry et al. (1997) state that 'perhaps 10% or so of Madagascar might still be covered with native vegetation' (p117), but neither provide a source for the claim nor proffer any analysis to substantiate it. Later, Ganzhorn et al. (2001) cite the Lowry et al. (1997) chapter as the source of their statement that habitat loss is estimated to exceed 90 percent.

In still other cases, the claim is made explicitly with reference to a source that provides only partial substantiation. For example, Lehman et al. (2005) make an assertion about 'the loss of 80–90% of forest habitats in Madagascar', citing Du Puy and Moat (1998) and Green and Sussman (1990) (in a companion article by the same authors the following year (Lehman et al., 2006), the same assertion appears verbatim, though without the second reference). As it happens, Du Puy and Moat (1998, p1) used Faramalala's (1988a, 1995) land cover maps to conclude that 'over 80% of the island has already been stripped of its native vegetation cover ... the majority of which is now very species-poor secondary grassland which is burnt annually and is subject to intense erosion'. Crucially, it must be noted that Faramalala (1988a) mapped secondary formations on the basis of untested assumptions about the genesis of a broad range of vegetation types thought to be degraded.

More recently, the Harper et al. (2007) study is being cited by many authors (e.g. Alnutt et al., 2008; Andreone et al., 2008; Barrett et al., 2010;

Durkin et al., 2011; Hannah et al., 2008; Johnson et al., 2011) as the source for their claims about the loss of 80–90 percent of the island's original (primary) forest (vegetation). This is particularly striking given that Harper et al. (2007) openly acknowledge controversy over the state of the island's vegetative mantle prior to the arrival of humans, and limit their conclusions to the five decades covered by the remote sensing data they analyzed.

The circulation of a claim about 90 percent forest loss is a striking example of the power of a received wisdom and of the fallibility of peer-reviewed science. One explanation is that authors simply repeat the 90 percent figure as a 'known fact', with a reference to a high-profile and much-cited article thrown in to support it, without actually checking to confirm whether this fact was indeed supported by data.

Yet, what our survey of the literature suggests is that there is a strong epistemic community of conservation scientists for whom this 'fact' forms part of motivating *raison d'être*, and as such becomes an unquestioned paradigm pervading the discourse. The most recent, and perhaps most authoritative, island-wide remote sensing-based study, Harper et al. (2007), acknowledges differing views on the question of forest loss statistics, yet persists in putting the 90 percent claim up front, and in pointing out the upper (but not the lower) bounds of forest loss statistics (see quote in Table 4.4). If they were not influenced by the dominant discourse, would they have phrased the discussion the same way? Their reinforcement of the conservationist dogma begs important questions of perspective and motivation. The authors are affiliated with the large conservation organization Conservation International. It is not insignificant that their analysis shows deforestation highest during the two decades of isolationist, socialist rule, with rates tapering off after 1990, when interventions by organizations such as their own boomed (see Chapter 7 by Kull). The article's conclusion may well be valid, but they bear verification to allay any potential concerns over the degree of objectivity of the analysis.

Amelot et al. (2012) point out two other instances in Harper et al. (2007) where oft-repeated ideas and theories from a conservation perspective probably unintentionally affect the statements and conclusions that are drawn. First, Amelot et al. investigate an inset map in Harper et al.'s Figure 1, which shows Madagascar as being covered by three forest zones: humid, dry, and spiny. Amelot and colleagues vividly illustrate the derivation of this simplified and misleading assessment. They trace this three-way forest classification back through its sources – through World Wide Fund for Nature (WWF) reports and botanic book chapters – to its origins in a complex map of climatic zonations they attribute to Cornet (1974). That pure climatic zonations became forest types through successive cartographic iterations reflects the discursive bias towards an 'original' (and potential) island-wide forest.

Second, Amelot et al. (2012) point out that the Harper study is used to justify claims about environmental change that it actually does not show.

The Harper study is cited in articles and policy documents that focus on slowing slash-and-burn cultivation in the eastern rainforest, yet the study itself clearly shows deforestation over the past two decades principally touched not the rainforest of the east, but the dry forest of the southwest (cleared for commercial corn production) and areas affected by urban demand for wood energy.

The interlinked influence of conservation interests and the dominant discourse

The persistence of the 90 percent claim can be traced to the paradigmatic dominance of certain ideas circulating in the administrative, policy, and scientific world about Madagascar's environment, and their correspondence with strong, foreign-funded conservation interests. The examples we show above are a manifestation of a broader phenomenon in Madagascar where what has been called a 'dominant discourse' influences the questions that are asked, the evidence that is seen, the stories that are told, and the actions that are taken. Dominant discourses are ways of understanding the world – shaped by the stories, metaphors, and language that we use – that gain their power from the influence of actors that initially promote them and that exercise power by shaping the realm of the possible (Fairhead and Leach 1998; Larson, 2011).

In Madagascar, it has long been clear that a certain discourse, a certain set of environmental narratives, dominates understandings of environmental change and the role of Malagasy farmers in that change (Jarosz, 1996; Kull, 2000; McConnell, 2002; Pollini, 2010; Amelot et al., 2012; Scales, 2011; Rakoto Ramiarantsoa et al., 2012). This discourse is rooted in a potent mix of fact with conjecture, spiced with the interests of its promoters – originally, colonial foresters and botanists, now conservationists – and made durable by their powerful position in the then colonial, now dirt-poor country with practically no tradition of rural social movements (whose voices might contest certain aspects).

The discourse paints Madagascar first and foremost as a hot spot of biological diversity, environmental degradation, and conservation action. It often relies on a narrative whose roots go directly back to the work of Perrier de la Bâthie and Humbert in the 1920s and 1930s. The story, as they framed it, begins with the original island-wide forest, and then blames the agricultural practices of Malagasy farmers as the primary cause of deforestation, together with colonial logging (see Chapter 5 by Scales). Perrier de la Bâthie and Humbert's stories had a great influence on contemporary writing; Madagascar became considered a type locality for the destruction of indigenous flora by fire and shifting agriculture. The narrative was reproduced nearly word-for-word by some scientists, in popular publications, and by development and environmental organizations throughout the environmental boom years of the 1980s and 1990s (Kull, 2000).

Scientific understanding of environmental change in Madagascar has continued to evolve. Yet for many scientists, as we show above, and in conservation and environmental organization documents, the Perrier–Humbert story remains the dominant narrative. This version of the story reaches the public through the media, the internet, travel guides, television documentaries, song lyrics (Emoff, 2004), environmentalists' writings, and agency documents, which use artistic license to further dramatize Madagascar's environmental degradation. They not only repeat the assertion of 90 percent forest loss due to slash-and-burn agriculture (for example on Wikipedia[14] and in *National Geographic News*[15]), but also build on tropes such as the blood-red, iron oxide-laden rivers that 'bleed' into the ocean, and the 'gangrenous wounds' of the island's *lavaka* erosion gullies (see examples reviewed in Kull, 2000, or Raharimahefa and Kusky, 2010 as a recent scientific example).

The persistence of the dominant narrative might not strike some people as unusual or problematic. After all, Madagascar *does* harbor a flora and fauna not shared with other places, native forest cover *has* declined in the past century in numerous regional contexts, and slash-and-burn agriculture certainly *is* one proximate cause of forest conversion. The discourse of exotic nature and environmental destruction, then, can be seen as necessary to justify conservation fundraising, policies, and actions, so it is unsurprising that it persists due to its compelling story line and its usefulness in gaining public and government support.

Yet, as Peet et al. (2011, p37) state, 'arguments over the apparently "given" facts and categories of ecology are always also arguments over social and political control of nature'. The dominant narrative and its exaggeration of forest loss contribute to strong conservation policies and actions that marginalize rural people, restrict their access to resources, and silence their viewpoints (see Chapter 14 by Kaufmann; Rakoto Ramiarantsoa et al., 2012). In addition, exaggerations, problematic assumptions and the use of outdated facts do a disservice to the credibility of science. They can undermine, and even contribute to the misuse of, scientific authority, serving to occlude other interpretations, forestall other topics of enquiry, and push to the sidelines important social debates about values and ethics (cf. Larson, 2011). It is important, then, that deforestation analyses return to careful, evidence-based approaches, in order to contribute to constructive debate about policy.

Lessons learned and the way forward

Our review of efforts to measure Madagascar's shrinking native forests results in four 'take home messages'. First, the science is complex and messy. Efforts to measure forest area, to determine historical and pre-historical forest area, and to assess changes over time are technically challenging and require more careful attention to detail than has heretofore been accorded.

Work has too often been motivated more by the assessment of new satellite sensors than by the careful repetition needed to build reliable knowledge.

Second, the dominant conservationist discourse influences interpretations of this complex and messy science. Power-laden ideas, such as that of the pre-human, island-wide forest, persist due to their correspondence with conservationist worldviews. They shape the questions that are asked, the interpretation of data, and the choice of statistics that are highlighted and repeated.

Third, the existing evidence documents a general trend of forest loss, though *not* of 90 percent as commonly cited. Increasingly rigorous comparisons of data derived from historical aerial photography and recent satellite imagery appear to be converging on an estimate that as much as half – but perhaps much less – of the most easily identified 'primary' forest types changed to other land covers during the latter half of the twentieth century. Changes in other forest types are quite varied, with some gaining and others losing. The available evidence does not support quantitative estimates at any time prior to the photographic record, and prudence demands that we avoid conjecture about forest cover dynamics during that time.

Fourth, there are specific ways in which forest cover analyses could be improved in order to support more evidence-based policy deliberations. Even if we succeed in defining forest for current purposes, we cannot change the definitions used in the past without redoing the prior analyses, and even if we were to repeat the prior analyses, the data we might use are different, leading to incompatible results. Perhaps the most certain way to judge land cover change would be to repeat the air photography conducted around 1950 and apply the same analytic methods (e.g. Kull, 2012). We must temper the habit of simply availing ourselves of the latest remote sensing technology in the belief that its superior qualities will lead to more reliable results and that it will become the standard to which future analyses will be compared. A number of concrete steps could be taken.

1. A systematic and fully documented reconciliation of the various remote sensing datasets should be undertaken. This would involve re-interpretation of at least some of the mid-century aerial photos, as well as of satellite imagery used in later studies. As part of this exercise, the topographic map series (at least the land cover information) could be digitized, with complete metadata, especially the dates of the photographs from which each sheet was derived. The exercise would require sharing access to all the imagery collected by different scientists. While this would require a significant investment of labor, the capacity exists in-country.
2. Rigorous validation should be conducted by an impartial third party.
3. The digital data and metadata from the exercise should be made freely available and further validation encouraged.
4. Careful classification of vegetation, allowing comparison with classification schemes used in prior studies, is absolutely crucial.

Madagascar has long been counted among the world's conservation hotspots due to its highly endemic and unusual wildlife and plants, and a perception that this natural heritage was under grave and imminent danger of disappearing in the face of a growing human onslaught. The Earth's biological riches developed over hundreds of millions of years, and avoiding further erosion of biodiversity by human activity is certainly among the most important challenges facing society today. In this struggle, it is entirely appropriate to focus attention on those areas of particular richness and where the threat of human impact is the greatest. At the same time, however, curbing human activity is never a simple undertaking, especially when those activities are directly linked to basic human needs, as they are in a place such as Madagascar. In this context, policy success depends on confronting the values and interests of different actors, and such confrontations depend, in part, on quality information. Scientists and practitioners alike must seek to build knowledge around the best evidence, avoiding conjecture and hyperbole. Journals should enforce such restraint, and should require that authors make data available on an ongoing basis so that the chaotic state of our knowledge about what has occurred can be redressed. The gold standard of knowledge building is replication, and the existing studies should be repeated to verify their results. Policy decisions are, in the end, about values, interests, and power, but good data speaks volumes.

Acknowledgements

Thanks to Ivan Scales for unearthing additional 90 percent references!

Notes

1 Details in Kull (2000, 2004). See also Koechlin et al. (1974), Burney (1997), and Dewar (1984)
2 Perrier de la Bâthie (1921, pp260–261) made sure to note that he meant not a pure dense forest across the island, but an island-wide 'forest flora' including tall shrublands and xerophyllous plants.
3 It should be noted that Guichon's main table contains several arithmetic errors, and that the totals should probably be 16,791,672 ha and 19,440,672 ha, not including and including savanna, respectively.
4 Oddly, the IEFN report's bibliography does not include Faramalala (1995). It does, however, include Faramalala (1988b); unfortunately we have not yet succeeded in obtaining a copy of the 1995 work, which may contain numeric estimates.
5 The date should be 1988. Mayaux et al. (2000) also incorrectly lists Faramalala's data source as Landsat Thematic Mapper (TM). Green and Sussman (1990) and Nelson and Horning (1993) also apparently incorrectly cite this thesis as dated 1981.
6 It is also possible that the study used a slightly different set of images, however the link provided in the article to the Supplementary information (p5) is faulty, making it impossible to verify.

7 The IEFN report (DEF et al., 1996) cites Nelson and Horning (1993) as having detected 5,809,000 ha. The difference with Nelson and Horning's own figure may have resulted from a misinterpretation of their 'Table 3'. The IEFN report apparently rounded the original estimate of 'rainforest' (34,167 km^2) to 3,417,000 ha of 'evergreen forest', then combined the original estimates of hardwood forest (6,697 km^2) and spiny forest (17,224 km^2) and rounded the resulting 23,921 km^2 to get 2,392,000 ha, and proceeded to ignore the 6,697 km^2 of grassland forest, perhaps misunderstanding this item in the table as simply grassland.
8 Dufils cites not Mayaux et al. (2000), but JRC (1999), yet this reference does not appear in the chapter bibliography. JRC is the home institution of Mayaux et al. – Dufils may have been working from a draft report.
9 The Humbert and Cours Darne (1965) maps depict 34 land cover classes, arranged in an array of floristic *séries* and elevational stages within three broad types (*humide, sec, littoral*). It is not possible to know which of these classes were combined to constitute the 'evergreen' estimate; while Dufils presented the same total for this category as appeared in the IEFN (DEF et al., 1996), that report provides no information on the aggregation methods. Meanwhile, the IEFN employed four phytogeographic zones (*Est et Sambirano, Centre, Ouest*, and *Sud*) within which they floristically differentiated a dozen major classes of forest. From these, Dufils appears to have selected a) *Forêts denses humides sempirvirentes de l'Est, du Sambirano et du Centre*, b) *Forêts sclérophylles des pentes occidentales et du Centre*, and c) *Forêts et fourrés sclérophylles de montagne et du Centre*. Finally, Mayaux et al. (2000) mapped dense humid and dense dry forests, along with mangroves and secondary forest complexes, from which Dufils selected the 'dense humid forest' class. The comparison across Dufils' harmonized 'evergreen' category from these three studies was imperfect. One example is the inclusion of highland forests (Forêts denses humides sempirvirentes du Centre) from the IEFN study, and the exclusion of the dense dry forests of Mayaux et al. (2000), some of which lie in the IEFN's 'centre' zone.
10 Dufils' (2003) 'Table 4.5' includes an apparently misplaced decimal indicating an implausible deforestation rate of 9.5 percent per year between the 1950s and the 1990s; this was probably intended to read 0.95 percent per year (which, it should be noted, is markedly lower than Green and Sussman's (1990) estimate of 1.5 percent per year for a subset of the area over the same period).
11 Dufils' (2003) estimate of the rate of deforestation is inflated somewhat by his choice of 1953 as the date represented in the Humbert and Cours Darne (1965) maps. While this represents the median of the period during which the aerial photos were acquired (1949–1957), our experience suggests that the bulk of the flights actually took place at the beginning of the period, as reflected in Green and Sussman's (1990) choice of the year 1950 in their calculations. The addition of three years to the calculation of a deforestation rate would slightly lower the resulting number.
12 Harper et al. (2007, pp2–3). They continue:

> In practice, this means that the canopy is at least 80% closed. 'Spiny forest and woodland' is primary vegetation dominated by closed-canopy trees or shrubs in the arid southern and south-western regions of Madagascar, sometimes as low as two meters in height in the extreme south. We did not include open-canopy areas, secondary formations or plantations in our estimates of forest and woodland areas. Lightly degraded primary forest and mature secondary forest may be indistinguishable from primary forest in Landsat imagery.

13 Unfortunately our ability to judge the quality of the results is hampered by the lack of cartographic documentation. Harper et al. (2007) probably employed many of the same scenes used by Faramalala (1988a) but it is difficult to know since their metadata are inaccessible.
14 http://en.wikipedia.org/wiki/Madagascar, accessed 21 May 2012. This oft-read site states 'Since the arrival of humans ... Madagascar has lost more than 90% of its original forest' (citing a WWF/National Geographic website as its source) and continues 'key contributors to the loss of forest cover include the use of coffee as a cash crop, illegal logging, and slash-and-burn activities'. The reference to coffee cultivation as a major cause of deforestation is not representative of the broader discourse (this reference is linked to a music scholar article, Emoff (2004), who in turn cites – indirectly – research published in Jarosz (1996)).
15 Stefan Lovgren, 'Madagascar creates millions of acres of new protected areas', *National Geographic News*, May 4, 2007, available at http://news.nationalgeographic.com/news/2007/05/070504-madagascar-parks.html (accessed 15 May 2012). Many other examples are easily accessible via internet searches of terms like '90 percent Madagascar forest'.

References

Agarwal, D. K., Silander, J. A. Jr., Gelfand, A. E., Dewar, R. E. and Mickelson, J. G. Jr. (2005) 'Tropical deforestation in Madagascar: analysis using hierarchical, spatially explicit, Bayesian regression models', *Ecological Modelling*, vol 185, pp105–131.

AGM (1969) *Atlas de Madagascar*, Bureau pour le Développement de la Production Agricole and Association des Géographes de Madagascar, Tananarive.

Allnutt, T. F., Ferrier, S., Manion, G., Powell, G. V. N., Ricketts, T. H., Fisher, B. L., Harper, G. J., Irwin, M. E., Kremen, C., Labat, J.-N., Lees, D. C., Pearce, T. A. and Rakotondrainibe, F. (2008) 'A method for quantifying biodiversity loss and its application to a 50-year record of deforestation across Madagascar', *Conservation Letters*, vol 1, pp173–181.

Amelot, X., Moreau, S. and Carrière, S. M. (2012) 'Des justiciers de la biodiversité aux injustices spatiales: l'exemple de l'extension du réseau d'aires protégées à Madagascar', in D. Blanchon, J. Gardin and S. Moreau (eds) *Justice et injustices environnementales*, Presses Universitaires de Paris Ouest, Paris.

Andreone, F., Carpenter, A. I., Cox, N., du Preez, L., Freeman, K., Furrer, S., Garcia, G., Glaw, F., Glos, J., Knox, D., Köhler, J., Mendelson, J. R. III, Mercurio, V., Mittermeier, R. A., Moore, R. D., Rabibisoa, N. H. C., Randriamahazo, H., Randrianasolo, H., Raminosoa, N. R., Ramilijaona, O. R., Raxworthy, C. J., Vallan, D., Vences, M., Vieites, D. R. and Weldon, C. (2008) 'The challenge of conserving amphibian megadiversity in Madagascar', *PLoS Biology*, vol 6, pp943–946.

Bakoariniaina, L. N., Kusky, T. and Raharimahefa, T. (2006) 'Disappearing Lake Alaotra: monitoring catastrophic erosion, waterway silting, and land degradation hazards in Madagascar using Landsat imagery', *Journal of African Earth Sciences*, vol 44, pp241–252.

Barrett, M. A., Brown, J. L., Morikawa, M. K., Labat, J.-N. and Yoder, A. D. (2010) 'CITES designation for endangered rosewood in Madagascar', *Science*, vol 328, pp1109–1110.

Blasco, F. (1965) 'Aperçu Geographique', in H. Humbert and G. Cours-Darne (eds) *Carte international du tapis végétal et des conditions écologiques à 1/1.000.000 (Notice de la Carte: Madagascar)*, Institut Français de Pondichéry (avec CNRS and ORSTOM), Pondicherry.

Bollen, A. and Donati, G. (2006) 'Conservation status of the littoral forest of southeastern Madagascar: a review', *Oryx*, vol 40, pp57–66.

Bond, W. J., Silander, J. A. Jr., Ranaivonasy, J. and Ratsirarson, J. (2008) 'The antiquity of Madagascar's grasslands and the rise of C4 grassy biomes', *Journal of Biogeography*, vol 35, pp1743–1758.

Bradt, H., Schuurman, D. and Garbutt, N. (1996) *Madagascar Wildlife: A Visitor's Guide*, Bradt Publications, Chalfont St Peter.

Brand, J. (1998) *Das Agro-Ökologische System am Ostabhang Madagaskars: Ressourcen- und Nutzungsdynamik unter Brandrodung*, PhD dissertation, Universität Bern, Switzerland.

Burney, D. A. (1987) 'Late Holocene vegetational change in central Madagascar', *Quaternary Research*, vol 20, pp130–143.

Burney, D. A. (1997) 'Theories and facts regarding Holocene environmental change before and after human colonization', in B. D. Patterson and S. M. Goodman (eds) *Natural Change and Human Impact in Madagascar*, Smithsonian Institution Press, Washington, DC.

Burney, D. A., Burney, L. P., Godfrey, L. R., Jungers, W. L., Goodman, S. M., Wright, H. T. and Jull, A. J. T. (2004) 'A chronology for late prehistoric Madagascar', *Journal of Human Evolution*, vol 47, pp25–63.

Burney, D. A., Robinson, G. S. and Burney, L. P. (2003) 'Sporomiella and the late Holocene extinctions in Madagascar', *Proceedings of the National Academy of Sciences of the United States of America*, vol 100, pp10800–10805.

CI, DEF, CNRE and FTM (1995) *Formations Végétales et Domaine Forestier National de Madagascar* (carte 1:1,000,000), Conservation International, Antananarivo.

CI, IRG, MINENVEF and USAID (2007) *Madagascar: Changement de la Couverture des Forêts Naturelles 1990–2000–2005*, Conservation International, Antananarivo.

Cornet, A. (1974) *Essai de Cartographie Bioclimatique à Madagascar: Notice Explicative*, no 54, ORSTOM, Tananarive.

Craul, M., Chikhi, L., Sousa, V., Olivieri, G. L., Rabesandratana, A., Zimmermann, E. and Radespiel, U. (2009) 'Influence of forest fragmentation on an endangered large-bodied lemur in northwestern Madagascar', *Biological Conservation*, vol 142, pp2862–2871.

DEF (Direction des Eaux et Forêts), Deutsche Forstservice GmbH, Entreprise d'Etudes de Développement Rural 'Mamokatra', and Foiben-Taosarintanin'I Madagasikara (1996) *Inventaire Ecologique Forestier National* (Report), Antananarivo.

Dewar, R. E. (1984) 'Extinctions in Madagascar: the loss of the subfossil fauna', in P. S. Martin and R. G. Klein (eds) *Quaternary Extinctions: A Prehistoric Revolution*, University of Arizona Press, Tucson.

Dufils, J.-M. (2003) 'Remaining forest cover', in S. M. Goodman and J. P. Benstead (eds) *The Natural History of Madagascar*, University of Chicago Press, Chicago.

Du Puy, D. J. and Moat, J. (1996) 'A refined classification of the primary vegetation of Madagascar based on the underlying geology: using GIS to map its distribution and to assess its conservation status', in W. R. Lourenço (ed.) *Biogéographie de Madagascar*, Editions de l'ORSTOM, Paris.

Du Puy, D. J. and Moat J. (1998) 'Vegetation mapping and classification in Madagascar (using GIS): implications and recommendations for the conservation of biodiversity', in C. R. Huxley, J. M. Lock and D. F. Cutler (eds) *Chorology, Taxonomy and Ecology of the Floras of Africa and Madagascar*, Royal Botanic Gardens, Kew.

Du Puy, D. J. and Moat, J. (1999) 'Vegetation mapping and biodiversity conservation in Madagascar Geographical Information Systems', in J. Timberlake and S. Kativu (eds) *African Plants: Biodiversity Taxonomy and Uses*, Royal Botanic Gardens, Kew.

Durbin, J. C. and Ratrimoarisaona, S. (1996) 'Can tourism make a major contribution to the conservation of protected areas in Madagascar?', *Biodiversity and Conservation*, vol 5, pp345–353.

Durkin, L., Steer, M. D. and Belle, E. M. S. (2011) 'Herpetological surveys of forest fragments between Montagne d'Ambre National Park and Ankarana Special Reserve, northern Madagascar', *Herpetological Conservation and Biology*, vol 6, pp114–126.

Elmqvist, T., Pyykönen, M., Tengö, M., Rakotondrasoa, F., Rabakonandrianina, E. and Radimilahy, C. (2007) 'Patterns of loss and regeneration of tropical dry forest in Madagascar: the social institutional context', *PLoS ONE*, vol 2, no 5, pp401–414.

Emoff, R. (2004) 'Spitting into the wind: multi-edged Malagasy environmentalism in song', in K. Dawe (ed.) *Island Musics*, Berg, New York.

Fairhead, J. and Leach, M. (1998) *Reframing Deforestation*, Routledge, London.

Faramalala, M. H. (1988a) *Etude de la Végétation de Madagascar à l'Aide de Données Spatiales*, PhD thesis, Université Paul Sabatier, Toulouse.

Faramalala, M. H. (1988b), 'Cartographie de la vegetation avec l'aide de satellite', in L. Rakotovao, V. Barre and J. Sayer (eds) *L'Equilibre des Ecosystèmes Forestiers à Madagascar: Actes d'un Séminaire International*, IUCN, Gland, Switzerland.

Faramalala, M. H. (1995) *Formations Vegetales et Domaine Forestier National de Madagascar*, Conservation International, Antananarivo.

Ganzhorn, J. U., Lowry, P. P. I., Schatz, G. E. and Sommer, S. (2001) 'The biodiversity of Madagascar: one of the world's hottest hotspots on its way out', *Oryx*, vol 35, pp346–348.

Goodman, S. M. and Benstead, J. P. (2005) 'Updated estimates of biotic diversity and endemism for Madagascar', *Oryx*, vol 39, pp73–77.

Grainger, A. (1984) 'Quantifying changes in forest cover in the humid tropics: overcoming current limitations', *Journal of World Forest Resource Management*, vol 1, pp3–63.

Green, G. M. and Sussman, R. W. (1990) 'Deforestation history of the eastern rain forests of Madagascar from satellite images', *Science*, vol 248, pp212–215.

Guichon, A. (1960) 'La superficie des formations forestières de Madagascar', *Revue Forestière Française*, no 6, pp408–411.

Guillaumet, J.-L. (1984) 'The vegetation: an extraordinary diversity', in A. Jolly, P. Oberlé and R. Albignac (eds) *Key Environments: Madagascar*, IUCN/Pergamon Press, Oxford.

Hannah, L., Dave, R., Lowry, P. P. I., Andelman, S., Andrianarisata, M., Andriamaro, L., Cameron, A., Hijmans, R., Kremen, C., MacKinnon, J., Randrianasolo, H. H., Andriambololonera, S., Razafimpahanana, A., Randriamahazo, H., Randrianarisoa, J., Razafinjatovo, P., Raxworthy, C. J., Schatz, G. E., Tadross, M. and Wilmé, L. (2008) 'Climate change adaptation for conservation in Madagascar', *Biology Letters*, vol 4, pp590–594.

Hannah, L., Rakotosamimanana, B., Ganzhorn, J., Mittermeier, R. A., Olivieri, S., Iyer, L., Rajaobelina, S., Hough, J., Andriamialisoa, F., Bowles, I. and Tilkin, G. (1998) 'Participatory planning, scientific priorities, and landscape conservation in Madagascar', *Environmental Conservation*, vol 25, no 1, pp30–36.
Hanski, I., Koivulehto, H., Cameron, A. and Rahagalala, P. (2007) 'Deforestation and apparent extinctions of endemic forest beetles in Madagascar', *Biology Letters*, vol 3, pp344–347.
Harper, G. J., Steininger, M. K., Tucker, C. J., Juhn, D. and Hawkins, F. (2007) 'Fifty years of deforestation and forest fragmentation in Madagascar', *Environmental Conservation*, vol 34, pp325–333.
Humbert, H. (1927) 'Principaux aspects de la végétation à Madagascar: la destruction d'une flore insulaire par le feu', *Mémoires de l'Académie Malgache*, Fascicule V.
Humbert, H. (1949) 'La dégradation des sols à Madagascar', *Mémoires de l'Institut de Recherche Scientifique de Madagascar*, vol D1, no 1, pp33–52.
Humbert, H. (1955) 'Les territoires phytogéographiques de Madagascar: leur cartographie', *Année Biologique*, vol 31, pp439–448.
Humbert, H. and Cours Darne, G. (1965) *Carte international du tapis végétal et des conditions écologiques à 1/1.000.000 (Notice de la Carte: Madagascar)*, Institut Français de Pondichéry (avec CNRS and ORSTOM), Pondicherry.
Hume, D. W. (2006) 'Swidden agriculture and conservation in eastern Madagascar: stakeholder perspectives and cultural belief systems', *Conservation and Society*, vol 4, no 2, pp287–303.
Ingram, J. C. and Dawson, T. P. (2005) 'Inter-annual analysis of deforestation hotspots in Madagascar from high temporal resolution satellite observations', *International Journal of Remote Sensing*, vol 26, no 7, pp1447–1461.
Ingram, J. C. and Dawson, T. P. (2006) 'Forest cover, conditions, and ecology in human-impacted forests, south-eastern Madagascar', *Conservation and Society*, vol 4, pp194–230.
Irwin, M. T., Johnson, S. E. and Wright, P. C. (2005) 'The state of lemur conservation in south-eastern Madagascar: population and habitat assessments for diurnal and cathemeral lemurs using surveys, satellite images and GIS', *Oryx*, vol 39, pp204–218.
Jarosz, L. (1996) 'Defining deforestation in Madagascar', in R. Peet and M. Watts (eds) *Liberation Ecologies*, Routledge, London.
Jenkins, M. D. (1987) *Madagascar: An Environmental Profile*, IUCN/UNEP/WWF, Gland, Switzerland and Cambridge.
Johnson, S. E., Ingraldi, C., Ralainasolo, F. B., Andriamaharoa, H. E., Ludovic, R., Birkinshaw, C. R., Wright, P. C. and Ratsimbazafy, J. H. (2011) 'Gray-headed lemur (*Eulemur cinereiceps*) abundance and forest structure dynamics at Manombo, Madagascar', *Biotropica*, vol 43, pp371–379.
Jolly, A., Oberlé, P. and Albignac, R. (1984) *Key Environments: Madagascar*, IUCN/Pergamon Press, Oxford.
Koechlin, J., Guillaumet, J.-L. and Morat, P. (1974) *Flore et Végétation de Madagascar*, J. Cramer, Vaduz.
Kull, C. A. (1998) 'Leimavo revisited: agrarian land-use change in the highlands of Madagascar', *Professional Geographer*, vol 50, pp163–176.
Kull, C. A. (2000) 'Deforestation, erosion, and fire: degradation myths in the environmental history of Madagascar', *Environment and History*, vol 6, pp421–450.

Kull, C. A. (2004) *Isle of Fire*, University of Chicago Press, Chicago.

Kull, C. A. (2012) 'Air photo evidence of historical land cover change in the highlands: wetlands and grasslands give way to crops and woodlots', *Madagascar Conservation and Development*, vol 7, pp144–152.

Laney, R. M. (2002) 'Disaggregating induced intensification for land-change analysis: a case study from Madagascar', *Annals of the Association of American Geographers*, vol 92, pp702–726.

Lanly, J. P. (1981) *Tropical Forest Resources Assessment Project (in the framework of the Global Environment Monitoring System – GEMS): Forest Resources of Tropical Africa*, FAO, Rome.

Larson, B. M. H. (2011) *Metaphors for Environmental Sustainability: Redefining our Relationship with Nature*, Yale University Press, New Haven.

Lehman, S. M., Rajaonson, A. and Day, S. (2005) 'Edge effects and their influence on lemur density and distribution in Southeast Madagascar', *American Journal of Physical Anthropology*, vol 129, pp232–241.

Lehman, S. M., Rajaonson, A. and Day, S. (2006) 'Lemur responses to edge effects in the Vohibola III classified forest, Madagascar', *American Journal of Primatology*, vol 68, pp293–299.

Lehtinen, R. M., Ramanamanjato, J.-B. and Raveloarison, J. (2003) 'Edge effects and extinction proneness in a herpetofauna from Madagascar', *Biodiversity and Conservation*, vol 12, pp1357–1370.

Leroy, J. E. (1978) 'Composition, origin and affinities of the Madagascar vascular flora', *Annals of the Missouri Botanical Garden*, vol 65, pp535–589.

Loh J., Randers, J., MacGillivray, A., Kapos, V., Jenkins, M., Groombridge, B., Cox, N. and Warren, B. (1999) *Living Planet Report*, WWF, Gland, Switzerland.

Lowry, P. P. (1986) *A Systematic Study of Three Genera of Araliaceae Endemic to or Centered in New Caledonia*, PhD dissertation, Washington University, St. Louis.

Lowry, P. P. I., Schatz, G. E. and Phillipson, P. B. (1997) 'The classification of natural and anthropogenic vegetation in Madagascar', in S. M. Goodman and B. D. Patterson (eds) *Natural Change and Human Impact in Madagascar*, Smithsonian Institution Press, Washington, DC.

Mayaux, P., Gond, V. and Bartholomé, E. (2000) 'A near-real time forest-cover map of Madagascar derived from SPOT-4 VEGETATION data', *International Journal of Remote Sensing*, vol 21, pp3139–3144.

McConnell, W. J. (2002) 'Madagascar: emerald isle or paradise lost?', *Environment*, vol 44, pp10–22.

McConnell, W. J. and Sweeney, S. P. (2005) 'Challenges of forest governance in Madagascar', *The Geographical Journal*, vol 171, pp223–238.

McConnell, W. J., Sweeney, S. P. and Mulley, B. (2004) 'Physical and social access to land: spatio-temporal patterns of agricultural expansion in Madagascar', *Agriculture, Ecosystems and Environment*, vol 101, pp171–184.

Mittermeier, R. A. (1986) *An Action Plan for Conservation of Biological Diversity in Madagascar*, World Wildlife Fund, Washington, DC.

Mittermeier, R. A., Louis, E. E., Richardson, M., Schwitzer, C., Langrand, O., Rylands, A. B., Hawkins, F., Rajaobelina, S., Ratsimbazafy, J., Rasoloarison, R., Roos, C., Kappeler, P. M. and Mackinnon, J. (2010) *Lemurs of Madagascar*, 3rd edn, Conservation International, Washington, DC.

Mittermeier, R. A., Myers, N., Gil, P. R. and Mittermeier, C. G. (1999) *Hotspots: Earth's Biologically Richest and Most Endangered Terrestrial Ecoregions*, Cemex, Conservation International and Agrupacion Sierra Madre, Monterrey, Mexico.

Myers, N. (1988) 'Threatened biotas: hot spots in tropical forests', *Environmentalist*, vol 8, pp187–208.

Myers, N., Mittermeier, R. A., Mittermeier, C. G., Da Fonseca, G. A. B. and Kent, J. (2000) 'Biodiversity hotspots for conservation priorities', *Nature*, vol 403, no 6772, pp853–858.

Nelson, R. and Horning, N. (1993) 'AVHRR-LAC estimates of forest area in Madagascar, 1990', *International Journal of Remote Sensing*, vol 14, no 8, pp1463–1475.

Norris, S. (2006) 'Madagascar defiant', *BioScience*, vol 56, no 12, pp960–965.

Peet, R., Robbins, P. and Watts, M. (2011) *Global Political Ecology*, Routledge, London.

Perrier de la Bâthie, H. (1921) 'La végétation Malgache', *Annales du Musée Colonial de Marseille*, vol Sér. 3, v. 9, pp1–266.

Perrier de la Bâthie, H. (1936) *Biogéographie des Plantes de Madagascar*, Société d'Editions Géographiques, Maritimes et Coloniales, Paris.

Pollini, J. (2010) 'Environmental degradation narratives in Madagascar: from colonial hegemonies to humanist revisionism', *Geoforum*, vol 41, pp711–722.

Quéméré, E., Amelot, X., Pierson, J., Crouau-Roy, B. and Chikhi, L. (2012) 'Genetic data suggest a natural pre-human origin of open habitats in northern Madagascar and question the deforestation narrative in this region', *Proceedings of the National Academy of Sciences*, vol 109, pp13023–13033.

Raharimahefa, T. and Kusky, T. M. (2010) 'Environmental monitoring of Bombetoka Bay and the Betsiboka Estuary, Madagascar, using multi-temporal satellite data', *Journal of Earth Science*, vol 21, pp210–226.

Rakoto Ramiarantsoa, H. (1995) *Chair de la Terre, Oeil de l'Eau: Paysanneries et Recompositions de Campagnes en Imerina (Madagascar)*, Éditions de l'Orstom, Paris.

Rakoto Ramiarantsoa, H., Blanc-Pamard, C. and Pinton, F. (2012) *Géopolitique et Environnement: Les Leçons de l'Expérience Malgache*, IRD, Marseille.

Rudel, T. K., Coomes, O. T., Moran, E. F., Achard, F., Angelsen, A., Xu, J. and Lambin, E. F. (2005) 'Forest transitions: towards a global understanding of land use change', *Global Environmental Change*, vol 15, pp23–31.

Sandy, C. (2006) 'Real and imagined landscapes: land use and conservation in the Menabe', *Conservation and Society*, vol 4, pp304–324.

Scales, I. R. (2011) 'Farming at the forest frontier: land use and landscape change in western Madagascar, 1896–2005', *Environment and History*, vol 17, pp499–524.

Vågen, T.-G. (2006) 'Remote sensing of complex land use change trajectories: a case study from the highlands of Madagascar', *Agriculture, Ecosystems and Environment*, vol 115, pp219–228.

Virah-Sawmy, M. (2009) 'Ecosystem management in Madagascar during global change', *Conservation Letters*, vol 2, pp163–170.

Volampeno, N. S. M., Masters, J. C. and Downs, C. T. (2010) 'A population estimate of blue-eyed black lemurs in Ankarafa Forest, Sahamalaza-Iles Radama National Park, Madagascar', *Folia Primatologica*, vol 81, pp305–314.

Whitehurst, A. S., Sexton, J. O. and Dollar, L. (2009) 'Land cover change in western Madagascar's dry deciduous forests: a comparison of forest changes in and around Kirindy Mite National Park', *Oryx*, vol 43, pp275–283.

Wilkie, D. S. and Finn, J. T. (1996) *Remote Sensing Imagery for Natural Resource Management: A Guide for First Time Users*, Columbia University Press, New York.

Wilmé, L., Goodman, S. M. and Ganzhorn, J. U. (2006) 'Biogeographic evolution of Madagascar's microendemic biota', *Science*, vol 312, pp1063–1065.

Yoder, A. D. and Nowak, M. D. (2006) 'Has vicariance or dispersal been the predominant biogeographic force in Madagascar? Only time will tell', *Annual Review of Ecology, Evolution, and Systematics*, vol 37, pp405–431.

5 The drivers of deforestation and the complexity of land use in Madagascar

Ivan R. Scales

Introduction

Deforestation plays a central role in Madagascar's environmental discourse. The 1984 national strategy for conservation and sustainable development, for example, warned that forest clearance would lead to 'brutal and apparently irreversible savannisation' (MEEF, 1984, p15), while a World Bank report (1996, p10) stated that:

> Madagascar has already lost 80 percent of its original forest cover, and the rest is under severe pressure for reasons that relate principally to poverty ... Traditional forms of itinerant agriculture, which are relied on by the poor because they have no incentives to intensify production, result in the burning of savanna and forests.

Forest clearance is thus painted as a one-way process of degradation, driven by poverty and population growth and inevitably leading to the permanent loss of forest (Scales, 2011).

Madagascar's deforestation narrative is simple and appealing. It presents a clear problem and an obvious solution – by reducing poverty and persuading Malagasy farmers to adopt different livelihoods, forest loss could be avoided and Madagascar's biodiversity protected. However, I argue in this chapter that policy has tended to assume the importance of population growth and poverty as drivers of forest clearance, without exploring the role of other factors shaping land use. It has also tended to ignore the role of other land uses, especially large-scale commercial agriculture.

In the first section of this chapter I take an historical approach to forest loss, looking at the land uses that have led to deforestation during the twentieth century. I show that a range of land uses, and not simply forest clearance agriculture by rural households, have led to changes in forest cover. In the second section I focus on the land use practices of rural households. While the received wisdom and conservation policy have tended to focus on the role of poverty as a driver of deforestation, I show that rural households make land-use decisions based on a complex range of factors.

Furthermore, rather than being the preserve of poor households, I argue that forest clearance can be practised by a wide range of households, rich and poor. This leads me to the third section, where I explore the broader political and economic factors that shape land use decisions. I show how policies and changes in the price of commodities have stimulated booms in the cultivation of cash crops and associated forest loss.

I finish the chapter by considering the future of rural land use and forest clearance and discussing the factors that have contributed to the failure of forest conservation policy. I conclude that, while rural households have played and continue to play a significant role in forest clearance, it is important not to ignore other land uses but more importantly the various environmental, cultural, political and economic factors that shape land use decisions.

Forest clearance and why the words we use to describe it matter

Before I begin there is an important semantic issue I must deal with. In Madagascar's environmental discourse, forest clearance by rural households is commonly referred to either as 'slash-and-burn' or by the Malagasy term *tavy*. Both of these are problematic. 'Slash-and-burn' is a highly emotive and pejorative term. The language is aggressive, painting a picture of wanton destruction. To see just how loaded it is, contrast it, for example, with other commonly used terms such as 'swidden', 'shifting cultivation' or less commonly used terms such as forest agrarian systems.

Another problem with the term is that it only describes part of the agricultural system – the cutting and burning of vegetation is only the first step of the process. What happens afterwards (for example what crops are planted, the order crops are planted in, whether the land is left completely fallow or not, how long it is left fallow for, whether the system is explicitly organised around rotation) is highly variable (Mazoyer and Roudart, 2006; Ruthenberg, 1976). As a result of these complexities and nuances, swidden agriculture has been described as 'not one system but hundreds or thousands of systems' (Brookfield and Padoch, 1994, p7) and one of the most complex forms of agriculture in the world (Thrupp et al., 1997).

The lack of consideration of the diversity of forest agrarian systems in Madagascar is most evident in the common and erroneous use of the term *tavy* to describe all forms of agriculture based on the cutting and burning of forest. *Tavy* refers specifically to a shifting system of forest clearance for the cultivation of rain-fed rice, as practised in the eastern rainforests of Madagascar. In the drylands of the west and southwest, for example, forest clearance is referred to as *hatsake* and involves the cultivation of maize rather than rice. In the northeast of the island, Betsimisaraka farmers refer to swidden cultivation as *jinja* (Sodikoff, 2012). And yet *tavy* has become shorthand for deforestation by rural households and Madagascar's environmental problems (Scales, 2011).

Finally, the term 'slash-and-burn' has become unhelpfully encumbered with negative connotations. Not only is it assumed to be irrational and unsustainable, it is automatically equated with the total and permanent removal of forest cover. In fact, there is evidence of forest agrarian systems in Africa dating back 10,000 years and in many parts of the tropics such systems have lasted thousands of years (Mazoyer and Roudart, 2006; Willis et al., 2004). The persistence of forest agrarian systems is down to the fact that they do not by nature involve permanent forest loss or declining soil fertility, as long as vegetation is given sufficient time to regenerate and soil fertility enough time to recuperate (Mazoyer and Roudart, 2006). This in turn is dependent on other demographic, economic and political factors (Angelsen, 1995; Cramb et al., 2009; Skole et al., 1994). Rather than being a short-term practice driven by poverty or stupidity, the cutting and burning of forest is a rational economic choice for farmers in the tropics (Ickowitz, 2006).

As William McConnell and Christian Kull showed in their chapter on estimating forest cover and rates of forest loss (Chapter 4), the choice of words and categories used when dealing with land use and land cover change has huge implications for what is measured and the 'facts' and figures produced about environmental change. There is no single universally accepted definition of 'deforestation', which can range from the total removal of forest cover to small changes in forest composition and structure (by selective logging for example). Often there is no distinction made between permanent and temporary conversions or between conversion and alteration (Angelsen, 1995). In this chapter, I use the word 'deforestation' to mean the total removal of forest cover (rather than selective logging for example). When discussing the land use practices of rural households, I avoid the term 'slash-and-burn' and use local terms such as *tavy, hatsake* and *jinja* or the more general term 'swidden agriculture'. 'Swidden' is a word of Scandinavian origin, meaning 'land clearing by burning', and refers to a system based on forest clearance using fire and employing a fallow phase longer than the cultivation phase of annual crops and dominated by woody vegetation (Mertz et al., 2009).

Lessons from the past: placing land use change and forest loss in Madagascar in an historical context

As Chapter 3 by Robert Dewar shows, deforestation is not a recent phenomenon in Madagascar. Ever since the first humans arrived on the island, they have modified its vegetation through the use of fire and there is no doubt that human action has led to considerable land cover change. However, our understanding of the drivers of land use and land cover change in Madagascar has been hampered by a narrow reliance on satellite imagery, which only became readily available in the 1970s. Very few studies on landscape change in Madagascar have gone further into the past. As I show in this chapter, taking a longer view of land use and land cover change can provide fascinating and useful insights into the drivers of forest loss.

In this section, I focus on forest lost during the twentieth century, especially during the colonial period. This is both because detailed archival sources allow us to reconstruct land use and land cover change and because the arrival of French colonialism in 1896 marked a dramatic change in the island's politics and economy, with significant implications for its forested landscapes. Looking at the twentieth century and the colonial period more specifically provides a deeper insight into the influence of broader political and economic factors shaping Madagascar's landscapes.

The impact of French colonial policy on Madagascar's forests was considerable and largely down to the government's desire to make the island profitable by encouraging the exploitation of natural resources such as timber and the cultivation of exportable cash crops (Jarosz, 1993; Randrianja and Ellis, 2009; Schlemmer, 1980). The land-use practices of rural Malagasy were therefore seen as a threat to Madagascar's forests, since swidden agriculture involved the clearing of valuable timber. Not only that, but 'traditional' livelihoods represented a barrier to the government's vision of progress. This was because, being largely subsistence households, they produced food for themselves rather than for markets, growing the types of crops they wanted to eat rather than the cash crops that might be exported (and taxed!). They also tended to keep their labour to themselves, rather than selling it to plantation owners or making it available for the government's large infrastructure projects such as roads or dams (see Chapter 6 by Scales for more on colonial policies and their impacts on the environment).

In order to achieve its plans, the French colonial government implemented a host of policies, including: i) the introduction of cash taxes to encourage farmers to switch from subsistence products to cash crops; ii) a system of forced labour to develop infrastructure; iii) the introduction of a system of private land tenure and land registration to replace local customary land tenure; and iv) the awarding of large concessions to (mostly French) individuals and companies, so that they could develop large plantations of cash crops (Randrianja and Ellis, 2009; Scales, 2011; Schlemmer, 1980).

My own research in the Menabe region of western Madagascar shows just how much impact such large-scale political and economic changes had on forests, and how important the colonial period was in terms of overall forest loss.[1] Prior to the arrival of the French colonial government in 1896, Menabe's economy was based primarily on extensive zebu pastoralism, with small amounts of swidden agriculture carried out mostly by slaves and some rice cultivation in the flood plains (Fauroux, 1980; Le Bourdiec, 1980; Schlemmer, 1980). The landscape was thus shaped largely by pastoralism rather than agriculture. However, once colonial control had been established over the region, the government rapidly went about trying to make the region profitable, with major implications for the region's dry-deciduous forest.

The government's aim was to increase the economic productivity of the region and boost tax income by focusing on agriculture. However, it was frustrated by the lack of willingness of rural Sakalava households, whose livelihoods were still primarily based on extensive cattle pastoralism, to undertake waged labour and participate in the cash economy (Fauroux, 1980; Scales, 2011). It therefore deployed a fairly standard set of policies in order to encourage/force them to change. By 1905, more than 50,000 hectares of concessions had been handed out, mostly in the fertile and seasonally flooded river valleys. One such concession was turned into a sisal plantation in the late 1950s and early 1960s, resulting in the clearance of 9,000 hectares of forest. This was a significant deforestation event, contributing 9.4 per cent of all the forest lost in Central Menabe between 1954 and 2005 (Scales, 2011).

Government policy resulted in a number of agricultural booms. The first involved irrigated rice, which was a relatively new technique in Menabe having been introduced to the region by Merina and Betsileo migrants during the course of the nineteenth century (Le Bourdiec, 1980). The second boom involved butter beans (*Phaseolus lunatus*), known as *pois du Cap* in French and lima beans in the United States. While the rice boom had been limited by the availability of irrigated land, the boom in butter bean cultivation saw the spread of agriculture onto the seasonally flooded alluvial soils around the region's main rivers. The large concessions awarded to expatriates underpinned a system of sharecropping, where concession owners offered the use of part of their land in return for a 50 per cent share of the crops produced. This meant they could convert large areas of the region's fertile river valleys to the cultivation of butter beans, with minimal effort, by relying on migrant labour.

Although the rice and butter bean booms were significant in the conversion of Menabe from a landscape based on extensive pastoralism to one based on the cultivation of crops, the expansion of agriculture (and therefore the conversion of forest to farmland) was limited by the availability of water. During the 1930s, a third agricultural boom occurred, this time in the cultivation of maize. This boom was particularly important in terms of landscape change due to the fact that unlike rice or butter bean, maize does not require irrigation or flooded alluvial soils. By cutting and burning trees and relying on the water provided by precipitation during the three-month long rainy season, high yields can be obtained on cleared land and maize can be grown in areas unsuitable for other crops. The maize boom therefore impacted the forest in a way that the previous agricultural booms had not.

Although swidden agriculture had existed in the region prior to the arrival of French colonialism, it had been primarily for subsistence. The 1930s thus saw a change in the underlying drivers of forest clearance from subsistence needs to the cultivation of maize as a cash crop grown for export. The records in the colonial archives show that maize exports passing through the port of Morondava went from 7,000 tonnes in 1935 to a peak of 29,500 tonnes in 1939 – more than a fourfold increase in four years. Approximately 45,000 hectares of forest were cleared during the maize boom. To put this into context, 33,000 hectares of forest were lost in Central Menabe between 1954 and 2005 (Scales, 2011). So more forest was lost in the four-year long maize boom than in the 51 years after 1954.

There were also long-term implications of the colonial government's policies. The possibility of sharecropping cash crops on forested land drew more migrants to the region, mainly from southern Madagascar. This not only contributed to deforestation, as migrants were allowed to clear forest by landowners in return for half of their crop, but it had a considerable impact on the region's demography and resulted in the establishment of new villages. The rapid demographic changes, and the resulting complex inter-ethnic dynamics, have had profound implications for forest use and management, not least because migrants often have a very different set of beliefs and attitudes when it comes to forest use (Scales, 2012).

There are important lessons to take away from the island's colonial environmental history. The first is that by taking our understanding further back in history, we get a better understanding of forest cover dynamics and the drivers of landscape change. The second is that colonial history shows us that other land uses, and not just swidden agriculture, have played a major role in forest loss during twentieth century. Large commercial plantations of cash crops, such as sisal, played a significant role in forest loss in many regions, not only through the direct clearance of forest but also by occupying the most fertile land and forcing subsistence cultivators into forest areas (Jarosz, 1993; Scales, 2011). Madagascar is currently experiencing dramatic political and economic changes, including increased trade liberalisation, foreign land acquisitions and a focus on cash crops (Cotula and Vermeulen, 2009). History warns us that these are likely to have significant impacts on forests.

Rural households and the complexity of land-use decisions

Household farms occupy a central position in the debate on tropical deforestation in Madagascar. The problem with the received wisdom regarding forest loss is that analyses of deforestation have tended to treat rural households as a homogenous group, characterising them as inefficient and destructive. The language used by policymakers to describe swidden agriculture is often damning. In Madagascar's Environmental Charter, published in 1990, it is labelled as 'a suicidal agricultural system based on fire' (MEP, 1990, p18). In this section, I show the range of factors that shape the land-use decisions of rural households and reveal the complexity of swidden agriculture. Rather than resorting to unhelpful and sweeping language, it is worth understanding the underpinning rationale of swidden agriculture.

The drivers of deforestation and the complexity of land use 111

Swidden agriculture: a risk averse, low labour and low capital system

The basic principle of swidden agriculture is to provide sunlight and a nutrient rich ash for crops. Once land is cultivated, fire is also used to clear weeds. The basic process is as follows. Smaller trees and shrubs are cut during the dry season and left to dry to create a large volume of fuel. Vegetation is cut so that canopies face the dominant wind direction to encourage rapid and intense burning (Figure 5.1) and reduce the chances of fire spreading to surrounding vegetation. Seeds may be planted before the rains arrive or after. This decision is crucial, especially in the semi-arid west and arid south, where the majority of rain falls in a short period between January and March. Later planting has the advantage that seeds are exposed to more rain, ensuring a higher proportion of germination, but it leaves less time for planting before the peak of the rainy season and means less land can be cultivated.

Although swidden agriculture is geared towards a main crop (rice in the eastern rainforests and maize in the south and west), it is usually followed by a secondary subsistence crop, with manioc (*Manihot esculenta*) being a

Figure 5.1 Forest recently cleared for *hatsake* (swidden cultivation) in western Madagascar (photograph taken in October prior to the start of the rainy season)

Source: photograph by Helen Scales.

common choice. As well as the primary and secondary crops, there are often fruit trees (for example bananas) and 'garden' crops such as melons and beans (Figure 5.2). Once land has been abandoned, there can be a number of post-clearance land uses, such as grazing for cattle and the conversion of shrubs to charcoal. It is important to note that forest clearance is rarely an exclusive activity for rural households, with families engaged in a range of other subsistence and income earning activities ranging from horticulture in household gardens, livestock rearing, the collection of non-timber forest products, the cultivation of high-value cash crops and also selling their labour (Casse et al., 2004).

The most advantageous aspects of swidden systems are their low labour and capital requirements. A man (vegetation cutting and burning is almost exclusively a male activity while weeding and harvesting tends to be carried out by women) can clear enough land to feed his family with little or no outside help. The nutrients released by the ash means that no additional and costly fertilizer inputs are required. In Menabe, maize production through *hatsake* can produce yields of over a tonne per hectare per year for two years on nutrient poor and rain-fed soils with no additional inputs or irrigation (Scales, 2011).

Figure 5.2 First year of cultivation following forest clearance (photograph taken in April following rainy season)

Source: photograph by Ivan R. Scales.

What happens to cleared land after the first agricultural cycle depends on the biogeographical conditions as well as the priorities of the household. Tandroy migrants in the dry forests of Menabe, for example, will grow maize as a cash crop until the soil is exhausted and then abandon the land. The goal is to use the cash income to buy zebu cattle and return south (Réau, 2002). This is therefore not a rotational system and the aim is rapid wealth creation. In the same region, as well as temporary migrants, there are settled villages that were established decades ago and have relied on swidden cultivation as the basis of their livelihoods. Burning forest releases enough nutrients for two to three years of maize cultivation, after which either manioc or groundnut are grown. After five years, soil nutrients are exhausted and land is left fallow. For these households, *hatsake* is, in principle at least, a shifting system.

Regardless of whether cultivation involves rotation or not, or whether it is primarily for subsistence or cash cropping, the key to understanding it is to see it as a form of agriculture that makes the most out of minimal inputs (Ickowitz, 2006; Mazoyer and Roudart, 2006). Seen through this lens, it is not surprising that so many households rely on it. It is a way of both ensuring food security, generating a surplus and, if crop prices are high, accumulating wealth. However, as forest clearance is outlawed, households are forced either to adopt different livelihood strategies or burn clandestinely.[2]

After the forest is burned: the social and ecological dynamics of cleared forest

What happens once land has been abandoned for cultivation, both in terms of ecological dynamics and other land uses, is poorly understood and in need of much more attention. As research in the eastern rainforests has shown, fallow use and vegetation dynamics can be complex. Betsimisaraka farmers, for example, have different names for different types of fallow, depending on species life form, species composition, particular stage in the cycle, height of vegetation and the agricultural potential of fallow (Styger et al., 2007). There are management strategies to go along with the different fallow types. So while policy documents refer to fallow land in the eastern rainforests simply as *savoka* (in the west this is referred to as *monka*), the reality is more complex. Farmers are aware of optimal fallow periods and can recognise when a fallow is ready to be farmed again. To produce good rice crops (>1.5 tonnes per hectare), a fallow period of at least three years is required for the first two cycles, followed by five years for the third cycle, eight years for the fourth cycle, 12 years for the fifth and 20 years for the sixth. The time needed for the soil to recover its nutrients thus increases with each cycle of cultivation (Styger et al., 2007).

The important question for an environmental management perspective is how swidden systems are affected by the shortening of fallow periods that tends to accompany increased demands on land. Research suggests that shortened fallow periods lead to nutrient loss while repeated burning tends to hinder the regeneration of native tree species and favours exotic and invasive shrubby and herbaceous species, although this depends on the frequency, intensity and seasonality of fires (Bloesch, 1999; De Wilde et al., 2012; Styger et al., 2007).

The cultural dimensions of forest clearance

While the decision to clear forest has major pragmatic dimensions, to simply reduce swidden agriculture to a calculated cost/benefit analysis on the part of rural Malagasy farmers would be to miss out important cultural dimensions. Chapter 14 (by Kaufmann) shows how Malagasy beliefs can shape attitudes towards nature and these beliefs can play an important role in how rural households relate to and use forests.

For rural households in Madagascar, forests are considered to be *tany fivelomana* – land where one can create a livelihood. As Christian Kull (2000a, p433) puts it: 'Malagasy farmers are not sacrificing nature for short-term needs, they are instead transforming nature to be of more use to them. It is a matter of perspective.' My research has shown that, in Menabe, the Sakalava perceive the forest in a range of ways – both material and spiritual, beneficial and potentially dangerous (Scales, 2012). The forest is considered to be materially beneficial because it provides a broad range of resources – building materials, medicines, fuel, meat, honey and wild tubers such as *ovy* (*Dioscorea* spp.) and *tavolo* (*Tacca leontopetaloides*) which are important foods, especially towards the end of the dry season when household food reserves are running low. The forest can also be cleared for farming.

However, forests are also considered to be dangerous places – especially for women and children – because of the presence of bandits (*dahalo*) and spirits. There are areas of *ala fady* (taboo forest), where people are not allowed to collect wood, clear forest and in some cases even travel through. These taboos mostly relate to links with the spiritual world, in particular ancestors.

In Malagasy cosmology, there is a strong link between nature and the supernatural. Sakalava households in the Menabe region of western Madagascar believe that the forest is inhabited by diverse spirits, who are wary of humans and can only be found deep in the forest. They are associated with old trees in the forest such as *kily* (*Tamarindus indica*) and *renala* (*Adansonia grandidieri*). Such trees often form the focus points of *zomba* (sacred places for traditional practices) and people will make offerings (usually honey, rum or tobacco) when rains are poor and harvests are looking bad. It is also important to make offerings prior to clearing forest to avoid ancestral retribution. So forest clearance is acceptable as long as the correct rituals are observed.

Such cultural and spiritual dimensions to forest clearance have been found elsewhere in Madagascar. Hume (2006), working in the eastern rainforests between Antananarivo and Toamasina, reports on the ritualised aspects of *tavy*. As in Menabe, prayers and offerings are made prior to clearing forest. Besimisaraka farmers practising *tavy* in the eastern rainforests believe that the soil must be 'hot' in order for seeds to germinate and produce strong seedlings. Such 'heat' can be produced by burning forest or adding fertiliser to the soil. The nutrient releasing aspect of swidden agriculture is thus articulated through the concept of heat. Furthermore, the practice of *tavy* is related closely to ethnic identity, with farmers believing that a part of themselves would be lost if they stopped practising *tavy*.

The underlying drivers of forest loss

In explaining the drivers of forest loss, I have mostly considered factors at the household level: the decisions made by farmers based on a set of household capabilities and priorities, environmental constraints, as well as cultural values. However, it is also important to consider how households respond to broader political and economic factors. Contrary to the received wisdom, forest clearance is not driven simply by the need to feed a family and a lack of alternatives.

The problem with many studies of the causes of tropical deforestation, in Madagascar and beyond, is that they tend to focus on a narrow range of obvious precursors to deforestation (such as an increase in the number of people in a forested area), without considering the other factors or mechanisms leading to such precursors or how households might actually respond to such conditions. The assumption is automatically that more people equals less forest, despite evidence from around the world that this is not necessarily the case (Fairhead and Leach, 1998; Geist and Lambin, 2002). Population growth in itself is not enough to lead to deforestation. In the context of abundant and unregulated resources – at a forest frontier, for example – population growth, especially through migration, can lead to rapid forest loss. However, there is a wealth of research, in Madagascar and elsewhere in the tropics, showing that population growth can act as a stimulus for landscape enhancement and intensification (Boserup, 1965; Kull, 1998; Tiffen et al., 1994). The link between poverty, economic growth and forest cover is equally complex. Conservation policy in Madagascar is often built on the assumption that poverty constrains livelihood choices and forces households to clear forest, and that reducing rural poverty will therefore automatically lead to reductions in forest loss. However, as Kull (2000a, p433) points out: 'Give the average Malagasy *tavy* farmer more money, and deforestation may just as well increase as they utilise better tools and pay for additional labour.'

In Madagascar, studies have identified a wide range of factors linked to forest loss, including cash cropping by poor migrants (Casse et al., 2004;

Réau, 2002); cash cropping by wealthy households using migrant labour (Minten and Méral, 2006; Scales, 2011); new roads opening up access to remote forest areas (Moser, 2008; Tidd et al., 2001); increases in international commodity prices (Casse et al., 2004; Scales, 2011); clearance for large-scale commercial plantations (Jarosz, 1993; Scales, 2011); and political and economic factors operating at a global level (Minten and Méral, 2006; Scales, 2011). Studies have found considerable variation in drivers over time (Scales, 2011), as well as differences between regions and at different spatial levels (Moser, 2008).

The other problem with analyses of deforestation to date is that as well as focusing on a narrow range of factors, they have often mixed direct and indirect factors in their analyses. Bromley (1999) uses the concept of intent to distinguish between proximate factors and underlying causes of deforestation. Deforestation does not occur by accident or neglect but for a specific purpose. According to Bromley, it is that purpose which is the underlying driver. For example, migration is often cited as a 'driver' of deforestation. But migration is itself the result of social, economic and political drivers. Similarly, roads are often cited as 'drivers' of deforestation, but while they facilitate the movement of people into (and crops out of) forested areas, they are not themselves 'drivers'. The drivers relate to why households moved to the forest frontier in the first place and the factors that result in specific land use choices over other possibilities. What happens when people actually move into forest areas depends on other drivers of land use decisions. Thinking about swidden cultivation specifically, this chapter has shown that it is a land use that can result from a wide range of drivers. This might seem like I am playing semantic games, but conceptually the distinction is crucial. To put it simply, if we want to understand the drivers of forest loss we must understand *why* people are in forested areas (i.e. not just *what* has facilitated their movement) and *why* they carry out forest clearance as opposed to other livelihood strategies. Too much research on deforestation has focused on the *where* and the *what* of forest clearance without thinking enough about the *why*.

So what drives the conversion of forest to other forms of land cover in Madagascar? The short answer is a diverse range of political, economic, cultural, demographic and environmental factors operating at a range of spatial levels. While this is perhaps unsatisfactory as an answer for shaping policy, it reflects both the complexity of land-use decisions and the dearth of sophisticated, historically informed, multi-level and multi-factor analyses of deforestation. Until more of these are carried out and incorporated into comparative studies and meta-analyses, it will not be possible to identity which factors, out of the diverse range of possible drivers, have been the most important in driving forest loss in Madagascar.

The role of international trade in forest loss: a case study from southwestern Madagascar

Although more sophisticated research on the drivers of deforestation is urgently required, recent studies have emerged that challenge the received wisdom and suggest that policymakers need to be much more aware of factors other than poverty and population growth. One such study in southwestern Madagascar, by Minten and Méral (2006), reminds us just how important forces outside Madagascar can be.

Between 1990 and 2000, the region around Toliara experienced a boom in maize cultivation, with a corresponding boom in the clearance of spiny forest. Many regional and national factors contributed to this boom: migration, a lack of secure land tenure rights and a lack of viable alternatives. However, as Minten and Méral (2006) have shown, the ultimate drivers of the boom were political and economic factors operating at the international level.

Up until the 1990s, maize was grown in the region primarily as a subsistence crop. Maize was traded, but mostly on local and regional markets. In the late 1980s, European Union (EU) policy radically changed the economic dynamics of maize cultivation. In 1989, the EU established a set of *Programmes d'Option Spécifiques à l'Eloignement et à l'Insularité* (POSEI). These programmes were designed to stimulate economic development in the outermost regions of the EU, including French Guiana, Guadeloupe, Martinique and Réunion.

In the case of Réunion, a small island and French overseas department approximately 700 km east of Madagascar, the government focused on developing its farming and agri-business sector. With support from POSEI, it offered tax breaks on imported grains in order to make animal feedstock cheaper and support meat production, particularly pig farming. The rapid expansion of pig farming in Réunion created a large demand for animal feed in the form of maize. Although maize was also imported from mainland France and Argentina, Madagascar benefited from the comparative advantage of low transport costs and the ability to satisfy orders at short notice due to its proximity. In order to benefit from the sudden demand, the Société de Production de Stockage et de Manutention des Produits Agricoles (SOPAGRI) was established by the Malagasy government in 1990. It played a central role in connecting rural farmers to the new international market through the construction of a storage silo in Toliara and the establishment of a collection system for maize.

As well as EU policy, other national and international political and economic factors played major roles. After 1980, Madagascar's economy became more open to international trade through liberal reforms to various sectors, including agriculture (Shuttleworth, 1989). This dramatic change was driven by the fact that Madagascar faced economic collapse, having borrowed heavily to fund a range of large-scale development projects.

Madagascar was forced to turn to the International Monetary Fund for emergency funds and in return had to accept conditions imposed as part of a Structural Adjustment Program (see also Chapter 7 by Kull for more on the debt crisis and its impacts). This included the removal of trade barriers and the devaluation of the Malagasy franc.

As a result of increased links to international agricultural commodity markets, maize prices increased 460 per cent between 1985 and 1998 (Casse et al., 2004). The impacts on southwestern Madagascar's maize sector were significant. Maize changed from being a household subsistence crop to an export commodity. Unclear land tenure facilitated the influx of migrants to the region, who were keen to profit from the maize boom. Unsurprisingly, the combination of 'free' land and abundant labour encouraged the expansion of forest clearance by rural households. The impact on the forest was equally dramatic. Between 1990 and 2000, over 200,000 tonnes of maize were exported from Toliara (including 160,000 tonnes to Réunion), leading to the clearing of between 30,000 and 50,000 hectares of forest (Minten and Méral, 2006).[3]

The maize export boom ended rather abruptly in 2000. The causes were once again outside Madagascar. To further assist its growing livestock sector, the government of Réunion built a cereal port and silo to handle imported maize. This allowed the handling of much larger, and therefore cheaper, shipments of maize. Malagasy maize, produced primarily by household farms and transported in smaller quantities on smaller ships, was more expensive than maize produced elsewhere through large-scale agriculture and transported on bigger ships. By 2001, the island was importing virtually all its maize from France and Argentina. However, the trade in maize has not stopped completely. While maize ceased to be an export commodity in southwestern Madagascar, the 1990's boom has had a lasting legacy, leaving in place a commodity network and infrastructure that now focuses on national markets, for example chicken-feed in Antananarivo.

Swidden agriculture in a changing world

Given the contribution of rural households to forest loss, it is not surprising that there have been considerable efforts to change their land use practices. This has ranged from laws banning the practice of burning vegetation (see Chapter 6 by Scales) through to attempts to involve local communities in forest management and the creation of alternative livelihoods (see Chapter 8 by Pollini *et al.*). It is fair to say that, to date, these have had a limited impact on rates of forest loss. While in other parts of the world, under a wide range of social and ecological conditions, swidden agriculture has been replaced by more intensive land uses, in Madagascar it has proved to be persistent (Pollini, 2009). The case studies covered in this chapter provide insights into why this is. It is also important to note that under certain conditions, swidden agriculture can be sustainable. With low population densities and long fallow periods, it can be a highly efficient and productive system and one that has a very long history.

What is the future for swidden agriculture in Madagascar? For possible pointers it is worth considering forest agrarian systems in other parts of the world. Ruthenberg (1976) believes that demographic and economic changes would inevitably result in household farms evolving from shifting cultivation to more permanent and intensive forms of agriculture. In this model, shifting cultivation is therefore the expression of a certain stage in the relationship between population density, technology and price relations. This rather teleological view is echoed by Angelsen (1995), who sees shifting cultivation as an early stage in the evolution of agricultural systems. The idea is that increasing population leads to a progressive shortening of fallow periods until land is permanently cultivated. Once land is permanently cultivated, intensification often occurs, facilitated by technological innovation (Mazoyer and Roudart, 2006).

However, others have argued that such models are overly deterministic. If swidden systems are initially diverse and adaptable, why should their development paths be so uniform? For example, studies of the *citemene* swidden system in Zambia have shown that there is great variety, both within and between communities, in how households respond to economic and political changes (Moore and Vaughan, 1994; Sharpe, 1990). In Vietnam, farmers have responded to population increases by developing a system of complex composite swiddens, incorporating permanent rice fields and cassava crops with subsystems of shifting cassava, home gardens, livestock and fishponds (Rambo and Tran, 2001). In many parts of the tropics, swidden agriculture is thus part of a strategy of flexibility, diversification and, ultimately, risk minimisation (Ickowitz, 2006).

There have been numerous attempts to encourage Malagasy households to change their cultivation methods, primarily based around the intensification of agriculture on fallow land (e.g. the 'Alternatives to Slash-and-Burn' programme discussed in Chapter 13 by Brimont and Bidaud). In western Madagascar, non-governmental organisations (NGOs), working in collaboration with the French *Centre de Coopération Internationale en Recherche Agronomique pour le Développement* (CIRAD), tried to introduce sorghum (*Sorghum bicolor*) to households, but uptake was poor due to the fact that it is not a staple food and there is no real market for it. It failed both as a subsistence crop and a cash crop. During interviews with farmers in the eastern rainforests, Hume (2006) found that farmers tended to rely on *tavy* because of a lack of work and other opportunities. Many also did not have faith in new systems of agriculture. Given the debates in the academic literature about the effectiveness of schemes such as the System of Rice Intensification (SRI), this is hardly surprising. Strategies in the east have often focused on promoting *tanimbary* (irrigated rice cultivation on terraced land) and improved rice varieties. However, this is expensive, requiring dams, infrastructure and technical training. Styger *et al.* (2007), also working in the eastern rainforests, has reported that government agencies and NGOs often encourage the cultivation of fruit trees to produce cash crops as a livelihood alternative. This ignores the reality of rural livelihoods, where the priority is to ensure basic food provision, and the precarious nature of cash cropping dependent on faltering infrastructure and uncertain markets.

The other major question about the future of Madagascar's forest frontiers is over the role that cash crops will play. The historical evidence from Menabe and the eastern forests, together with the recent evidence from southwestern Madagascar, has shown the importance of international markets in stimulating forest loss. In 2008, spikes in the price of basic agricultural commodities, including maize, were largely related to a rise in the price of petroleum – food production is increasingly dependent on fossil fuels for mechanisation, chemical inputs and transport of bulk commodities around the world. Another key factor was increased demand for feedstock for the production of biofuels (FAO, 2008).[4] As well as being one of the principle global staples, maize is one of the main crops converted to bioethanol. As the EU and US, and increasingly China and India, promote biofuels for reasons of both energy security and climate change mitigation, demand for maize is likely to increase (Murphy, 2010). This presents us with the possibility, and bitter irony, of eastern rainforests being conserved through carbon offset and trading schemes (see Chapter 13 by Brimont and Bidaud) at the same time as western dry forests and southern spiny forests being cleared to cultivate maize to be converted into biofuels, both in the name of decarbonising the economies of industrialised nations. This reminds us that Madagascar's forests are increasingly linked to global political and economic processes.

When thinking about the future, it is vital to remember that, contrary to the narrow received wisdom, farming is not static. Christian Kull's (1998) work in the highlands of Madagascar shows that under certain conditions, Malagasy farmers are perfectly willing and able to change their practices from extensive to more intensive forms of agriculture. His research has shown that the trajectory of land use change depends on a diverse range of factors, including population changes, state policies, market incentives, climate variations and access to water and land. His work also shows how farmers have a range of choices, reflecting opportunities that are conditioned by ecological, political, economic and cultural factors. Some factors, such as diminishing water availability, constrain choices. Others, such as the growth of urban vegetable markets, can increase choices. In Leimavo (in the southern highlands), population growth constrained the amount of pasturelands and helped reduced water availability, thereby reducing land available for rice paddies. As grazing and the cultivation of rice became less of an option, many households focused on profitable orange crops to be sold in urban markets.

Perhaps the most important question is how climate change might affect land use and forest clearance. Madagascar's rainfall is highly spatially and temporally variable in comparison to similar regions around the world (Dewar and Richard, 2007). Drought conditions have long been a driver of migration from the south of Madagascar (Casse et al., 2004; Réau, 2002). It is possible that Madagascar's climate will become warmer, droughts will become more common and rainfall less predictable over the next three decades, especially in the semi-arid west and arid south (Tadross et al., 2008). If this is the case, there is likely to be more pressure on forests as households migrate to forest frontiers in search of cultivatable land and are forced to extensify production due to decreasing yields.

Conclusions

Environmental discourse and policy in Madagascar have been based on the view that deforestation is principally carried out by rural households and driven by population growth and poverty. This narrative ignores the contribution of other land uses to forest loss. Even more crucially, it is based on a narrow understanding of the factors that shape the land use decisions of rural households. It is true that poor households often rely on forest clearance agriculture and can have a mistrust of different methods proposed by donors and NGOs. Given the precarious nature of rural life, households rely on tried and tested methods. On a practical level, the burning of vegetation releases nutrients to improve otherwise poor soil. The technique also requires a minimum input of time, effort and money, allowing a household to farm a hectare of land with labour supplied entirely from the household. It is even capable of generating considerable profits when crop prices are high.

However, to reduce the land use choices of households to a simple tallying up of costs and benefits misses the point. Agriculture is at heart a social process. It has major cultural, political and economic dimensions. In Madagascar, forest clearance has been related to identity, ancestral beliefs and taboos, and to desires to acquire wealth and status. Any attempt to work with local households, be it for conservation or development (or both), must start with a better understanding of the complexity of rural land-use choices. Any attempt to reduce forest loss in Madagascar must also engage with other land uses and the broader political and economic factors that shape land use.

In terms of research on deforestation, the diversity in household agricultural systems, particularly regarding responses to socio-economic and environmental conditions, is an area in need of further investigation. Research outside Madagascar has shown that different social groups can have diverse agricultural practices and distinct priorities, which may influence the response to socio-economic change (Carr, 2005; Kunstadter, 1988). There are often important differences between established farmers who have been in an area for a considerable period of time and migrant farmers who have recently moved into an area (Sponsel et al., 1996). While this is a simplistic dichotomy, these two broad groups can have very different priorities and agricultural methods, resulting in different impacts on the landscape. The relationship (and possible conflict) between these users is an important issue in need of more attention. Rural household farms are complex entities and rather than simplifying the relationship between them and the environment, researchers and policymakers need to understand how they react to changing conditions and the relationships that they have with landscapes. Finally, climate change and the emerging global political economy of food production are likely to have major impacts on land use. Both research and policy need to do more to consider how agriculture might respond to such changes and the impact this will have on land cover and biodiversity.

Acknowledgements

I would like to thank Bob Dewar and Christian Kull for their helpful comments.

Notes

1 The case study material is taken from Scales (2011).
2 For an in-depth analysis of the conflict over fire between the state and rural households, see Kull (2000b; 2002a, b; 2004).
3 Estimates of forest loss are based on the fact that one hectare of cleared forest produces on average two tonnes of maize a year for two years before land starts to become exhausted of nutrients and is abandoned.
4 There is considerable debate over the exact causes of the 2008 increase in food prices and related food crisis. Rising demand from an expanding global population; increases in the price of petroleum; competing demand for crops from the biofuel industry; and speculation on global agricultural commodity markets have all been proposed as possible drivers.

References

Angelsen, A. (1995) 'Shifting cultivation and "deforestation": a study from Indonesia', *World Development*, vol 23, pp1713–1729.
Bloesch, U. (1999) 'Fire as a tool in the management of a savanna/dry forest reserve in Madagascar', *Applied Vegetation Science*, vol 2, pp117–124.
Boserup, E. (1965) *The Conditions of Agricultural Growth: The Economics of Agrarian Change under Population Pressure*, Aldine, Chicago.
Bromley, D. W. (1999) 'Deforestation: institutional causes and solutions', in M. Palo and J. Uusivuori (eds) *World Forests, Society, and Environment*, Kluwer Academic Publishers, Dordrecht.
Brookfield, H. and Padoch, C. (1994) 'Appreciating agrodiversity: a look at the dynamism and diversity of indigenous farming practices', *Environment*, vol 36, pp37–45.

Carr, D. L. (2005) 'Forest clearing among farm households in the Maya Biosphere Reserve', *The Professional Geographer*, vol 57, pp157–168.
Casse, T., Milhøj, A., Ranaivoson, S. and Randriamanarivo, J. R. (2004) 'Causes of deforesation in southwestern Madagascar: what do we know?', *Forest Policy and Economics*, vol 6, pp33–48.
Cotula, L. and Vermeulen, S. (2009) 'Deal or no deal: the outlook for agricultural land investment in Africa', *International Affairs*, vol 85, pp1233–1247.
Cramb, R. A., Colfer, C. J. P., Dressler, W., Laungaramsri, P., Le, Q. T., Mulyoutami, E., Peluso, N. L. and Wadley, R. L. (2009) 'Swidden transformations and rural livelihoods in Southeast Asia', *Human Ecology*, vol 37, pp323–346.
De Wilde, M., Buisson, E., Ratovoson, F., Randrianaivo, R., Carriere, S. M. and Lowry II, P. P. (2012) 'Vegetation dynamics in a corridor between protected areas after slash-and-burn cultivation in south-eastern Madagascar', *Agriculture, Ecosystems and Environment*, vol 159, pp1–8.
Dewar, R. E. and Richard, A.F. (2007) 'Evolution in the hypervariable environment of Madagascar', *Proceedings of the National Academy of Sciences*, vol 104, pp13723–13727.
Fairhead, J. and Leach, M. (1998) *Reframing Deforestation: Global Analysis and Local Realities – Studies in West Africa*, Routledge, London.
FAO (2008) *The State of Food Insecurity in the World 2008*, Food and Agriculture Organization of the United Nations, Rome.
Fauroux, E. (1980) 'Les rapports de production Sakalava et leur évolution sous influence coloniale (Région de Morondava)', in R. Waast, E. Fauroux, B. Schlemmer, F. Le Bourdiec, J. P. Raison and G. Ganday (eds) *Changements Sociaux dans l'Ouest Malgache*, ORSTOM, Paris.
Geist, H. J. and Lambin, E. F. (2002) 'Proximate causes and underlying driving forces of tropical deforestation', *BioScience*, vol 52, pp143–150.
Hume, D. W. (2006) 'Swidden agriculture and conservation in eastern Madagascar: stakeholder perspectives and cultural belief systems', *Conservation and Society*, vol 4, pp287–303.
Ickowitz, A. (2006) 'Shifting cultivation and deforestation in tropical Africa: critical reflections', *Development and Change*, vol 37, pp599–626.
Jarosz, L. (1993) 'Defining and explaining tropical deforestation: shifting cultivation and population growth in colonial Madagascar (1896–1940)', *Economic Geography*, vol 69, pp366–379.
Kull, C. A. (1998) 'Leimavo revisited: agrarian land-use change in the highlands of Madagascar', *Professional Geographer*, vol 50, pp163–176.
Kull, C. A. (2000a) 'Deforestation, erosion, and fire: degradation myths in the environmental history of Madagascar', *Environment and History*, vol 6, pp423–450.
Kull, C. A. (2000b) 'Madagascar's burning: the persitent conflict over fire', *Environment*, vol 44, pp8–19.
Kull, C. A. (2002a) 'Empowering pyromaniacs in Madagascar: ideology and legitimacy in community-based natural resource management', *Development and Change*, vol 33, pp57–78.
Kull, C. A. (2002b) 'Madagascar aflame: landscape burning as peasant protest, resistance, or a resource management tool?', *Political Geography*, vol 21, pp927–953.
Kull, C. A. (2004) *Isle of Fire: The Political Ecology of Landscape Burning in Madagascar*, University of Chicago Press, Chicago.

Kunstadter, P. (1988) 'Hill people of Northern Thailand', in J. Sloan Denslow and C. Padoch (eds) *People of the Tropical Rain Forest*, University of California Press, Berkeley.

Le Bourdiec, F. (1980) 'Le développement de la riziculture dans l'ouest Malgache', in G. Sautter, R. Waast, E. Fauroux, B. Schlemmer, F. Le Bourdiec, J. P. Raison and G. Dandoy (eds) *Changements Sociaux dans l'Ouest Malgache*, ORSTOM, Paris.

Mazoyer, M. and Roudart, L. (2006) *A History of World Agriculture from the Neolithic Age to the Current Crisis*, Earthscan, London.

MEEF (1984) *Stratégie Malgache pour la Conservation et le Développement Durable*, Ministère de l'Environnement, des Eaux et Forêts, Antananarivo.

MEP (1990) *Chartre de L'Environnement*, Ministère de l'Economie et du Plan, Antananarivo.

Mertz, O., Padoch, C., Fox, J., Cramb, R. A., Leisz, S. J., Lam, N. T. and Vien, T. D. (2009) 'Swidden change in Southeast Asia: understanding causes and consequences', *Human Ecology*, vol 37, pp259–264.

Minten, B. and Méral, P. (2006) *International Trade and Environmental Degradation: A Case Study on the Loss of SpinyF in Madagascar*, World Wild Fund For Nature, Antananarivo.

Moore, H. L. and Vaughan, M. (1994) *Cutting Down Trees: Gender, Nutrition, and Agricultural Change in the Northern Province of Zambia, 1890–1990*, James Currey, London.

Moser, C. M. (2008) 'An economic analysis of deforestation in Madagascar in the 1990s', *Environmental Sciences*, vol 5, pp91–108.

Murphy, S. (2010) 'Biofuels: finding a sustainable balance for food and energy', in G. Lawrence, K. Lyons and T. Wallington (eds) *Food Security, Nutrition and Sustainability*, Earthscan, London.

Pollini, J. (2009) 'Agroforestry and the search for alternatives to slash-and-burn cultivation: from technological optimism to a political economy of deforestation', *Agriculture, Ecosystems and Environment*, vol 133, pp48–60.

Rambo, A. T. and Tran, D. V. (2001) 'Social organization and the management of natural resources: a case study of Tat Hamlet, a Da Bac Tay ethnic minority settlement in Vietnam's northwestern mountains', *Southest Asian Studies*, vol 39, pp299–324.

Randrianja, S. and Ellis, S. (2009) *Madagascar: A Short History*, Hurst, London.

Réau, B. (2002) 'Burning for zebu: the complexity of deforestation issues in western Madagascar', *Norwegian Journal of Geography*, vol 56, pp219–229.

Ruthenberg, H. (1976) *Farming Systems in the Tropics*, Clarendon Press, Oxford.

Scales, I. R. (2011) 'Farming at the forest frontier: land use and landscape change in western Madagascar, 1896 to 2005', *Environment and History*, vol 17, pp499–524.

Scales, I. R. (2012) 'Lost in translation: conflicting views of deforestation, land use and identity in western Madagascar', *The Geographical Journal*, vol 178, pp67–79.

Schlemmer, B. (1980) 'Conquête et colonisation du Menabe: une analyse de la politique Gallieni', in R. Waast, E. Fauroux, B. Schlemmer, F. Le Bourdiec, J. P. Raison and G. Ganday (eds) *Changements Sociaux dans l'Ouest Malgache*, ORSTOM, Paris.

Sharpe, B. (1990) 'Nutrition and commercialisation of agriculture in Northern Province', in A. Woods (ed.) *The Dynamics of Agricultural Policy and Reform in Zambia*, Iowa State University Press, Ames.

Shuttleworth, G. (1989) 'Policies in transition: lessons from Madagascar', *World Development*, vol 17, pp397–408.

Skole, D. L., Chomentowski, W. H., Salas, W. A. and Nobre, A. D. (1994) 'Physical and human dimensions of deforestation in Amazonia', *BioScience*, vol 44, pp314–322.

Sodikoff, G. (2012) *Forest and Labor in Madagascar: From Colonial Concession to Global Biosphere*, Indiana University Press, Bloomington.

Sponsel, L. E., Bailey, R. C. and Headland, T. N. (1996) 'Anthropological perspectives on the causes, consequences, and solutions of deforestation', in L. E. Sponsel, T. N. Headland and R. C. Bailey (eds) *Tropical Deforestation: The Human Dimension*, Columbia University Press, New York.

Styger, E., Rakotondramasy, H. M., Pfeffer, M. J., Fernandes, E. C. M. and Bates, D. M. (2007) 'Influence of slash-and-burn farming practices on fallow succession and land degradation in the rainforest region of Madagascar', *Agriculture, Ecosystems and Environment*, vol 119, pp257–269.

Tadross, M., Randriamarolaza, L., Rabefitia, Z. and Zheng, K. Y. (2008) *Climate Change in Madagascar: Recent Past and Future*, World Bank, Washington, DC.

Thrupp, L. A., Hecht, S. and Browder, J. (1997) *The Diversity and Dynamics of Shifting Cultivation: Myths, Realities, and Policy Implications*, World Resources Institute, Washington, DC.

Tidd, S. T., Pinder, J. E. and Ferguson, G. W. (2001) 'Deforestation and habitat loss for the Malagasy Flat-Tailed tortoise from 1963 through 1993', *Chelonian Conservation and Biology*, vol 4, pp59–65.

Tiffen, M., Mortimore, M. and Gichuki, F. (1994) *More People, Less Erosion: Environmental Recovery in Kenya*, Wiley, Chichester.

Willis, K. J., Gillson, L. and Brncic, T. M. (2004) 'How "virgin" is virgin rainforest?', *Science*, vol 304, pp402–403.

World Bank (1996) *Madagascar Poverty Assessment*, World Bank, Washington, DC.

Part 3
The politics of biodiversity conservation and environmental management

6 A brief history of the state and the politics of natural resource use in Madagascar

Ivan R. Scales

Introduction

Since the 1980s, environmental policy in Madagascar has undergone rapid expansion, with the island becoming a focus of global conservation attention (see Chapter 7 by Kull). However, the politics of natural resource use on the island have a much longer history. It is important to understand how this history unfolded, not only because of its role in laying the foundations for contemporary environmental legislation, but also because it raises major questions about the relationship between the state, natural resource use and environmental change.

In this chapter, I provide an overview of the key events of the pre-colonial, colonial and early post-colonial period relating to Madagascar's environmental politics. I pay particular attention to the evolution and role of the state, defined here as a nation or territory considered as an organized political community under one government (Oxford English Dictionary). While this is a very loose definition, encompassing a broad range of systems of political organization, social scientists tend to agree that states differ significantly from earlier forms of socio-political organization, not only in their scale but also in the scope of their activities. This includes maintaining general order, exerting control over fiscal matters (taxing and spending) and the protection of sovereignty through military force (Kottak, 1977).

From the seventeenth century onwards, it is possible to see such forms of socio-political organization on the island in the form of distinct kingdoms. During the nineteenth century a process of unification began, albeit incomplete and strongly resisted. It is not until the colonial period that it becomes possible and meaningful to talk of Madagascar as a single 'nation state', in other words 'a complex array of modern institutions involved in governance over a spatially bounded territory which enjoys monopolistic control over the means of violence [i.e. control of the military and police]' (Johnston *et al.*, 2001, p534).

It is important to note that it is impossible to isolate state politics from processes occurring both 'above' and 'below' the state. Chapter 7 (Kull) and Chapter 9 (Corson), for example, show just how significant geopolitics and international relations have been in shaping both conservation ideology and practice and demonstrate the growing importance of international non-governmental organizations (NGOs) in shaping policy. Meanwhile, Chapter 8 (Pollini *et al.*) and Chapter 14 (Kaufmann) reveal the importance of local customary rules in controlling access to land and regulating the use of natural resources. However, the state occupies a central place in environmental politics and thus merits considerable attention (Box 6.1).

Before I begin this history of state-based environmental politics, there is an important caveat. While it is possible to talk of state power, the 'state' is never a monolithic or homogenous entity. For example, there are key differences between ministers sat in offices in the capital and the field officers in charge of monitoring and enforcing rules. There are often tensions between ministries pursuing different goals, for example agricultural development and economic growth versus forest protection. As Kull (2004, p25) reminds us, the state is both 'vertically' and 'horizontally' diverse and made up of a 'complex, ever-changing set of institutions and practices, full of competing personal, institutional, and political agendas representing different parts of civil society'. As this chapter is designed to provide an overview of state-level environmental politics, I necessarily gloss over many of these differences. However, heterogeneity in state-based politics will become evident in subsequent chapters. I also refer readers to Christian Kull's (2004) *Isle of Fire: The Political Ecology of Landscape Burning in Madagascar* and various papers (e.g. Kull, 2000, 2002a, 2002b) which provide an in-depth analysis of the island's complex multi-level environmental politics. Finally, I refer readers interested in knowing more about Madagascar's political and economic history to Randrianja and Ellis' (2009) *Madagascar: A Short History* and Mervyn Brown's (2002) *A History of Madagascar*.

Box 6.1 Environmental politics and different forms of power

Environmental politics can take a wide range of forms and encompass a diverse set of institutions. Politics can be broadly thought of as 'the practices and processes through which power, in its multiple forms, is wielded and negotiated' (Paulson *et al.*, 2005 p28), with power defined as the potential to control, command or direct the actions of others (Allen, 1997). There are of course many ways a state can exercise power. The most obvious is through control of law – both in terms of creating legislation and enforcing rules. However, power is often exercised in other ways, for example through the ability to influence environmental knowledge. Chapters 4 (McConnell and Kull), 5 (Scales) and 7 (Kull), for example, show how discourse has played a key part in the politics of natural resource use and biodiversity conservation. A discourse can be defined as the process by which people form and share knowledge about the world around them. Discourses reflect particular sets of beliefs and ways of looking at the world that enable people 'to interpret bits of information and put them together into coherent stories or accounts' (Dryzek, 1997, p8). A good example of environmental discourse in action can be seen in Chapter 4 (McConnell and Kull), which shows how the idea of a pristine island-wide forest existing before human arrival has come to dominate environmental thinking in Madagascar, leading to the oft-repeated 'fact' that Madagascar is 80–90 per cent deforested. Chapter 5 (Scales) shows how policy is also dominated by the idea that deforestation is driven primarily by population growth, poverty and a lack of options. This is despite considerable evidence that challenges both these received wisdoms.

Stories, or 'narratives', such as an island-forest being destroyed by human action are seductive and persistent because they help to simplify very complex realities, presenting a clear problem (and therefore solution) around which to organize policy. Together, they contribute to the central idea that 'pristine nature' is under threat from rural Malagasy and that they must therefore be excluded from 'wilderness', for example through the expansion of protected areas ('fortress conservation'). Such discourses are important in environmental politics because they help to define which forms of natural resource are acceptable and which are not, according to particular worldviews. Chapter 11 (Scales) shows how ideas of pristine nature have driven (mostly unsuccessful) efforts to integrate conservation and development through nature tourism, while attempts to derive financial benefits from consumptive uses of natural resource (e.g. timber exploitation) have largely fallen out of fashion. Such policy changes often have less to do with pragmatism and more to do with changes in western environmental ideas (see Chapter 7 by Kull for a discussion of the interplay between western environmentalism and Malagasy conservation).

Pre-colonial kingdoms, natural resources and forest politics

It is difficult to reconstruct much of Madagascar's pre-colonial history, especially prior to the seventeenth century, as there is little written material (Kull, 2004; Randrianja and Ellis, 2009). From the second half of the seventeenth century, the accounts of European explorers and traders provide us with detailed descriptions of society and politics in different parts of Madagascar. The most famous of these is Etienne de Flacourt's (1658[1995]) *Histoire de la Grande Isle Madagascar*.

According to early European visitors, before the eighteenth century the majority of Malagasy engaged in subsistence-based fishing, cultivation or herding, sometimes combined with other craft activities such as blacksmithing and weaving (Randrianja and Ellis, 2009). Political groupings were thus small and highly localized. The first large-scale, socio-political organizations began to form on the island in the seventeenth century, with the rise of a variety of kingdoms (Randrianja and Ellis, 2009). The most powerful were the Sakalava, a group of more or less independent monarchies that originated in south-central Madagascar in the sixteenth century and gradually expanded westward to the coast and then northwards between the seventeenth and nineteenth centuries to cover almost one-third of the island (Feeley-Harnik, 1978; Kent, 1968). The Sakalava kingdoms were characterized by large-scale herding of zebu cattle as well as slave trading (Chazan-Gillig, 1991; Fauroux, 1977, 1989; Goedefroit, 1998).

The rise of the Sakalava dynasties is arguably the single most important political revolution in the history of the island:

> By marrying their religious powers to their commercial acumen and military might, the Sakalava kings permanently altered the form of the island's social organization, changing fundamentally the moral values of the small communities into which people had been divided into for the previous thousand years.
>
> (Randrianja and Ellis, 2009, p99)

The kingdoms possessed considerable economic, political and military strength and their expansion saw large population movements, as smaller groups were incorporated and moved towards new economic centres. By the eighteenth century, the Sakalava were involved in an extensive international trade network made possible by links with Indian and Swahili traders (Campbell, 2005). The Sakalava traded slaves, zebu and rice in return for muskets and gunpowder. The arms purchased by the Sakalava helped fuel further slaving forays into the Betsileo and Imerina regions of the central highlands. Some slaves were kept as agricultural labour (carrying out swidden cultivation of forest areas) but most were surplus to the largely pastoral economy and were therefore sold (Fauroux, 1980; Schlemmer, 1980).

There were other powerful kingdoms at this time, most notably the Betsimisaraka and Imerina. By the eighteenth century, the central highlands had become densely populated. In contrast to the Sakalava kingdoms, highland monarchs based their systems on and derived power from intensive rice cultivation, with kings providing patronage and protection in return for payment in various forms (Randrianja and Ellis, 2009). Such agricultural systems depended on the ability to organize collective labour for communal work on dams and irrigation. From an environmental management perspective, this had important consequences, as increasingly sophisticated political structures allowed much larger, ambitious and intensive agricultural schemes and major landscape change. Only with the larger and more complex socio-political systems of the eighteenth century did the transformation of the central highlands into an intensively cultivated landscape become possible.

The key to the power of the various kingdoms during the seventeenth and eighteenth centuries was control over the supply of natural resources and slaves, and the trading of both with foreign (mostly European) traders. While various groups in Madagascar were exporting commodities and slaves before the arrival of Europeans, the eighteenth century saw the increasing incorporation of Madagascar into the European-dominated world economic system. Connection to global markets gave kingdoms access to modern weapons, and with it the ability to expand and maintain territorial control (Brown, 2002; Campbell, 2005; Randrianja and Ellis, 2009).

With growing royal power came a greater desire and means to control natural resources, especially the hardwood timber provided by the island's forests. Prior to this, there was most likely no state regulation of burning, with people using fire freely to manage pastures and prepare fields for cultivation, and managing fires through evolving traditions and community-based rules and institutions to resolve conflict (Kull, 2004). The earliest recorded forest legislation dates to the end of the eighteenth century, when the Imerina King Andrianampoinimerina (who reigned between 1797 and 1810) ordered his subjects to look after the island's forests, banning the cutting of live firewood and burning of forest and allowing charcoal fires only outside forested areas (Kull, 2004; Raik, 2007; Ramanantsoavina, 1973). This was one of the first-known attempts in precolonial Africa of state protection and management of wooded regions (Gade, 1996). King Andrianampoinimerina's reign also saw the rapid development of agriculture through politically directed communal works (Randrianja and Ellis, 2009).

The nineteenth century saw significant political developments, particularly regarding the island's relationship with western nations. During the course of the nineteenth century, the Indian Ocean was a major area of tension and conflict between Britain and France, as the two powers sought to expand their spheres of influence and control key trade routes. Major naval battles between 1809 and 1811 saw Britain taking control of both Mauritius and Réunion. With the defeat of Napoleon in 1815, the balance of power shifted in Britain's favour. Britain returned Réunion to France but kept Mauritius. The governor of Mauritius played a significant role by recognizing King Radama I as the King of Madagascar to help strengthen ties between Britain and Madagascar. In 1817, King Radama I signed a treaty with Britain that formalized political and economic arrangements between the two powers. The treaty created a new form of government and new institutions in Madagascar, much more akin to what Europeans considered a 'modern' state. This period also saw the growing influence of missionaries, who not only spread Christianity but also helped to write down and formalize the Malagasy language. Madagascar began to be recognized diplomatically as a nation by Europe and the United States of America (Kull, 1996). King Radama I agreed to ban slavery in return for compensation in the form of military supplies from Britain. This led to a major increase in the military strength of the Imerina kingdom, giving it a significant advantage over other groups (Randrianja and Ellis, 2009). However, neither King Radama I nor any of his successors managed to achieve authority over the whole of the island.

In 1861, King Radama II passed a law taking possession of Madagascar's forests and any land without claim (Healy and Ratsimbarison, 1998; Henkels, 2001). This was subsequently supported by two key pieces of legislation. The first was the Code of 101 Articles, published in 1868, which banned various forest use practices and forbade human settlement in forested areas (Henkels, 2001; Kull, 1996; Raik, 2007):

> One may not clear the forest by fire with the goal of cultivating rice fields, corn or other crops. One who clears by fire a new terrain or expands those which exist already, that person will be put in irons.
> (Ratovoson, 1979 cited in Henkels, 2001, p2)

This was followed in 1881 by the Code of 305 articles, published by Queen Ranavalona II. Article 91 declared that all forests and 'uninhabited' land belonged to the state, while Article 101 to 106 prohibited forest burning, the production of charcoal in forested areas, the cutting of large trees, as well as settlement in forested areas, with punishments as severe as ten years in irons (Kull, 2004). While forest legislation was severe, it is noteworthy that neither King Andrianampoinimerina's proclamations in the late eighteenth century nor the Code of 305 articles in 1881 regulated grassland fires (Kull, 2004).

It might be tempting to look at the developments in environmental policy during the nineteenth century through the lens of modern western environmentalism and attribute them to a conservation ethic. The reality was probably more pragmatic. Dez (1968), for example, believes these measures were related to securing supplies of construction wood and ensuring public security and defence – rulers preferred to keep people out of forests where they could keep an eye on them, as forests were often hiding places for bandits and potential invaders. For example, the *Tantaran'ny Andriana* (a transcription of oral histories and major historical source of the period) reports one of King Andrianampoinimerina's declarations: 'However it is forbidden for people to come to forge clandestinely arms in the forest because they can prepare a rebellion' (Ratovoson, 1979 cited in Henkels, 2001, p2). Furthermore, swidden cultivation did not benefit the state or traders in the Malagasy elite and threatened valuable stocks of timber. The rights to exploit the forests of the east coast were sold to a variety of foreign interests, with a trade in hardwoods such as rosewood (*Dalbergia* spp.), as well as non-timber forest products such as rubber, honey and gum copal – a tree resin used in the production of paints, varnishes and perfumes (Campbell, 2005).

The last two decades of the nineteenth century saw another dramatic shift in the island's politics, linked to the broader international dynamics of European colonialism and more specifically the colonial machinations of Britain and France. The 1870s had seen the start of the European 'Scramble for Africa'. France began adopting a more aggressive stance towards British-influenced Madagascar, invading the island in 1883. At the end of the war between France and the Merina Kingdom of Madagascar, Diego Suarez (Antsiranana) was ceded to France. Madagascar's fate was finally sealed in 1884 by events elsewhere in Africa, when King Leopold II of Belgium annexed a large part of central Africa that became Congo Free State (which in turn became Belgian Congo and is now the Democratic Republic of Congo). This precipitated the Berlin Conference in 1884–1885, where Africa was divided between European countries. As well as carving out the African continent, this also involved exchanging existing territories and Britain, in order to obtain the Sultanate of Zanzibar (in present day Tanzania), ceded all claims to Madagascar to France, leaving the French with a free hand over the island. In 1895, the French army landed in Mahajanga and marched to the capital, Antananarivo. The colonial era was about to begin.

The colonial state, environmental management and resistance

The first three decades following the arrival of French colonialism in 1896 was a period of tremendous political and economic upheaval in Madagascar, as the colonial government sought to exert control over the island and radically re-orientate its economy towards the production of exportable (and taxable) commodities. While the colonial period officially began in 1896, colonization was met with fierce resistance in many places (Feeley-Harnik, 1984; Kaufmann, 2001, Middleton, 1999; Randrianja and Ellis, 2009). In Menabe, for example, control over key settlements such as Morondava was not established until 1902 and even then large areas of the west remained outside the government's sphere of influence (Scales, 2011).

It is hard to overstate the impact that the colonial government had on the island's politics, economics, culture and landscape. The only effective way of implementing policies in Madagascar was by government action and 'the history of colonization becomes inseparable from the history of the state in Madagascar' (Randrianja and Ellis, 2009, p159). At the core of the government's policies was 'the foreign imposition of radically different ideas and practices on a subject population' (Feeley-Harnik, 1984, p7), which involved not only the violent quelling of resistance but also the destruction of institutions, identities and livelihoods. This often had significant implications for the island's natural resources, especially its forests, as the new economic ventures busied themselves with the pillage of natural resources (Randrianja and Ellis, 2009, pp158–159). The colonial government undertook large-scale timber exploitation, created vast plantations of exotic tree species and also introduced crops such as sisal, coffee and tobacco, transforming the landscape of large parts of the island (Jarosz, 1993; Sodikoff, 2005).

France's expectations and intentions with regards to its African colonial possessions were made abundantly clear when, in 1900, the parliament in Paris passed a law requiring its colonies to be economically self-sufficient. The phrase used was *mise en valeur* (i.e. to make something productive/profitable). This placed great pressure on colonial governors to radically change the land use practices of rural households, which were principally geared towards production for subsistence. Their response was a 'three-pronged' strategy of land enclosure, state concessions and forced labour (Sodikoff, 2012).

The first governor of Madagascar, General Joseph Simon Gallieni, based his strategy for economic self-sufficiency on awarding large concessions to a small number of foreign companies. This required the introduction of a system of private land tenure and land registration to replace local customary land tenure (Scales, 2011). In 1926, the government passed a law claiming any unoccupied or un-enclosed land for the state (Healy and Ratsimbarison, 1998). However, the colonial government soon realized that a shortage of labour and a lack of infrastructure were holding back its attempts to develop an export-oriented and large-scale agricultural economy.

From the state's perspective, peasant agriculture and pastoralism were problematic, as these households produced food largely for themselves. Not only did this mean that they were not growing cash crops or trading in exportable commodities, their labour was also not available for activities such as working on plantations or infrastructural projects. The government was often frustrated by the lack of willingness of households to undertake waged labour and participate in the cash economy, accusing them of idleness and backwardness. For example, a regional governor working in Menabe stated in his 1909 annual report that:

> In groups of three, four or five families, the Sakalava travel through the forest, following their whims ... They find an area in the middle of nowhere and build themselves vile little huts, where they will quite happily live, and when we need them to pay their taxes or require their labour, we won't be able to find anyone.[1]

The challenge for the state was thus to 'encourage' rural Malagasy households to become 'modern' and productive members of the colonial economy. With this in mind, General Gallieni introduced *l'impot moralisateur* (moralizing tax) to be paid in cash and designed to force people into the cash economy. The idea was that by having to pay such taxes, not only would Malagasy households be obliged to undertake money earning activities, but in doing so would also develop a new profit-motivated 'work ethic'. Ethnic stereotypes (Box 6.2) played their part, with the Betsileo and Merina of the highlands generally seen as industrious and described by one regional governor as having 'the souls of agriculturalists' and being a model for 'the peasant movement which we seek'.[2] By contrast, the Tandroy and Sakalava tended to be seen as 'work-shy vagabonds' (Scales, 2011, 2012). In an effort to change the land use practices of rural households, the colonial government also introduced a policy of minimum cultivation per village, hoping that this would boost the production of cash crops as well as encourage households to establish more permanent settlements and stop living what it considered to be 'backwards' migratory lifestyles. The settlement of rural Malagasy into larger and more permanent villages also facilitated tax collection and labour recruitment (Scales, 2011; Sodikoff, 2012). These policies exemplified the French *mission civilisatrice* (civilizing mission) and the desire not only to make its colonies profitable but also to impart a set of moral values on the Malagasy population.

After World War I, colonial officials were able to direct their attention toward the island's infrastructure and, in 1926, the government initiated a vast public works campaign to build railways, bridges, roads and ports (Sodikoff, 2005). Over the course of the next decade, the colonial administration received 730 million francs, to be repaid with interest (Randrianja and Ellis, 2009). The island's infrastructural developments required a significant labour force and, despite General Gallieni's earlier efforts, a lack of available labour continued to hinder the administration's grand plans. Various forced labour schemes had been developed during the first two decades of the colonial period. In 1926, these were consolidated and codified by Governor-General Marcel Olivier in a programme called Service de la Main-d'Oeuvre des Travaux d'Interet Général (SMOTIG), which required every able-bodied man between the ages of 16 and 60 to work for the state for up to three months per year for a period of two years (Randrianja and Ellis, 2009; Sodikoff, 2005, 2012). Concession owners and private industrialists were able to 'borrow' workers from public works (Sodikoff, 2012). Labourers were expected to feed themselves, having to rely on the support of family and extended kin. Although it was banned, swidden agriculture was generally tolerated as it helped to feed workers and keep wages low (Sodikoff, 2012). Both public works and agricultural concessions were thus often accompanied by large-scale forest clearance (Scales, 2011; Sodikoff, 2012).

Box 6.2 'Tribes', ethnicity and the politics of identity

The eighteenth and nineteenth centuries were important in the political history of Madagascar because they saw the start of many 'ethnic' identities, i.e. the emergence of distinct groups with shared ancestry, their own political arrangements and generally tied to particular geographical areas. However, the concept of ethnicity must be treated with caution. The official 'ethnic groups' used in contemporary discourse were in fact largely defined during the colonial period (Covell, 1987; Larson, 1996). This was part of a broader strategy of indirect rule, or 'ethnic administration' (Randrianja and Ellis, 2009), where chiefs were appointed and identified with specific territories around carefully defined ethnic labels (Covell, 1987; Rakotondrabe, 1993). By the 1930s, the government had classified the island's population into 18 official 'tribes' and drawn maps to delineate their geographical boundaries (Randrianja and Ellis, 2009). Ethnic identity thus became categorically and geographically fixed.

Such fixed boundaries are problematic in that they treat ethnic identity as something concrete and permanent. Not only that, but both colonial administrators and modern-day policymakers and conservation organizations have been guilty of attaching certain behavioural traits to different groups (Scales, 2012). According to Randianja and Ellis (2009, p221):

> The unchanging existence of such [ethnic] groups has been taken as a fact by far too many writers on Madagascar. The reality of ethnicity is historical above all: it is never static, being permanently inscribed neither in the human genome nor even on the landscape.

There is now a considerable body of work demonstrating the fluidity of ethnic identity in Madagascar (see, for example, Astuti, 1995; Eggert, 1981; Larson, 1996; Poyer and Kelly, 2000). Rather than something attributed simply to shared ancestry or geographical location, it is common for ethnic identity to be understood as something achieved through performance and livelihood activities, with adherence to taboos playing an important part in this process (Brown, 2004). As Chapter 14 discusses in greater depth, policy risks running into trouble when it starts treating culture, taboos, local rules and ethnic identity as unchanging.

As well as agricultural and infrastructural policy, the colonial government paid particular attention to the island's forest resources. The *Service des Eaux et Forêts* (Water and Forestry Service) was created in 1896 and its policies focused on three key areas: i) reforestation and afforestation programmes often using introduced exotic species; ii) the establishment of forestry reserves and protected areas; and iii) the control of the burning practices and forest use of Malagasy. In 1897, General Gallieni launched a policy to plant trees on the highland plateau and eastern escarpment, instructing district forestry officers to stop swidden agriculture and limit pasture fires (Kull, 2004). Reforestation was largely based on plantations of fast growing, non-native species such as eucalyptus (*Eucalyptus* spp.), pine (*Pinus* spp.) and acacia (*Acacia decurrens*) and, by 1960, over 200,000 hectares had been planted (Kull, 1996; Raik, 2007). Gallieni also established an experimental garden and forestry school for Malagasy near the capital, with a focus on the propagation of both indigenous and exotic species (Sodikoff, 2005). While the colonial administration made attempts to involve Malagasy in the Water and Forestry Service, positions tended to be beyond the reach of the majority of Malagasy, most of whom were illiterate (Sodikoff, 2005). The lack of Malagasy involvement is a theme that has subsequently echoed through years, for example in community-based forest management (see Chapter 8 by Pollini *et al.*) and nature tourism (see Chapter 11 by Scales).

It is important to note that the government's forestry practices were often contested, even within the administration. On the one hand, many believed the island's forests to be a huge and underexploited resource. One regional governor remarked that: 'To preserve intact forests would be a serious mistake for their futures. Those who have travelled through Madagascar's forests have been struck by the number of dead trees, representing lost income from unexploited forests.'[3] On the other hand, timber exploitation came under criticism from many French naturalists, such as Henri Jean Humbert, who accused concession owners of destroying fragile forests and disrupting biological equilibrium (Sodikoff, 2005).

The colonial government's priorities clearly lay in exploiting the island's abundant natural resources. However, as well as this utilitarian view of nature, the colonial period also saw the emergence of more protectionist viewpoints and the establishment of the first areas dedicated to the protection of the islands' flora and fauna. The Académie Malgache, founded in 1902 by General Gallieni, played a key role in banning the killing of lemurs (Kull, 1996). The French National Committee for the Protection of Colonial Wildlife was established in Paris in 1923 and, in 1925, decided to create protected areas in its colonies (Peters, 1999; WCMC, 1992). In 1927, the government strengthened forest protection legislation and established Madagascar's first protected areas in the form of ten Réserves Naturelles Intégrales managed by the Forestry Service and chosen to represent a range of the island's forest types (Kull, 1996; Sodikoff, 2005). These constitute some of the earliest protected areas in Africa.[4] Decrees of 11 June 1939 and 3 January 1952 created two more reserves (WCMC, 1992). Legislation in 1958 (Decree No. 58-07 of 28 October 1958) created Madagascar's first national park, Montagne d'Ambre, as well as more reserves (WCMC, 1992; see also Chapter 10 by Virah-Sawmy et al.).

In terms of the control of land use, the swidden and grassland burning practices of rural Malagasy were seen as a serious threat both to the rational exploitation of forests and to the island's flora and fauna, so much so that during the 1940s and 1950s, the French colonial government discussed the possibility of moving up to 500,000 people from forested areas (Kull, 2004). The colonial period saw the establishment of a series of laws to control the use of fire for forest clearance and the management of pastures. In 1930, the government passed Article 36, which prohibited all forest fires and other forms of deforestation (Montagne, 2004). By 1932, those convicted of burning land without authorization faced six months in prison and a fine of 200 francs (Sodikoff, 2012). The colonial state thus criminalized burning for both economic and ideological reasons – to protect valuable timber resources and because burning was seen as backwards and irrational (Kull, 2004). From the colonial perspective, the banning of swidden cultivation also had the advantage of removing the means of subsistence of many rural Malagasy and thus pushing them towards waged employment in crop plantations and timber concessions (Jarosz, 1993; Sodikoff, 2012). As Sodikoff (2005, p408) argues:

> Whereas the state's ideological justification of forced labor for public works lay in inculcating Malagasy people to a 'work ethic', the colonial forest service sought to instill a 'conservation ethic' among loggers, miners, and Malagasy rice farmers who practiced slash-and-burn agriculture. What was 'good' for the Malagasy person and the Malagasy forest was good for the empire.

Despite increasingly stringent legislation, forest clearance and pasture fires continued, with efforts to control burning hindered by a lack of foresters to enforce rules (Kull, 2000, 2004). The relationship between the colonial government and farmers became increasingly tense, with the livelihood practices of Malagasy agriculturalists deemed to be irrational and destructive and thus criminalized. Forests became sites of conflict, often serving as refuges against the brutality of colonial taxes and forced labour (Sodikoff, 2005). For example, in western Madagascar the dry-deciduous forest became a place to hide away from the colonial administration (Scales, 2011). By the end of the colonial period, the relationship between the state and large parts of the island's rural population has become increasingly antagonistic, with fire not just an environmental management tool but also a symbol of protest and resistance (Kull, 2000, 2002b).

'Independence', the First Republic and a 'break' from France

Madagascar gained its independence from France in 1960. However, independence was initially more in principle than in practice. President Philibert Tsiranana and the government of the First Republic left colonial economic policies largely unaltered and membership of *La Zone Franc* left exchange rates and monetary policy in the hands of the French Treasury (Barrett, 1994). France's continued influence was felt in other diverse ways: French was enshrined in the constitution as an official language; French consuls sat as members of provincial assemblies that decided provincial budgets; the French army had bases on the island; French officers worked in Madagascar's military academy; the head of Malagasy military intelligence was a French officer; and the island of Sainte Marie enjoyed double status as both French and Malagasy (Randrianja and Ellis, 2009). The government continued to rely on colonial era taxes (for example poll and cattle taxes) for much of its revenue, and French-owned businesses and concessions continued to dominate the economy (Barrett, 1994).

In terms of environmental ideology and policy, there were important continuities with the colonial era. The national constitution of 1959 stated that all individuals should protect, improve and use the natural resources of Madagascar in a responsible fashion (Ramanantsoavina, 1973). The focus on protected areas continued, with Madagascar's second national park (Isalo) established in 1962 (Peters, 1999). Five additional categories of protected areas were created in addition to strict nature reserves: national parks, special reserves, classified forests, afforestation zones and non-hunting reserves (Kull, 1996). Like the previous colonial administration, the government of the First Republic also focused considerable attention on reforestation and afforestation (Kull, 2004; Raik, 2007). A presidential decree in 1962 required all men to plant 100 seedlings per year or otherwise be forced to pay a fine, but this law proved unpopular and difficult to enforce and was abolished in 1972 (Kull, 1996).

Although there were many similarities with the colonial mindset, the establishment of the first post-colonial government saw a shift in anti-fire legislation, with a relaxation of the rules and a 'paradoxical mix of heavy-handed force and politically pragmatic tolerance' and fire seen as a 'necessary evil' (Kull, 2004, p226). The first piece of post-independence legislation on forest clearing and vegetation fires was Ordonnance 60-127 of 1960. It differed from colonial rules in that it contained no general ban on fires but created a distinction between different types of fires (forest clearance, cleaning fires, pasture fires and wildfires) and prohibited or restricted different types of fires in certain types of vegetation (Kull, 2004).

As well as some expansion in protected areas, the First Republic saw a burst of conservation activity in other spheres. Madagascar became a member of the International Union for Conservation of Nature (IUCN) in 1961.

The first conservation project to be established in the country was a joint World Wild Fund for Nature (WWF) and IUCN effort aimed at preserving the aye-aye, *Daubentonia madagascariensis* (Kull, 1996). The international conservation presence grew in 1966 when the United States Agency for International Development (USAID) and Coopération Suisse (Swiss Aid) began their environmental activities. In 1970, Madagascar hosted the International Conference on the Conservation of Natural Resources, which led to the Ministère des Eaux et Forêts publishing *Protection de la Nature* (MEEF, 1972). This was the first attempt at a national conservation plan, although it had few concrete proposals and was little more than a manifesto.

Conclusions

This chapter has briefly explored the relationship between the state, environmental policy and natural resource use in Madagascar. A common thread has been the desire of the state to regulate particular activities, especially the use of fire to clear forest and manage grasslands. Such legislation pre-dates the colonial government and reveals the tension between different interest groups. This process was particularly evident during the colonial period, when the state sought to exert not only political and economic control – through changes in land tenure and a whole raft of taxes for example – but also ideological control. The result was a potent mix, combining: i) a desire to exploit Madagascar's abundant natural resources for profit; ii) a 'civilizing mission'; and iii) a western environmental ethic, based on the idea of wilderness, that manifested itself most clearly in the strict nature reserves set aside to protect the island's flora and fauna. The result was a transformation of the Malagasy landscape, as the French government sought to order it according to western principles, with strict areas for the protection of wildlife, forestry reserves for the 'rational' exploitation of timber and new villages where migrant agriculturalists and pastoralists could settle and produce the export-oriented cash crops of a profitable colony. Taxes were imposed not just to fill the coffers of the colonial administration but also to help foster a new work ethic.

By the time independence arrived in 1960, much of the scene had been set for the environmental politics of the remainder of the twentieth century. The island of Madagascar possessed a nation-state (albeit with continuing involvement of the ex-colonial power). All forested areas were claimed by the state and vast areas of land had been privatized, thereby ignoring customary land tenure and land use rights. A system of protected areas had started to take shape, with areas of land set aside specifically for the purpose of protecting the island's flora and fauna. There were links to the world of international biodiversity conservation through various treaties and associations. Perhaps most importantly, there was environmental legislation with a strong anti-fire stance that criminalized livelihood practices such as forest clearance and pasture burning. The state and much of Madagascar's rural population had become locked into a tense confrontation they have yet to emerge from.

Notes

1 Gouvernement de Madagascar et Dépendances, Cercle de Morondava. Rapport politique, administrative et Economique, Année 1909. Archives D'Outre-Mer, Aix-en-Provence, Aix 2 D 172 B.
2 Région de Morondava, Rapport Annuel, Année 1941. Archives D'Outre-Mer, Aix-en-Provence, Aix 2 D 176.
3 Province de Morondava, Rapport Economique 1920. Archives D'Outre-Mer, Aix-en-Provence, Aix 2 D 175.
4 Parc National Albert (now Virunga National Park) in Belgian Congo (now Democratic Republic of Congo) created in 1925 was the first national park in Africa.

References

Allen, J. (1997) 'Economies of power and space', in R. Lee and J. Wills (eds) *Geographies of economies*, Arnold, London.
Astuti, R. (1995) '"The Vezo are not a kind of people". Identity, difference and "ethnicity" among a fishing people of western Madagascar', *American Ethnologist*, vol 22, pp464–482.
Barrett, C. B. (1994) 'Understanding uneven agricultural liberalisation in Madagascar', *The Journal of Modern African Studies*, vol 32, pp449–476.
Brown, M. (2002) *A History of Madagascar*, Markus Wiener, Princeton.
Brown, M. L. (2004) 'Reclaiming lost ancestors and acknowledging slave descent: insights from Madagascar', *Comparative Studies in Society and History*, vol 46, pp616–645.
Campbell, G. (2005) *An Economic History of Imperial Madagascar, 1750–1895: The Rise and Fall of an Island Empire*, Cambridge University Press, Cambridge.
Chazan-Gillig, S. (1991) *La Société Sakalave*, Karthala, Paris.
Covell, M. (1987) *Madagascar: Politics, Economics and Society*, Frances Pinter, London.
de Flacourt, E. (1658[1995]) *Histoire de la Grande Isle Madagascar*, Karthala, Paris.
Dez, J. (1968) 'Un des problèmes du développement rural: la limitation des feux de végétation', *Terre Malgache Tany Malagasy*, vol 4, pp97–123.
Dryzek, J. S. (1997) *The Politics of the Earth: Environmental Discourses*, Oxford University Press, Oxford.
Eggert, K. (1981) 'Who are the Mahafaly? Cultural and social misidentifications in southwestern Madagascar', *Omaly Sy Anio*, vols 13–14, pp149–176.
Fauroux, E. (1977) 'La formation sociale Sakalava dans les rapports marchands: pour l'introduction de la dimension historique dans les études d'anthropologie économique', *Cahiers des Sciences Humaines*, vol 14, pp71–81.
Fauroux, E. (1980) 'Les rapports de production Sakalava et leur évolution sous influence coloniale (Région de Morondava)', in R. Waast, E. Fauroux, B. Schlemmer, F. Le Bourdiec, J. P. Raison and G. Ganday (eds) *Changements Sociaux dans l'Ouest Malgache*, ORSTOM, Paris.
Fauroux, E. (1989) *Le Boeuf et le Riz dans la Vie Economique et Sociale Sakalava de la Vallée de la Maharivo*, ORSTOM, Paris.
Feeley-Harnik, G. (1978) 'Divine kingship and the meaning of history among the Sakalava of Madagascar', *Man*, vol 13, pp402–417.
Feeley-Harnik, G. (1984) 'The political economy of death: communication and change in Malagasy colonial history', *American Ethnologist*, vol 11, pp1–19.

Gade, D. W. (1996) 'Deforestation and its effect in highland Madagascar', *Moutain Research and Development*, vol 16, pp101–116.
Goedefroit, S. (1998) *A L'Ouest de Madagascar: Les Sakalava du Menabe*, Karthala, Paris.
Healy, T. M. and Ratsimbarison, R. (1998) 'Historical influences and the role of traditional land rights in Madagascar: legality versus legitimacy', in M. Barry (ed.) *Proceedings of the International Conference on Land Tenure in the Developing World with a Focus on Southern Africa*, University of Cape Town, pp365–377.
Henkels, D. (2001) 'A close up of Malagasy environmental law', *Vermont Journal of Environmental Law*, vol 3, pp1–16.
Jarosz, L. (1993) Defining and explaining tropical deforestation: shifting cultivation and population growth in colonial Madagascar (1896–1940), *Economic Geography*, vol 69, pp366–379.
Johnston, R. J., Gregory, D., Pratt, G. and Watts, M. (2001) *The Dictionary of Human Geography*, Wiley-Blackwell, Oxford.
Kaufmann, J. C. (2001) 'La question des raketa: colonial struggles with prickly pear cactus in southern Madagascar, 1900–1923', *Ethnohistory*, vol 48, pp87–121.
Kent, R. K. (1968) 'Madagascar and Africa: the Sakalava, Maroserana, Dady and Tromba before 1700', *Journal of African History*, vol 9, pp517–546.
Kottak, C. P. (1977) 'The process of state formation in Madagascar', *American Ethnologist*, vol 4, pp136–155.
Kull, C. A. (1996) 'The evolution of conservation efforts in Madagascar', *International Environmental Affairs*, vol 8, pp50–86.
Kull, C. A. (2000) 'Madagascar's burning: the persistent conflict over fire', *Environment*, vol 44, pp8–19.
Kull, C. A. (2002a) 'Empowering pyromaniacs in Madagascar: ideology and legitimacy in community-based natural resource management', *Development and Change*, vol 33, pp57–78.
Kull, C. A. (2002b) 'Madagascar aflame: landscape burning as peasant protest, resistance, or a resource management tool?', *Political Geography*, vol 21, pp927–953.
Kull, C. A. (2004) *Isle of Fire: The Political Ecology of Landscape Burning in Madagascar*, University of Chicago Press, Chicago.
Larson, P. M. (1996) 'Desperately seeking "the Merina" (Central Madagascar): reading ethnonyms and their semantic fields in African identity histories', *Journal of Southern African Studies*, vol 22, pp541–560.
MEEF (1972) *La Protection de la Nature à Madagascar: Document 2866*, Ministère de l'Environnement, des Eaux et Forêts, Antananarivo.
Middleton, K. (1999) 'Who killed "Malagasy Cactus"? Science, environment and colonialism in Southern Madagascar (1924–1930)', *Journal of Southern African Studies*, vol 25, pp215–248.
Montagne, P. (2004) *Analyse Rétrospective du Transfert de Gestion à Madagascar et Aperçu Comparatif des Axes Méthodologiques des Tranfserts de Gestion sous loi 96–025 et sous Décret 2001–122*, Consortium RESOLVE – PCP – IRD, Antananarivo.
Paulson, S., Gezon, L. and Watts, M. (2005) 'Politics, ecologies, genealogies', in S. Paulson and L. Gezon (eds) *Political Ecology Across Spaces, Scales and Social Groups*, Rutgers University Press, New Brunswick.
Peters, J. (1999) 'Understanding conflicts between people and parks at Ranomafana, Madagascar', *Agriculture and Human Values*, vol 16, pp65–74.

Poyer, L. and Kelly, R. L. (2000) 'Mystification of the Mikea: constructions of foraging identity in Southwest Madagascar', *Journal of Anthropological Research*, vol 56, pp163–185.

Raik, D. B. (2007) 'Forest management in Madagascar: an historical overview', *Madagascar Conservation and Development*, vol 2, pp5–10.

Rakotondrabe, T. D. (1993) 'Beyond the ethnic group: ethnic groups, nation state and democaracy in Madagascar', *Transformation*, vol 22, pp15–29.

Ramanantsoavina, G. (1973) *Note sur la Politique et l'Administration Forestières à Madagascar*, Ministère du Developpement Rural, Antananarivo.

Randrianja, S. and Ellis, S. (2009) *Madagascar: A Short History*, Hurst, London.

Scales, I. R. (2011) 'Farming at the forest frontier: land use and landscape change in western Madagascar, 1896 to 2005', *Environment and History*, vol 17, pp499–524.

Scales, I. R. (2012) 'Lost in translation: conflicting views of deforestation, land use and identity in western Madagascar', *The Geographical Journal*, vol 178, pp67–79.

Schlemmer, B. (1980) 'Conquête et colonisation du Menabe: une analyse de la politique Gallieni', in R. Waast, E. Fauroux, B. Schlemmer, F. Le Bourdiec, J. P. Raison and G. Ganday (eds) *Changements Sociaux dans l'Ouest Malgache*, ORSTOM, Paris.

Sodikoff, G. (2005) 'Forced and forest labor regimes in colonial Madagascar, 1926–1936', *Ethnohistory*, vol 52, pp407–435.

Sodikoff, G. (2012) *Forest and Labor in Madagascar: From Colonial Concession to Global Biosphere*, Indiana University Press, Bloomington.

WCMC (1992) *Protected Areas of the World: Vol. 3 – Afrotropical: A Review of National Systems*, World Conservation Union, Gland, Switzerland and Cambridge.

7 The roots, persistence, and character of Madagascar's conservation boom

Christian A. Kull

Introduction

In the span of 30 years, Madagascar has been transformed from a forgotten, isolationist republic to an emblem of biodiversity and environmental crisis. Even Disney has cashed in on the island's natural image with its 2005 eponymous animated film. Behind this transformation lies a complex, multi-million-dollar effort that has linked together international conservation organizations, multilateral agencies, bilateral donors, the Malagasy government, and many passionate individuals in an effort to protect the island's flora and fauna. Conservation spending in Madagascar by the World Wide Fund for Nature (WWF) increased more than ten-fold between 1983 and 1993 (Figure 7.1). Other actors, such as American and Swiss development aid (Figure 7.2), showed similar explosions in spending. This level of activity has – with occasional hiccups from political crises – been maintained until today. The multi-year National Environmental Action Plan (NEAP), running from 1990 to 2009, mobilized almost half a billion dollars. While the post-NEAP future is in question due to the impacts of a national political crisis and the global economic downturn on conservation funding, conservation organizations such as the WWF, Conservation International (CI), and the Wildlife Conservation Society (WCS) are as active as ever, joined by a host of smaller actors and an entire generation of Malagasy professionals and students.

What is behind this transformation? Naturalists have long raved about the island's distinct flora and fauna – Philibert Commerson reported on it after a visit in 1773, BBC naturalist David Attenborough first filmed there in 1961 – so this interest alone is not sufficient to explain the conservation boom. Did environmental degradation pass some critical threshold? Can we attribute it to the international wave of environmental consciousness characterizing the late twentieth century? What role does Madagascar's situation as a poor, indebted, third-world nation play? This chapter tells the story of the conservation boom and its perpetuation. It demonstrates how geopolitics, political, and economic ideologies, environmental discourses, specific institutional logics, and motivated actors came together to write the story of conservation in Madagascar.

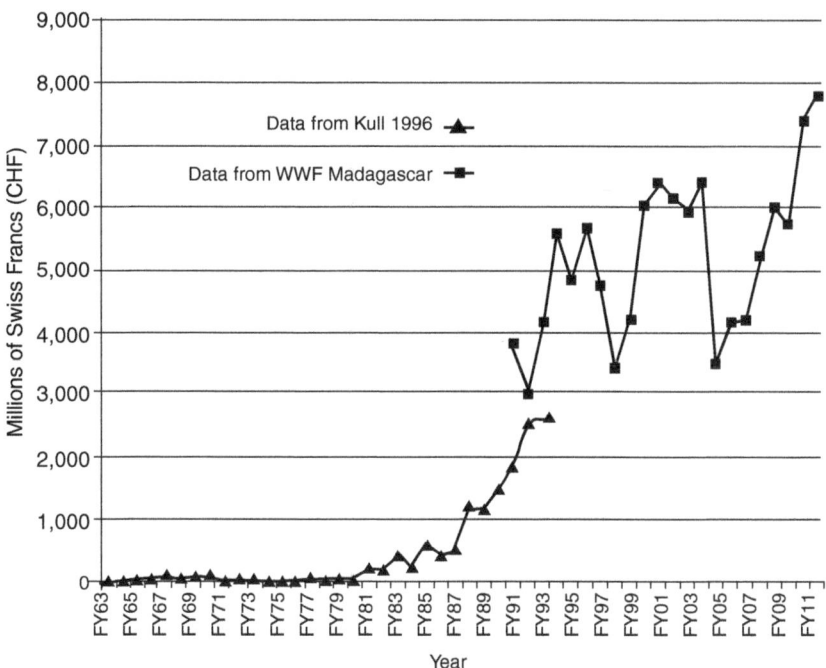

Figure 7.1 Annual expenditures by WWF in Madagascar – currently celebrating 50 years of work on the island – are illustrative of the conservation boom and its persistence

Notes: WWF, while the oldest and largest, is only one of many actors investing in conservation on the island. Note also that strong WWF expenditures in the past few years, since the 2009 political crisis, reflect its ability to seek alternative funding through its global network at a time when much traditional bilateral and multilateral donor environmental funding has dried up. Many other conservation actors have struggled to maintain funding and activities in the current political situation.

Sources: FY63–FY93 are from WWF International as reported in Kull (1996); FY91–FY2012 were kindly provided by WWF Madagascar (Richard Hughes and Zo Rakotonomenjanahary). Note that differences in accounting procedures result in inconsistent data between the two series (the 1962–1993 data, for example, only includes those funds passing through the Swiss headquarters of WWF International). Note also that fluctuations in foreign exchange rates strongly affect the figures.

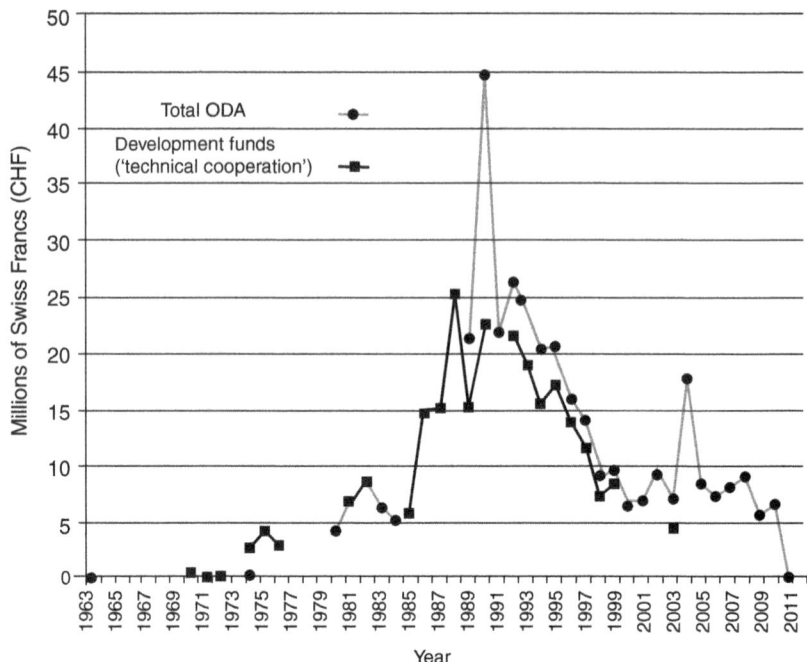

Figure 7.2 Bilateral aid from Switzerland to Madagascar, 1963–2011. Swiss aid in Madagascar has a five-decade history representing nearly half a billion francs of investment, often in the agricultural, forestry, and rural development sectors

Notes: from 1987 to 1995, Switzerland participated in the initial development and implementation of NEAP. At this point, Madagascar was always among the top three countries in Africa receiving Swiss aid, with funding stabilizing at 15–20 million Francs per year (except for a 1990 peak related to 21 million francs of debt service). Funding declined from 1996 in direct response to the unresolved assassination of Walter Arnold, a Swiss aid worker. While some projects continued through to final closure at the end of 2012, the Swiss Agency for Development and Cooperation (SDC) closed their Antananarivo office in 1996 and Madagascar was no longer considered a priority country (Jaberg, 2011).

Source: SDC annual reports, available at www.sdc.admin.ch/en/Home/Documentation/ Publications/Annualreports, accessed June 5, 2013. Note that reporting formats in these reports have not been consistent over the years, hence some years do not have data for either ODA (a combined figure for Overseas Development Assistance that includes development projects, humanitarian aid, loans, and grants) or specifically development funds. Some early figures were converted from U.S. dollars at average exchange rates for that year.

The chapter begins by outlining the important events in Malagasy environmental and conservation history leading up to the 'boom' in conservation marked by the creation of NEAP. Then I trace the perpetuation of the boom through the three distinct political regimes and environmental programs of the 1990s and 2000s. After these historical sections (Figure 7.3 provides a summarized timeline), I reflect on five important factors that have caused, shaped, and facilitated the conservation boom.

Early conservation in Madagascar

Today's environmental efforts have numerous antecedents. Here I summarize a few important ones that set the scene for the ongoing boom (for more detail, see Chapter 6 by Scales). I begin in the nineteenth century, for while matters of resource management surely occupied the minds and activities of people on the island long before then, they are poorly documented. Events of the nineteenth century are significant to conservation history for three reasons. First, a proliferation of contact with Europeans set the stage for long-term influences – from missionaries to colonists to global trade and environmental politics. Second, the early explorers that brought stories of the island's exotic natural beauty back to Europe contributed to the romantic myth of a wild Africa that lies at the roots of the conservation movement. Finally, this period includes several oft-cited precedents for forest conservation policies, including a ban on cutting live firewood issued by King Andrianampoinimerina, and forest burning prohibitions in the 1881 Code of 305 Articles.[1]

France conquered Madagascar in 1896 and controlled the island until 1960. The colonial period is an important precursor to today's power relations and resulted in many enduring administrative, legal, and social structures (see Figure 1.2, Chapter 1, and Chapter 6). For one, the colonial government intervened strongly in a variety of natural resource management sectors, both in the interest of creating a profitable colony (e.g. protecting valuable logging timber from peasant fires) and reflecting nascent scientific interest in conservation (e.g. establishing the first nature reserves in remote areas in the 1920s). Roads and train lines were built, the Agricultural and Livestock Services supported plantation crops, fought pests, and sought to intensify rice production, and a Forest Service oversaw the exploitation and protection of native hardwoods and widespread planting of eucalypts, pines, and other exotic trees (Kull, 2004). Today, linguistic, commercial, personal, and geopolitical links with France (including nearby Réunion) persist strongly through business channels, a strong French role in training institutions, and a large elite diaspora.

The First Republic of President Philibert Tsiranana began with independence in 1960 and is important to our story in three ways. First, the government replaced colonial rules with a fairly complete and long-reaching environmental legislation that reflected a strong agenda to develop the

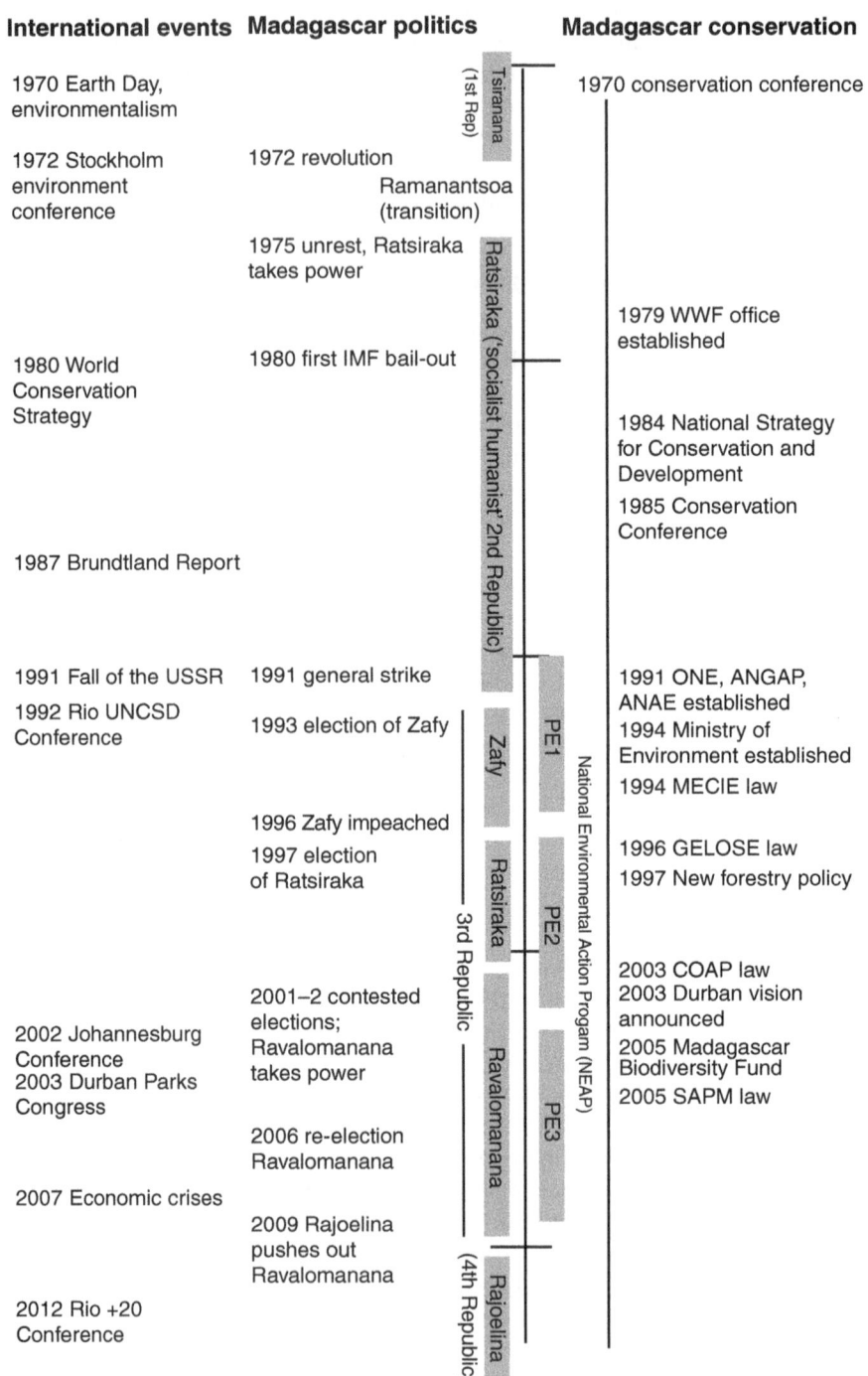

Figure 7.3 Timeline of key events in Madagascar politics and conservation, 1970–2012

nation through increased, modernized agricultural production twinned with environmental protection. This included land tenure laws stressing investment in cultivated land, rules that forbade forest fires but made room for economically necessary pasture fires, policies for reforestation that required all men to plant 100 seedlings a year, new protected areas categories, and hunting restrictions on endangered species. The second important thing is that the enforcement of this hopeful set of laws was meagre, reflecting limits in the reach of the state and underlying tensions with affected parties that persist to this day (Kull, 1996, 2004). Third, this period is an important precedent to the conservation boom, in that non-French foreign involvement in Madagascar through development and conservation agencies began at this time. The WWF began working on the island in 1963 with a project aimed at preserving the endangered aye-aye (*Daubentonia madagascariensis*), a lemur. The United States Agency for International Development (USAID) began its activities with a 1966 loan for railway improvements and Switzerland began to develop its Malagasy aid program at the same time.

The 1970s and 1980s: the run up to the boom

In 1970, the Malagasy government, working in close collaboration with the International Union for Conservation of Nature (IUCN), hosted an International Conference on the Conservation of Nature and its Resources, co-sponsored and attended by the leaders of institutions such as the WWF, the Food and Agriculture Organization of the United Nations (FAO), France's overseas research office (Orstom), and the Paris Museum of Natural History. The high-level meeting took place during a major global wave of environmental interest and was justified with reference to the island's scientific importance and the spectre of species extinctions. Calvin Tsiebo, Vice President of the Republic, stated in his opening remarks: 'Unfortunately, our incomparable natural heritage, this unique natural capital, is gravely endangered. According to the specialists, few areas of the world suffer from such grand and rapid degradation' (IUCN, 1972, p9).

The conference was a milestone in Malagasy conservation; it put the issues on the front page of the newspapers.[2] Tsiebo closed the conference by calling for a variety of actions, including new protected areas and, most importantly, increased involvement and financial support from international and national organizations.

At the same conference, however, Etienne Rakotomaria, then Director of Scientific Research, questioned the dominance of foreign scientists:

> We have touched on three problems – forest reserves, education, and the role of foreign scientists. In all three spheres we have seen international organizations negotiate with Frenchmen in the name of Madagascar but systematically exclude the Malagasy from our own

concerns ... in the future, however, you will find that negotiations must take place only with our government's representatives. Scientists will only be allowed to work here if they arrange reciprocal benefits for Malagasy colleagues. The people in this room know that Malagasy nature is a world heritage. *We are not sure that others realize that it is our heritage.*

(Jolly, 1980, p7, emphasis added)

Rakotomaria's statement is relevant today: the conservation boom is still largely driven by outsiders. But it particularly reflects the post-colonial tensions felt in Madagascar at the time. The Tsiranana government and armed forces were full of French advisors, foreigners controlled 80 percent of the economy, and university instructors were predominantly French. University students instigated protests that were joined by the masses and which led to Tsiranana handing over power to a military transition government lead by General Ramanantsoa. The 1972 revolution was seen as a second independence from continued French domination. During the tumultuous period that followed, France was asked to close its military base at Diego Suarez, Madagascar quit the *Zone franc* (the French-backed monetary system in its ex-colonies), most Western technical assistants and scientists were banned, and conservation efforts stalled (Covell, 1987).

The following three years were characterized by power struggles, riots, and a presidential assassination. In 1975, Admiral Didier Ratsiraka, minister of foreign affairs under Ramanantsoa, took power. By the end of that year, he had put in place a new Constitution and been elected President. Thus began the Second Republic, characterized at its inception by a commitment to nationalization, 'scientific socialism', 'humanist Marxism', and a lack of environmental concern (Jolly, 1990).

Like other developing countries, Madagascar followed the advice of wealthy country lenders and international financial institutions and borrowed heavily from commercial banks (which were cash rich from oil-exporting countries following the 1970s energy crisis) to invest in education, the military, transportation, communications, and industrial development. Irresponsible lending, ill-advised borrowing, and a global recession led to a rapidly deepening crisis of deficits, debts, and inflation. By 1980, a billion dollar external debt meant that Madagascar had little choice but to seek its first bail-out from the International Monetary Fund (IMF). The condition for IMF assistance was a program of 'structural adjustment'. Prescribed for developing countries around the world in order to receive IMF loans, structural adjustment consisted of macroeconomic reforms such as government austerity and balanced budgets, currency devaluation to improve exports, and reduction of trade barriers. By 1986, the World Bank had become a struggling Madagascar's dominant source of funding, with attendant pressure for further policies such as liberalization and privatization

(Covell, 1987; Mukonoweshuro, 1994). Such structural adjustment policies were known as the 'Washington Consensus', reflecting their advocacy by the headquarters of the IMF, the World Bank, and the U.S. Treasury Department. They were underpinned by an emerging 'neoliberal' ideology. Neoliberalism seeks to orient economic and political governance in line with classical liberal theory, including faith in the market and civil society, and hostility to the state. A strong ideological influence in governments around the world over the past three decades, neoliberalism promotes free-market policies such as trade liberalization, privatization of state assets, outsourcing of state services, and opening of markets.

It is within this political and economic context that conservation activities resumed. Slowly but surely, the debt crisis forced a geopolitical rapprochement of the isolationist Second Republic with foreign influences. Doors were also re-opening through connections in the conservation community. In 1979, the WWF established an official representation in Antananarivo under the direction of Barthélémi Vaohita, a long-time conservation activist and a strong advocate for environmental education. As a good public speaker and a friend of President Ratsiraka, Vaohita was important in aiding progress in Malagasy conservation.[3] While the WWF continued to focus on species conservation and protected areas, it also initiated an awareness campaign aimed at decision-makers and the public. Foreign research was invited again and, in 1983, a council was created, under the guidance of the Jersey Wildlife Preservation Trust and Yale, Duke, and Washington universities, which facilitated the granting of research permissions.[4]

By the mid-1980s, the momentum created at the 1970 conference resumed. The 1980s was a decade when Malagasy governmental opinion shifted from 'outright denial that the environment could affect human welfare, to being one of the leading countries in at least the rhetoric of sound policy' (Jolly, 1990, p121). In 1984, Madagascar adopted a National Strategy for Conservation and Development, signed by every government minister. Called for in the World Conservation Strategy (IUCN/WWF/UNEP, 1980), Madagascar was the first major nation in the Afrotropics to prepare such a document. The strategy stresses public awareness and environmental education, behavioural changes with respect to the environment, technical competency, program evaluation, and local participation. What precipitated this groundbreaking document? According to one conservationist, 'it was kind of a Malagasy thing', and 'few expatriates were involved'; yet another supposed that WWF pressure did have an influence.[5]

A second milestone International Conference on Conservation and Development was held in November 1985. Joseph Randrianasolo, then Minister of Livestock, Water, and Forests, declared that 'before, people only spoke of the beauty and scientific interest of our flora and fauna. This time we are speaking of our people, and how to manage our resources to be self-sufficient in food and fuelwood' (Jolly, 1990, pp119–120).

The linkage of conservation to human welfare in the newly minted concept of 'sustainable development' was a critical factor in bringing the government on board. Rooted in the World Conservation Strategy and enthroned by the Brundtland Report (WCED, 1987), sustainable development responded in part to a concern by poor countries that environmental rhetoric would impede their desperate goal of economic development.

The 1985 conference was attended by many international agencies, government members, as well as by Prince Philip, the International President of the WWF. It is said that he confronted President Ratsiraka with the statement 'your nation is committing environmental suicide', an event that is touted as a major milestone in Malagasy conservation awareness. One conservation agent I interviewed called it the key event in his career, as he was deeply impressed by the intense external interest in his nation's natural heritage.[6] During the conference, Minister Randrianasolo established the first new protected area since the 1960s (Beza-Mahafaly), and requested financial and technical assistance to implement the 1984 National Strategy.

Several programs were initiated in the aftermath. These included, for example, soil conservation and forest management projects financed by the World Bank, Switzerland, and Norway. The WWF and American universities were asked to assist in the management of protected areas with USAID and United Nations Educational, Scientific and Cultural Organization (UNESCO) funding. An important review of the protected areas system was commissioned, which stimulated plans for expansion and integrated conservation and development projects (Nicoll and Langrand, 1989; Hannah, 1992).[7] Finally, the WWF was asked to develop an environmental education program, which by 1992 had reached most school districts.

Despite high expectations, however, the successes of the mid-1980s were limited. The Forest Service, responsible for the protected areas, was caught between structural adjustment budget austerity requirements and an overwhelming number of projects (Schmid, 1993). As a result, the government, strongly pushed by the World Bank, asked international donors to help design a more effective action plan for the environment. The World Bank, under its new President Barber Conable, was seeking to demonstrate its green credentials after years of harsh environmental criticism. It introduced, encouraged, and largely funded NEAPs across Africa, pushing them as unofficial structural adjustment conditions for the receipt of country-operating budget loans in the wake of the debt crisis (Dorm-Adzobu, 1995; Lindemann, 2004). In each country, the NEAP took on its own character, leading to different kinds of conversations and results. Madagascar's was particularly prominent, given the antecedents described above and external conservation interest (World Bank, 1988; Falloux and Talbot, 1993; Hufty and Muttenzer, 2002; Sarrasin, 2007; Pollini, 2011).

On the Malagasy side, political support was not immediate from all parts of government. In one recounting:

> Many of the influential Malagasy were preoccupied with the country's urgent economic problems ... they simply had other priorities than the environment ... However ... on the basis of the alarming estimates of the costs of the environmental degradation the Prime Minister joined the Director of Planning as a sponsor of the NEAP. The President initially remained on the sidelines ... [but] was obliged to enter the arena after the showing of an excellent series of televised environmental episodes produced by *Radio Télévision Malgache* ... intended to strengthen public opinion in favour of the NEAP. This coincided with the start of the President's re-election campaign ... Happily, the development of the NEAP came at a good time, and the President adopted it and became an ardent supporter of the NEAP.
>
> (Falloux and Talbot, 1993, pp34–35)

Another source claimed that the 1989 appointment of Victor Ramahatra as Prime Minister was pushed by the World Bank, thus ensuring support and good coordination of the NEAP.[8] Malagasy cooperation in environmental action was never an explicit condition for World Bank aid, yet 'which government can ignore the strong wishes of its foreign donors?'[9]

The NEAP was developed by Malagasy government with strong technical guidance and financial support from the World Bank, U.S. and Swiss bilateral aid agencies, the WWF, the United Nations Development Programme (UNDP), and UNESCO (Sarrasin, 2007). UNESCO and the UNDP had, through their Man and the Biosphere programme, been playing an important role in environmental efforts on the island in the 1980s, including sponsoring another conference in Toamasina in 1988 (Maldague et al., 1989). The Swiss, at the time, included Madagascar as a focal country for their development aid, focussing on soil conservation and farming systems. The Americans were increasingly becoming involved specifically through the environmental sector. France, despite ongoing ties and an active development aid program, was absent in this original formulation (Andriamahefazafy and Méral, 2004; Freudenberger, 2010).

Booming conservation: from 1990 to today

This section tells the story of conservation in Madagascar since 1990, highlighting key themes and putting events into broader political and historical context. The NEAP served as an umbrella for most conservation activity in the 1990s and 2000s. Over $100 million was committed for the initial years of the NEAP by foreign donors in an accord signed in January 1990 in Paris, and the national assembly put the plan into law as the *Charte de l'Environnement* later that year. The NEAP was developed with an unusually long 15–20 year vision. Promotional materials announced that it would promote sustainable development by raising living standards, better resource management, and conservation of nature, and touted principles of dialogue, benefits to local communities, and continuity. Emphasis, however, was on the environment. Priority programs included rural and urban environmental management, institutional support, mapping, environmental education and training, and, in particular, biodiversity protection through the addition of 400,000 ha to the protected areas system (World Bank, 1988; Kull, 1996; Freudenberger, 2010).

Conservation activities in Madagascar accelerated rapidly as the NEAP got underway, ushering in an era of multi-million dollar projects. Madagascar became an 'El Dorado' for conservation,[10] witnessing an incredible surge in conservation activity (Hough, 1994; Kull, 1996). USAID, which had sponsored only a smattering of projects in Madagascar, introduced no less than six major conservation projects in the years 1988–1992, and took the lead in sponsoring the biodiversity component of the NEAP. The U.S. Peace Corps arrived in September 1993, and sent volunteers to work in environmental projects. This was also the era of Debt-for-Nature swaps, whereby conservation donors pay off a portion of a country's foreign debt in exchange for an agreement that the country will finance local conservation activity. CI, the WWF, and USAID facilitated at least seven swaps beginning in 1989. In 1991, the government gazetted brand-new Ranomafana National Park, with the help of a broad consortium of largely American sponsors. This first new National Park in 30 years symbolized the resurgent activity.

In 1991, conservation activities were dealt minor setbacks as the opposition to authoritarian President Ratsiraka, frustrated by the lack of economic progress, organized a general strike that paralyzed the economy. A new constitution in August 1992 marked the beginning of the Third Republic. On February 10, 1993, a general election replaced Ratsiraka with Albert Zafy, leader of the opposition coalition, in a smooth transition. Zafy's government, led by Prime Minister Francisque Ravony, was largely preoccupied with economic and political problems, but did not interfere with the NEAP. This is unsurprising, given the context of government bankruptcy, an economy stricken by the political crisis, and the unity of major donors behind a sustainable development agenda (remember, the UN Conference on Environment and Development (UNCED) took place in Rio in 1992). Conservation actors had unusually high-level access to government officials and convinced them to continue conservation actions in order to attract foreign funding. For example, the WWF's *Chargé du Programme*, Sheila O'Connor, was reputed to have had the ear of the Prime Minister.[11] It was in this context – political uncertainty, economic difficulties, and the near-free reign of conservation organizations – that the first phase of the NEAP was implemented. Known as the first environment program, or PE1, it ran from 1990 to 1996.

PE1, integrated conservation and development under Zafy

The PE1 had three major thrusts. The first was the creation of institutional structures to carry out the NEAP. In 1991, the government established an

Office National pour l'Environnement (ONE) to coordinate the activities of the NEAP. It also created the *Association Nationale d'Actions Environnementales* (ANAE) to focus on soil management and rural development, and the *Association National pour la Gestion des Aires Protégées* (ANGAP) to oversee protected areas management. ANGAP was created under a neoliberal spirit as a parastatal agency to facilitate the temporary management of protected areas by international non-governmental organizations (NGOs) with the idea of building up local capacity to take over.[12] Finally, in 1994, a new Ministry of the Environment was created, but had a weak mandate.

The creation of new institutions – designed under a dominant neoliberal ideology of shrinking the state and outsourcing certain functions – was not without tensions at the time as well as long-term consequences (Andriamahefazafy and Méral, 2004). For example, the *Direction des Eaux et Forêts* (the Forest Service) was historically responsible for protected areas and resisted transferring authority to ANGAP.[13] The Forest Service had a proud tradition in the colonial administration as a highly professional agency, but had fallen apart during 1980s austerity measures. The NEAP simultaneously sought to circumvent it and to re-build its capacity (Montagne and Ramamonjisoa, 2006; Freudenberger, 2010).

A second, related thrust of PE1 was the modernization of the country's environmental legislation. A significant milestone was the passage in 1994 of the first environmental impact assessment legislation, called MECIE (for *Mise en Compatibilité des Investissements avec l'Environnement*). A few years later, facilitated by a Swiss project, Madagascar also passed a full new set of forestry legislation (Montagne and Ramamonjisoa, 2006).

The third and most visible thrust of PE1 was to solidify and expand the country's network of protected areas. Donors sponsored over a dozen Integrated Conservation and Development Projects (ICDPs), which sought to twin conservation efforts inside the park or reserve with development efforts in the surrounding villages, with the idea that the latter efforts would reduce human pressures on the protected areas. Some illustrative examples included Ranomafana National Park, funded by USAID via a consortium of American universities; Mananara-Nord, a UNESCO biosphere reserve funded by the World Bank and the UNDP; Masoala peninsula, funded by USAID and implemented by the WCS together with the development organization CARE; and the Marojejy region, operated by the WWF initially through German funding (Kull, 1996).

USAID was effectively the key driver of NEAP activities during PE1, as it and the World Bank were the main funding sources (Méral, 2012). USAID's offices were classified as a 'major mission' and had a large staff presence, while the World Bank had its key staff based in Washington (Freudenberger, 2010). The USAID mission was downgraded in 1994 and 1996 due to Madagascar not meeting structural adjustment commitments, but support to environmental programs continued due to a Congressional earmark for biodiversity (Medley, 2004; Corson, 2010), so USAID maintained its key role throughout the rest of the NEAP. Its strong presence was also aided by the continuity in the coordination of its environment programs by the same person, Lisa Gaylord (Freudenberger, 2010).

A key trend in PE1 was the inexorable shift of emphasis from the broad goal of sustainable natural resource management (including biodiversity conservation), to a narrow focus on biodiversity first and foremost. This tension was already visible in earlier 'conservation and development' conferences, and is manifested in the fate of the three institutions that were created – ONE, ANGAP, and ANAE. Of the three, ANAE, which focussed more on soil conservation and rural development, was the weakest and eventually ceased being funded (Andriamahefazafy and Méral, 2004). The move towards a biodiversity focus was aided by the structural and political constraints on USAID funding (Corson, 2010). Furthermore, the Swiss emphasis on sustainable forestry and rural development was reduced when Swiss aid demoted Madagascar from 'focal country' status (in response to the unresolved assassination of one of its contractors in 1994).

In the closing two years of PE1, there was a sudden rush – reflecting global trends – towards community-based natural resource management. The main NEAP donors (France was now involved too) sponsored a series of workshops and expert missions that promoted ideas from a community-management paradigm, including from University of Wisconsin's Land Tenure Center and from the French research agency Cirad (Weber, 1995; Montagne and Ramamonjisoa, 2006). The result was that policymakers developed a law, called GELOSE for *Gestion Locale Sécurisée*, facilitating the transfer of specific resource management rights and responsibilities to community associations. It was passed in 1996 (for greater detail see Chapter 8 by Pollini et al.; Pollini and Lassoie, 2011).

The return of Ratsiraka, PE2, and regional approaches

After little progress in the management of Madagascar's economic crisis, Albert Zafy was forced out of the Presidency in 1996 (Marcus, 2004). Reflecting frustration with his efforts to centralize authority, the National Assembly impeached him for exceeding his constitutional powers. This political crisis, which ended with the re-election of former dictator Didier Ratsiraka in early 1997, did not result in strikes or violence, but provided a measure of political uncertainty during the transition from PE1 to PE2.

The second environmental program ran from 1997 to 2002. Evaluators of PE1 had found that the ICDP approach had delivered little conservation benefit from poorly targeted development activities around protected areas. They proposed for PE2 a broader, more strategic, more comprehensive regional approach with significant emphasis on decentralization and participation (Gezon, 2000; Freudenberger, 2010; Pollini, 2011). USAID projects focussed on 'eco-regions' that linked corridors of protected areas with the regions around them. For instance, the U.S. sponsored Landscape

Development Interventions project (LDI) focussed on two major forest corridors in eastern Madagascar, undertaking a bewildering diversity of initiatives including alternatives to slash-and-burn agriculture, remote rural health, natural product commodity chains, ecotourism, market road-building, local irrigation systems, farmers cooperatives, and participatory planning structures. One of the impacts of this regional focus was the territorial entrenchment of different actors – the Americans in Fianarantsoa and Moramanga, the Swiss in the Menabe, and the Germans in the Vakinankaratra (Moreau, 2008; Méral, 2012).

Several projects sought to promote the co-management of natural resources, as initiated through GELOSE. Yet, for some, GELOSE was seen as too legalistic and cumbersome. They developed a simpler alternative, called GCF (*Gestion Contractualisée des Forêts*; for forest management contracts), by decree in 2001. Unlike GELOSE, GCF can only be applied to lands controlled by the Forest Service, and it requires no tenure allocation and no negotiation with a municipality. At one point, these competing approaches became emblematic of rivalries between donors, sometimes over-simplified as tensions between French adherents of GELOSE and American sponsors of GCF.[14] In the end, over 450 local management contracts using one or the other legislation were reported in the period 1997–2006 (see Chapter 8 by Pollini et al.; Montagne and Ramamonjisoa, 2006; Montagne et al., 2007).

In the meantime, much work was continuing in Antananarivo on refining the institutional, technical, and legislative basis for environmental management. Aid and conservation agencies provided support, for instance, to the Forest Service, by developing a satellite-based fire monitoring tool.[15] Likewise, a Multi-Donor Secretariat was established by a number of the multilateral and bilateral agencies and conservation organizations, with a mission to coordinate their activities (Lindemann, 2004; Freudenberger, 2010).

PE3, Ravalomanana, and the Durban Vision

The transition from PE2 to PE3 was long and confused, due to another round of political crises. The results of Presidential elections in December 2001 were contested by partisans of the mayor of Antananarivo, Marc Ravalomanana, who claimed he had won an outright majority in the first round of voting. Months of street protests and tensions followed. Rapidly recognized diplomatically by the Americans, Swiss, and Norwegians, Ravalomanana's control of the island was assured in July 2002 when France facilitated the exile of President Ratsiraka. Ravalomanana, a self-made entrepreneur and businessman, brought a brash new attitude to the Presidency: a results-oriented, top-down management style; openness to non-French contacts and investors (the U.S., South Africa, China, and other Asian countries); and hostility to the old Franco-Malagasy establishment (Marcus, 2004; Rakoto Ramiarantsoa, 2008).

While bilateral funders and NGOs quickly resumed their work, the World Bank delayed implementation of PE3, awaiting guarantees that the new government would demonstrate its commitment. In the event, Ravalomanana's government did so in spades. It announced a strict ban on burning, launched awareness and repression campaigns (jailing several slash-and-burn farmers), linked municipal (*commune rurale*) budgets to performance measures such as fire prohibition, and, most dramatically, announced an ambitious goal of tripling protected areas in five years. As a result, the start of PE3 was messy and uncoordinated (Pollini, 2011; Méral 2012).

President Ravalomanana's aggressive agenda, which also included large mining projects, land tenure reforms facilitating foreign investment, agribusiness deals, and road building (Rakoto Ramiarantsoa, 2008; Rakoto Ramiarantsoa et al., 2012), changed the tenor of the third phase of the NEAP. While there was a general continuity in the key actors, in their projects, and in their intervention zones, PE3 (2003–2009) was marked by three major departures from earlier efforts: i) a resurgent focus on the protected areas system; ii) a more assertive state and less room for local participation; and iii) efforts to achieve conservation through economic tools. I address each in turn.

First, the protected areas system – and its expansion – became the central focus of conservation action. In 2003, the government passed new protected areas legislation called COAP (*Code des Aires Protégées*), which outlined different categories of protection and prohibited most human resource use within protected areas. Then, President Ravalomanana announced at the World Parks Congress in Durban, South Africa, that Madagascar would triple its protected areas in five years. The urgency of this 'Durban Vision' unleashed a flurry of activity and debate (see Chapter 9 by Corson for more detail). To simplify, the debate pitted preservationist interests, led by U.S.-based conservation organizations CI and WCS (whose stars were ascendant under Ravalomanana) against diverse actors whose interests were threatened by the rapid, top-down expansion of strict conservation zones. This included not only government ministries, mining, and forestry interests, but also many within the environment and development sectors who preferred sustainable use approaches and saw the hard-core approach as undermining relationships with poor rural communities (such as French[16] and German bilateral aid agencies, some local NGOs, UNESCO, and to a lesser extent the WWF and UNDP).

The impasse was only broken through after the IUCN (the sponsor of the Durban conference) sent two major missions, led by Grazia Borrini-Feyerabend, which proposed that the less restrictive IUCN categories 5 and 6 (protected landscapes and sustainable use areas) be used for the new protected areas. The result was a new policy in 2005, called SAPM for *Système des Aires Protégées de Madagascar*, which aligns the protected areas system with the IUCN categories and gives its overall management to the Ministry of Environment and Forests, but gives responsibility for individual reserves to actors such as Madagascar National Parks (formerly ANGAP), regional governments, NGOs, or private actors. Furthermore, modifications to the COAP legislation in 2008 allowed for participatory approaches and some natural resource use in the more flexible park categories[17] (Duffy, 2006; Freudenberger, 2010; Pollini, 2011; Corson, 2012).

Second, during PE3 and Ravalomanana's presidency the state became more assertive, sometimes in partnership with conservationists, while the participatory, community-based management spirit that had grown during PE2 was dealt a number of setbacks, despite many intentions and much rhetoric to the contrary (for detail see Chapter 8 by Pollini et al. and Chapter 9 by Corson). It should be noted, however, that on the ground, many other actors sought to maintain strong co-management components to their projects. Government assertiveness was aided by some long-overdue administrative changes that clarified rivalries among different sectors and institutions, reducing the influence of the para-statal agencies created in PE1. Most crucially, in 2003 a newly merged Ministry of Environment, Water, and Forests took over coordination of the NEAP (from ONE) and later of the protected areas system (from ANGAP and the Forest Service). In principle, such centralization facilitates control and coordination, but this particular ministry has also been seen as an 'institution having as a mission to obey international environmental objectives' (Rakoto Ramiarantsoa et al., 2012, p253).

Third, environmental initiatives during PE3 took a strong turn towards 'neoliberal' conservation approaches, seeking to harness financial or economic tools to make environmental protection last. In 2005, the WWF and CI collaborated with the Malagasy government to establish a 'Madagascar Biodiversity Fund'. With funding from the World Bank, France, and Germany, the endowment accumulated US$25 million by the end of 2010. The goal is to have an operating budget capable of sustaining the protected areas system. At the same time, increasing efforts have been made to harness Payments for Environmental Services (PES) approaches. Such approaches were already discussed in the lead-up to PE2, but – aside from an innovative pilot project at Makira – were dropped due to uncertainty about their efficacy (Freudenberger, 2010). More recently, however, market-based approaches have climbed the global agenda. The Malagasy government approved its first carbon purchases in 2005, and several PES projects have been set up, including four focussed on carbon, three on biodiversity, and one on watershed protection. Funding comes from sources ranging from conservation NGOs or the World Bank (via its Biocarbon Fund), to private sector actors such as Air France, Mitsubishi, and Pearl Jam (for more detail see Chapter 13 by Brimont and Bidaud; see also Méral et al., 2011; Méral, 2012).

After NEAP

The NEAP ended in 2009 with somewhat of a whimper, overtaken by the events of yet another political crisis. Ravalomanana's heavy-handedness in the environmental sector contributed (among many other complaints, tactical errors, and geopolitical machinations) to growing dissatisfaction with his regime. After his troops shot at street protesters, the tide turned against him and he was forced into exile in March 2009, with power over a 'transitional authority' going to young rival Andry Rajoelina. Rajoelina has clung to power for three years now (these words are written in 2012). Numerous donors suspended their non-humanitarian funding in 2009 in protest, including the U.S., the World Bank, the EU, the African Development Bank, the IMF, and the UNDP (Rakoto Ramiarantsoa et al., 2012). While the U.S. has continued to maintain its distance from what it calls an 'illegitimate regime', the World Bank resumed partial funding of critical programs (including the environmental sector) in June 2011, and proposed in Februray 2012 to resume full relations.

Explaining the boom and its persistence

This overview of the history of conservation in Madagascar shows that the events of the past decades have a variety of antecedents and driving forces. In this section, I discuss five important and interlinked factors that have driven and shaped the conservation boom and its persistence.

Madagascar's environment, real and imagined

Conservation action in Madagascar is motivated by people's perceptions that the island's biodiversity is both unique and particularly threatened. The highly endemic flora and fauna has inspired naturalists for decades (see Chapter 2 by Ganzhorn et al.).[18] Their observations of the rapid disappearance of the island's natural forests, habitat of much of this natural heritage, impelled them to action. On top of these empirical observations, several 'received wisdoms' have shaped how Madagascar's environment is talked and written about. These include contested ideas of an original island-wide forest and oft-repeated but exaggerated figures such as the loss of 90 percent of forests (see Chapter 4 by McConnell and Kull; Kull, 2004), as well as theories about the causes of degradation (such as the 'spiral of degradation' that linked population growth, poverty, and environmental destruction – World Bank, 1988) that tend to ignore other driving forces (see Chapter 5 by Scales). Together, these real and imagined components of Madagascar's environment have constructed the island as a global conservation priority. The island is now one of the 'hottest' of the biodiversity hotspots (Ganzhorn et al., 2001), an image endlessly reproduced through television and the media.

Global environmentalism

Conservation in Madagascar was undoubtedly shaped by the broader context of global (or, perhaps more honestly, Euro-American) environmentalism. With antecedents in concerns over pollution and species losses caused by industrial activities, the environmental movement first captured widespread public attention in the period around 1970, leading to the UN Conference on the Human Environment in Stockholm in 1972. Resurgent from the mid-1980s, the emphasis turned to crisis buzzwords such as acid rain, desertification, rainforests, biodiversity, and global warming. Attention moved towards the new idea of 'sustainable development', pushed by high-profile documents such as the Brundtland Report (WCED, 1987), which sought to reconcile economic growth in both industrialized and poor countries with environmental conservation. This contributed to a 'greening of aid' (Adams, 1990). Sustainable development arguably reached its zenith in the 1992 UNCED, held in Rio de Janeiro. Global environmentalism has continued to evolve, with two trends reflecting different manifestations of neoliberal ideas: the move towards decentralization and participation in the 1990s; and another move towards market-based instruments in the 2000s (reflected in the UN's Rio+20 Conference and its theme of the 'Green Economy'; Carrière et al., 2013). Throughout this long history, global conservation has felt tensions between more 'preservationist' views that seek to preserve portions of nature from human influence, and more 'sustainable use' views that draw less stark lines between humans and the wild.

The story of conservation in Madagascar clearly reflects this evolving global context. In the 1970s, interest in the island's nature dovetailed with a global interest in 'ecology'. The late 1980s focus on sustainable development and the greening of aid were a critical element in the conservation boom, as they led to a dramatic increase in available funding. Madagascar was an 'object of prestige' for environmentally minded donors.[19] More recently, conservation activities on the island reflect, as we have seen, the trends towards community- and market-based approaches, and the tensions between preservationist and sustainable use philosophies.

Indigenous environmentalism

While foreign interest and funding is undoubtedly a prime motor for the conservation boom, it is also important to recognize the role of Malagasy individuals in advocating for the protection of the island's natural heritage. Had key decision-makers not been on board, it would have been easy for the government to make it more difficult for foreign groups to undertake their activities.[20] In addition, the influence of the small but dedicated Malagasy scientific and intellectual community should not be underestimated. Key figures, for example, might include Leon Rajaobelina, Ambassador to the U.S. in the 1980s, Minister of Finance during the elaboration of the NEAP, and now regional Vice President at CI. His son, Serge, has taken up his mantle, founding the NGO Fanamby in 1997 (see also Kull, 1996). The implication of Malagasy actors has increased through each stage of the NEAP (Andriamahefazafy and Méral, 2004).

While some observers have suggested that 'the Malagasy perspective on the natural forest per se can best be characterized as indifferent' (Freudenberger, 2010, p89), their perspective might rather be seen as *different*, reflecting the cultural lenses and socio-economic interests of rural farmers (Keller, 2008; Scales, 2012). Malagasy environmental attitudes differ widely, depending on social context, urban vs. rural location, and what is meant by 'the environment'. What is sure is that three decades of booming environmental action on the island – including WWF's awareness programs[21] – has resulted in the mainstreaming of green discourses in urban society and the creation of a cadre of field workers, park staff, graduates, and professionals whose careers and identity are linked to the conservation boom. Surprisingly missing, however, is any broad-based movement or activism by rural residents related to land, environment, and livelihood issues.

Politics and economics: the foreign role

Foreign government agencies and international NGOs play a preponderant role in conservation in Madagascar (Duffy, 2006; Méral, 2012; Rakoto Ramiarantsoa et al., 2012). The amount of development aid disbursed to Madagascar in the past decades has been similar in magnitude to the operating budget of the central government. The NEAP involved nearly half a billion dollars of funding over 20 years. When donors such as the World Bank and USAID went green in the late 1980s, the country had little choice but to follow.[22] As Alison Jolly (1990, p121) stated, 'the country depends on Bank-funded projects ... Madagascar is too poor and too much in debt to do otherwise'.[23] This dependency arises from complex historical factors including colonialism and its political and economic legacy, ill-advised borrowing in the 1970s, and structural adjustment policies in the 1980s.

The timing of the conservation boom has much to do with the convergence of Madagascar's debt crisis, its political re-opening to the outside world, and the global rise of sustainability discourse. Geopolitical strategic issues also played a role. For instance, the increase in American aid to Madagascar coincided with threats posed by instability in South Africa to strategic mineral supplies (Hannah, 1992). The island's transition to democracy, exemplified in the 1993 elections and the relative stability of that decade, helped attract aid from major donors who saw the island as a 'good pupil'. In this context, the influence of conservation organizations was facilitated by the lack of a strong lobby for mining or logging interests.[24] The result was a close relationship between conservation NGOs and the bilateral aid sector (Duffy, 2006).

The strong role of foreign funders and international environmental organizations was compounded by the weakness of the state and the civil society (Freudenberger, 2010). The contrast with, say, another conservation prize – the Brazilian Amazon – could not be stronger. There, the federal state is better resourced and has a strong agenda, and social movements are vocal in defending a variety of interests. In contrast, no civil society groups participated in the development and running of Madagascar's NEAP (Lindemann, 2004). As an interviewee stated when asked why the conservation boom took place, 'it is cheaper and easier to work in Madagascar than, for example, Brazil'.[25] In the end,

> eco-power remains in the hands of big environmental NGOs specialized in being financial intermediaries [between donors and the field]. They orient the aid programs, giving priority to biodiversity conservation and climate change adaptation. These two types of action are re-packaged as poverty eradication.
> (Rakoto Ramiarantsoa et al., 2012, p256)

Rivalries, ideologies, and lobbies within the conservation effort

While it is crucial to appreciate the overwhelming foreign role in the conservation boom, it is also necessary to understand the ideological tensions and geopolitical rivalries that play out between the different actors (Méral, 2012). While nascent conservation activities up to the late 1980s were designed by a small, cooperative group of individuals, the higher stakes in the years that followed led to more competition and conflict.[26]

An illustrative tension has been between strict nature conservation goals and a broader focus on sustainable natural resource management. The 1970 and 1985 conferences and the 1988 NEAP documents were framed in terms of sustainable resource management in the context of a poor population seeking social and economic development. In part, this framing assured Malagasy government interest. Yet while there have been efforts focussed on soil conservation, sustainable farming systems, and rural development, the lion's share of attention and funding has gone to biodiversity conservation (Pollini, 2011).

A major impetus for this tendency has been the structure of American aid. For complex reasons, the domestic political landscape in the United States caused biodiversity protection to dominate USAID's environment portfolio. As a result, this agency has essentially adopted the protected areas approach of its conservation NGO partners (WWF, CI, WCS) (Andriamahefazafy and Méral, 2004; Medley, 2004; Corson, 2010; Freudenberger, 2010). Under the separate economic development portfolio, a major contribution of American aid was the 2000 African Growth and Opportunity Act, which lowered tariffs on imports from countries such as Madagascar, and contributed to the expansion of its textile industry.

The American approach, 'modernist' in the sense that it separates nature conservation on the one hand from economic (industrial) development, on the other, differs from the strong rural development tradition in the aid programs of other NEAP partners, such as France, Switzerland, and Germany. It also reflects a somewhat different conception of what constitutes 'nature', with American approaches dominated by a wilderness ideology and a preservationist model inspired by Yellowstone National Park, whereas continental European approaches are oriented more towards sustainable use, incorporating rural farm landscapes like in the French *Parcs naturels régionaux* (Marcus and Kull, 1999; Carrière and Bidaud, 2012; Méral, 2012).

While certainly not absolute in any sense, these ideological tensions played out – as we saw earlier – in an initial minimal French involvement in the NEAP, in the conflict between GELOSE and GCF participatory models of conservation, and in the debates over the post-Durban approach to expanding the protected areas. These ideological (or cultural) tensions are at times entwined with institutional and geopolitical rivalries. Some French, for example, perceive of the environment as a Trojan Horse for 'Anglo-Saxon' influence on the island (Moreau, 2008).

France provided less than three percent of development assistance in the category 'environment' between 1990 and 2003; the big environmental donors were the U.S. (32 percent), the World Bank (20 percent), Switzerland and Germany (15 percent each), the EU (six percent), and the UNDP (four percent). Instead, France funded rural development, fishing, cotton, irrigated rice, agro-ecology, and livestock (Andriamahefazafy and Méral, 2004; Méral, 2012). Its environmental influence came through advisory positions in government agencies (see Pollini, 2011 for an autobiographical view) and through large, long-term research collaborations through its research agencies.

With the rise and fall of the Ravalomanana regime, the French–American rivalry gained a geopolitical aspect. Ravalomanana offended the established Franco-Malagasy elite and cultivated Anglophone links, even establishing English as an official language. France was slow to recognize his Presidency in 2002. After the 2009 coup that deposed him, rumours abounded of French help or opportunism (Deltombe, 2012), while America was the quickest and most vocal in shutting down its programs in protest at what it called an illegitimate regime. Given the strong foreign role in conservation in Madagascar, the ideological and geopolitical tussles of the main actors – particularly ex-colonial master France and chief environmental financier America – have shaped the course and character of the 20-year-long conservation boom.

Conclusion

Madagascar has long exerted a particular attraction to nature-lovers, due to its peculiar flora and fauna found nowhere else. It is through this naturalist's lens that many foreigners have viewed the island – despite its other attractions, such as its musical traditions or cultural landscapes of rice terraces and red-brick houses. The booming efforts at nature conservation from the late 1980s until now were based in this particular view of the island, and have reinforced it.

The conservation boom was ultimately caused by a combination of the island's special biological characteristics, the degradation of this natural heritage, the dominant discourses of environmental crisis that amplified the speed and effects of this degradation, the expanding reach and evolving ideas of the global environmental movement, and, most crucially, the political–economic influence of bilateral and multilateral institutions in a desperately poor, post-colonial country. The timing of this boom was due to a global boom in the late 1980s of environmental activism (extending, via sustainable development, into the corridors of the World Bank) coinciding with Madagascar's financial crisis and political re-opening. The efforts of numerous passionate individuals contributed all along the way.

Conservation activity has come a long way in the 20 years of the NEAP. Yet, new challenges call for attention today. Two large mining projects are underway, with several more planned. Agricultural investors seek land concessions for cash crops. European and American aid donors are crippled by economic crises. Asian investors have an increased influence in business and in development projects. And a 'shadow state' of networked elites plays a nefarious role in profiting from activities in direct contradiction to environmental goals, such as illegal logging (Duffy, 2006; Pollini, 2011).

The results of the conservation boom – protected areas, legislation, institutions, and more (see Freudenberger, 2010) – are simultaneously appreciated and contested by many of the stakeholders. Conservationists may rightly be proud of progress such as the expansion of the protected areas, but also frustrated at the many failures and obstacles along the way – trees are still being cut, after all, even in parks. Advocates for poor rural Malagasy residents can appreciate the development efforts that have been undertaken around protected areas, the occasional employment opportunities for rural communities, and the recognition given to them through co-management initiatives, but may still be frustrated at restrictions on rural ways of life. This tension has never been resolved in a satisfactory way; the 'eco-power' of the conservation lobby continues to be 'confronted with problems of legality and legitimacy' (Rakoto Ramiarantsoa et al., 2012, p256).

Many of the world's flagship protected areas were tenuous and contested affairs at first. Madagascar nature reserves are certainly no exception, and one can ask whether over the longer term its parks will succeed. These protected areas and other initiatives certainly have positive implications for nature conservation, but their sustainability depends on broader social and economic factors. Most important at this point, now that so much energy has focussed on lemurs, chameleons, and endemic flora, is to resolve the political, economic, and governance challenges of the island nation, and to focus on the sustainable management of all landscapes, focussing on those who make a livelihood from them. After all, the original aims of the NEAP were not just conserving the natural heritage, but also developing human resources, raising living standards, and promoting sustainable development through improved resource management.

Acknowledgements

Certain portions of this chapter build on Kull (1996). Thanks to the participants in the Madagascar sessions at the 2007 AAG Conference in San Francisco who encouraged me to write an update, and to Barry and Ivan for the opportunity to do so.

Notes

1 These precedents, often cited to justify conservation policies, are often removed from their context (see Kull, 2004).
2 Source: anon-e (code refers to anonymous interviews conducted during 1994 with conservation and development professionals. The letter code identifies the individual).
3 anon-g.
4 anon-h.
5 anon-c; anon-j.
6 anon-f.
7 anon-j.
8 anon-a.
9 anon-d.
10 anon-a.
11 anon-c.
12 anon-j.
13 anon-e.
14 Moreau, 2008; interviews, Antananarivo, 2003.
15 Interview, Andy Keck, Jari-Ala program, 2006.
16 Ironically, in 2003, France moved the environment portfolio of its development assistance from the Ministry of Foreign Affairs (which, for example, had placed experts on matters such as decentralized resource management into government advisory positions) to the *Agence Française de Développement* which is more like a development bank and which decided to put all its money into the Biodiversity Foundation, effectively supporting a harder-core position (Méral, 2012 and pers. comm., October 26, 2010).
17 P. Méral, pers. comm., October 26, 2010.
18 anon-a; anon-g.
19 anon-n.
20 anon-j.
21 anon-c.
22 anon-a; anon-d; anon-k; anon-o.
23 anon-a; anon-d; anon-k.
24 anon-g; this changed in the 2000s.
25 anon-g. Similar comment also made by anon-a.
26 anon-h.

References

Adams, W. M. (1990) *Green Development*, Routledge, London.
Andriamahefazafy, F. and Méral, P. (2004) 'La mise en oeuvre des plans nationaux d'action environnementale: un renouveau des pratiques des bailleurs de fonds?', *Mondes en Développement*, vol 32.3, no 127, pp29–44.
Carrière, S. M. and Bidaud, C. (2012) 'Enquête de naturalité: représentations scientifiques de la nature et conservation de la biodiversité', in H. Rakoto Ramiarantsoa, C. Blanc-Pamard and F. Pinton (eds) *Géopolitique et Environnement*, IRD Editions, Montpellier.
Carrière, S. M., Rodary, E., Méral, P., Serpantié, G., Boisvert, V., Kull, C. A., Lestrelin, G., Lhoutellier, L., Moizo, B., Smektala, G. and Vandevelde, J.-C. (2013) 'Rio+20, biodiversity marginalized', *Conservation Letters*, vol 6, pp6–11.
Corson, C. (2010) 'Shifting environmental governance in a neoliberal world: USAID for conservation', *Antipode*, vol 42, no 3, pp576–602.
Corson, C. (2012) 'From rhetoric to practice: how high profile politics impeded community consultation in Madagascar's new protected areas', *Society and Natural Resources*, vol 25, pp336–351.
Covell, M. (1987) *Madagascar: Politics, Economics, and Society*, Frances Pinter Publishers, New York.
Deltombe, T. (2012) 'La France, acteur-clé de la crise malgache', *Le Monde Diplomatique*, March.
Dorm-Adzobu, C. (1995) *New Roots: Institutionalizing Environmental Management in Africa*, World Resources Institute, Washington, DC.
Duffy, R. (2006) 'Non-governmental organisations and governance states: the impact of transnational environmental management networks in Madagascar', *Environmental Politics*, vol 15, no 5, pp731–749.
Falloux, F. and Talbot, L. M. (1993) *Crisis and Opportunity*, Earthscan, London.
Freudenberger, K. S. (2010) *Paradise Lost? Lessons from 25 Years of USAID Environment Programs in Madagascar*, International Resources Group, United States Agency for International Development, Washington, DC.
Ganzhorn, J. U., Lowry, P. P. I., Schatz, G. E. and Sommer, S. (2001) 'The biodiversity of Madagascar: one of the world's hottest hotspots on its way out', *Oryx*, vol 35, no 4, pp346–348.
Gezon, L. L. (2000) 'The changing face of NGOs: structure and *communitas* in conservation and development in Madagascar', *Urban Anthropology*, vol 29, no 2, pp181–215.
Hannah, L. (1992) *African People, African Parks*, USAID, Biodiversity Support Program, and Conservation International, Washington, DC.
Hough, J. L. (1994) 'Institutional constraints to the integration of conservation and development: a case study from Madagascar', *Society and Natural Resources*, vol 7, no 2, pp119–124.
Hufty, M. and Muttenzer, F. (2002) 'Devoted friends: the implementation of the convention on biological diversity in Madagascar', in P. G. Le Prestre (ed.) *Governing Global Biodiversity*, Ashgate, London.
IUCN (1972) *Comptes rendus de la Conférence internationale sur la Conservation de la Nature et de ses Ressources à Madagascar, Tananarive 7–11 Octobre, 1970*, International Union for Conservation of Nature, Gland, Switzerland.

IUCN/UNEP/WWF (1980) *World Conservation Strategy*, International Union for Conservation of Nature, Gland, Switzerland.
Jaberg, S. (2011) '50 ans d'aide suisse à Madagascar, et après?', *Swissinfo.ch*, September 14, 2011.
Jolly, A. (1980) *A World Like Our Own: Man and Nature in Madagascar*, Yale University Press, New Haven.
Jolly, A. (1990) 'On the edge of survival', in F. Lanting (ed.) *Madagascar: A World Out of Time*, Aperture, New York.
Keller, E. (2008) 'The banana plant and the moon: conservation and the Malagasy ethos of life in Masoala, Madagascar', *American Ethnologist*, vol 35, no 4, pp650–664.
Kull, C. A. (1996) 'The evolution of conservation efforts in Madagascar', *International Environmental Affairs*, vol 8, no 1, pp50–86.
Kull, C. A. (2004) *Isle of Fire*, University of Chicago Press, Chicago.
Lindemann, S. (2004) *Madagascar Case Study: Analysis of National Strategies for Sustainable Development*, IISD, Environmental Policy Research Centre, Freie Universität Berlin.
Maldague, M., Matuka, K. and Albignac, R. (1989) *Environnement et Gestion des Ressources Naturelles dans la zone Africaine de l'Ocean Indien*, UNESCO, Paris.
Marcus, R. R. (2004) *Political Change in Madagascar: Populist Democracy or Neopatrimonialism by Another Name?*, Institute for Security Studies, Pretoria.
Marcus, R. R. and Kull, C. A. (1999) 'Setting the stage: the politics of Madagascar's environmental efforts', *African Studies Quarterly*, vol 3, no 2, pp1–8.
Medley, K. E. (2004) 'Measuring performance under a landscape approach to biodiversity conservation: the case of USAID/Madagascar', *Progress in Development Studies*, vol 4, no 4, pp319–341.
Méral, P. (2012), 'Économie politique internationale et conservation', in H. Rakoto Ramiarantsoa, C. Blanc-Pamard and F. Pinton (eds) *Géopolitique et Environnement*, IRD Éditions, Marseille.
Méral, P., Froger, G., Andriamahefazafy, F. and Rabearisoa, A. (2011) 'Financing protected areas in Madagascar: new methods', in C. Aubertin and E. Rodary (eds) *Protected Areas, Sustainable Land?*, Ashgate, Burlington, VT.
Montagne, P. and Ramamonjisoa, B. (2006) 'Politiques forestières à Madagascar: entre répression et autonomie des acteurs', *Économie Rurale*, vol 294–295, July–October, pp9–26.
Montagne, P., Razanamaharo, Z. and Cooke, A. (2007) *Tanteza: Le Transfert de Gestion à Madagascar: Dix Ans d'Efforts*, CIRAD, Antananarivo.
Moreau, S. (2008) 'Environmental misunderstandings', in J. C. Kaufmann (ed.) *Greening the Great Red Island: Madagascar in Nature and Culture*, Africa Institute of South Africa, Pretoria.
Mukonoweshuro, E. G. (1994) 'Madagascar: the collapse of an experiment', *Journal of Third World Studies*, vol 11, no 1, pp336–368.
Nicoll, M. E. and Langrand, O. (1989) *Madagascar: Revue de la Conservation et des Aires Protégées*, World Wide Fund for Nature, Gland, Switzerland.
Pollini, J. (2011) 'The difficult reconciliation of conservation and development objectives: the case of the Malagasy Environmental Action Plan', *Human Organization*, vol 70, no 1, pp74–87.
Pollini, J. and Lassoie, J. P. (2011) 'Trapping peasant communities within global governance regimes: the case of the GELOSE legislation in Madagascar', *Society and Natural Resources*, vol 24, no 8, pp814–830.

Rakoto Ramiarantsoa, H. (2008) 'Madagascar au XXIe siècle: la politique de sa géographie', *EchoGéo*, Number 7.

Rakoto Ramiarantsoa, H., Blanc-Pamard, C. and Pinton, F. (2012) *Géopolitique et Environnement: Les Leçons de l'Expérience Malgache*, IRD Éditions, Marseille.

Sarrasin, B. (2007) 'Le plan d'action environnemental Malgache: de la genèse aux problèmes de mise en œuvre: une analyse sociopolitique de l'environnement', *Revue Tiers Monde*, vol 190, pp1–20.

Scales, I. R. (2012) 'Lost in translation: conflicting views of deforestation, land use and identity in western Madagascar', *The Geographical Journal*, vol 178, pp67–79.

Schmid, S. (1993) 'Sauvegarde des forêts naturelles et développement rural à Madagascar: un premier bilan des actions en cours', *Cahiers d'Outre-mer*, vol 46, no 181, pp35–60.

WCED (1987) *Our Common Future*, Oxford University Press, Oxford.

Weber, J. (1995) 'L'occupation humaine des aires protégées à Madagascar: diagnostic et éléments pour une gestion viable', *Natures Sciences Sociétés*, vol 3, no 2, pp157–164.

World Bank (1988) *Madagascar Environmental Action Plan*, World Bank, USAID, Coop. Suisse, UNESCO, UNDP, WWF, Washington, DC.

8 The transfer of natural resource management rights to local communities

Jacques Pollini, Neal Hockley, Frank D. Muttenzer and Bruno S. Ramamonjisoa

Introduction

Over the past 30 years, a wide range of conservation approaches have been implemented in Madagascar, including anti-fire policies and protected area gazetting (see Chapter 7 by Kull and Chapter 10 by Virah-Sawmy *et al.*); Integrated Conservation and Development Projects (ICDPs) in the early 1990s; decentralization and community-based natural resource management (CBNRM) and conservation through the transfer of resource management rights to local communities since the late 1990s; and market-based approaches during the last decade (see Chapter 13 by Brimont and Bidaud). CBNRM through management transfers is still seen as playing an important role in forest management, underpinning most of the new protected areas gazetted since 2003 (see Chapter 10 by Virah-Sawmy *et al.*).

This chapter tells the story of resource management transfers in Madagascar. It reviews the ideals behind the policies and shows that these ideals were quickly perverted by the agencies in charge of their implementation. Looking at the social, environmental and economic impact of management transfers, we reveal conflicting agendas that might now require renegotiation if community conservation is to work in Madagascar. We show that the Malagasy case is consistent with a general pattern observed worldwide in the implementation of decentralized natural resource management policies, and question the feasibility of such policies within the current global environmental regime.

The chapter reviews a series of case studies and builds on the field experience of the authors, who have all been involved in the design, implementation and/or evaluation of management transfers in Madagascar. We use a series of snapshots, taken in various locations in Madagascar from various angles by authors belonging to diverse academic disciplines (geography, anthropology, economics, agronomy, natural resources, law), to tell the story of management transfers as they happen on the ground and in a holistic way.

The birth and perversion of CBNRM policies
Historical and international context

Before the eighteenth century, the majority of Malagasy were engaged in subsistence-based cultivation, hunting, fishing and gathering (see Chapter 6 by Scales). They developed community institutions to manage these activities as well as the transformation of landscapes through the use of fire and the large-scale conversion of natural ecosystems to agricultural land. However, the formation of the Merina state during the nineteenth century and French colonization between 1896 and 1960 progressively reduced community control (Bertrand *et al.*, 2009). The French colonial administration declared that the Malagasy forest belonged to the state and created a forestry service that regulated access to forests and limited resource extraction (Bertrand *et al.*, 2009; see also Chapter 6 by Scales).

Following the economic collapse of the late 1970s (see Chapter 7 by Kull), state control over access to natural resources became weak or non-existent. According to Bertrand *et al.* (2009), in some areas decades of state intervention had weakened customary institutions to the point that when the state collapsed, nothing remained to control access to natural resources. This often led to a situation of open access (Weber, 1995), or what Hardin (1968) called the 'tragedy of the commons'. Similar histories have been observed in many developing countries, where land tenure rights often became more rather than less centralized after independence from European powers (Dressler *et al.*, 2010). Thus, conservation and management of forests and other lands often continued to follow the 'fortress conservation' approach originally pioneered in the national parks of the USA (Brockington *et al.*, 2008).

During the 1980s and 1990s, two global trends prepared the ground for the emergence of CBNRM in Madagascar and elsewhere. The first was a general trend towards the decentralization of government. This was driven by changes in development theory, where the idea of local participation in the development process became popular (Chambers, 1983), as well as pragmatic realization of the limited capacity of many African states to assert de facto control of resources. The second trend was the recognition that 'fortress conservation' could have severe implications for local livelihoods, such as loss of access to natural resources, and that this could undermine its legitimacy and effectiveness (Brandon and Wells, 1992). Some responses, such as ICDPs, aimed to compensate for the local costs of resource preservation, but did not substantially change tenure or power relations. Others (including CBNRM, community-based conservation and community forest management) were conceived as devolving power and rights over resources to local communities, improving management and livelihoods as a result (Dressler *et al.*, 2010).

CBNRM is usually defined as the devolution of rights to make management decisions and capture benefits, in relation to resources located on community lands (Dressler *et al.*, 2010). The strategy is based on the assumption that local communities are best placed to protect natural resources and manage the benefits arising out of their extraction, because they possess deep knowledge of the ecosystems within which they live, and because the dependence of their livelihood on these ecosystems would motivate sustainable management rules and long-term conservation efforts.

In Madagascar, CBNRM occupied centre stage in the conservation agenda during the first National Environmental Action Plan (NEAP) (between 1990 and 1995, see Chapter 7 by Kull). Several participatory forest management projects were conducted during the early 1990s and a series of workshops were organized to capitalize upon these experiences. This led to the creation of *Gestion Locale Sécurisée* (GELOSE or secure local management), supported by a new law issued in 1996 (and subsequent implementation decrees). To address perceived limitations of GELOSE law, *Gestion Contractualisée des Forêts* (GCF or joint forest management) was introduced in 2000 through an implementation decree of the forest law of 1997.

The GELOSE legislation

The GELOSE law provides a legal framework to transfer resource management rights from the state to local communities through a tripartite agreement between i) forestry services; ii) a new institution referred to as the *Communauté de Base* (COBA), created for this purpose and supposed to represent the local community; and iii) the municipal government (*commune*), which is the most local decentralized institution with elected leaders (Figure 8.1). These agreements, or management contracts, are signed for a three-year period, then evaluated and renewed for ten more years if successful, and evaluated again before being turned into definitive agreements. They concern the management of a specific resource, not necessarily the ecosystem where these resources are found, and are established with support from non-governmental organizations (NGOs) or other external stakeholders (Figure 8.1) for the preparation of technical and official documents (e.g. management plans, rule book, zoning, resource maps and the contracts themselves). They require the hiring of an environmental mediator and the partial securitization of land tenure, but these two provisions were not universally implemented.

GELOSE's false assumptions

One of the key assumptions made by the promoters of GELOSE was that granting resource management authority to local communities would put an end to the problem of 'open access' (Bertrand *et al.*, 2009) where there were no rules controlling natural resource use, or, where rules existed, to conflicts between local customary rules and national legislation (referred to as

(a) In theory …

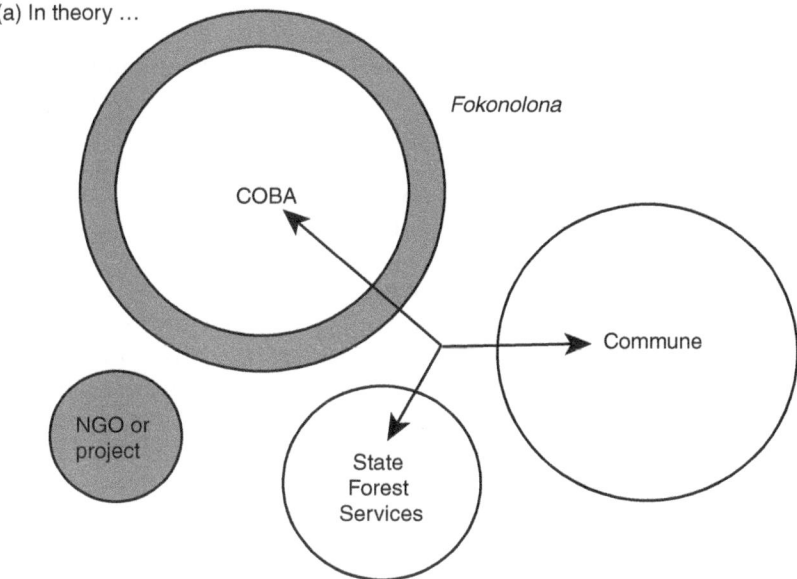

… And (b) in practice

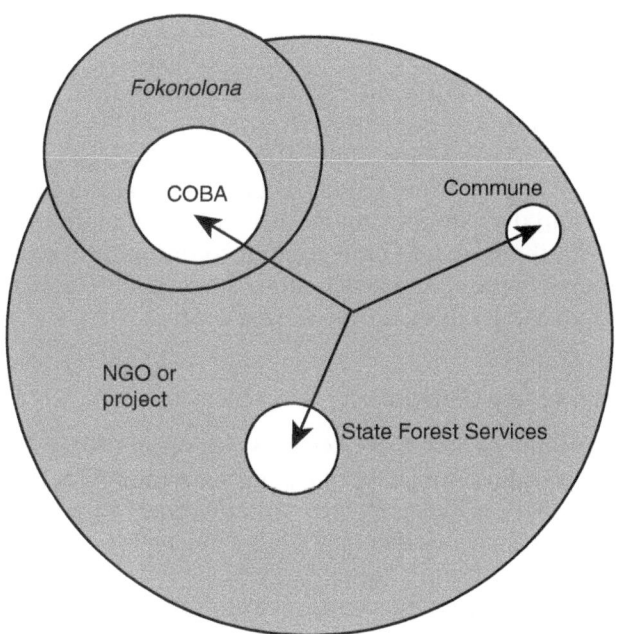

Figure 8.1 Key stakeholders in early stages of GELOSE management transfers

legal dualism). Muttenzer (2010) has argued that this assumption was overly simplistic. Rather than a binary of either 'open access' or legal dualism, the situation on the ground is often a sophisticated hybrid of regulations involving multiple social groups that settled in forest areas at different periods, as well as staff from the state forest service who still have regulatory power but whose management decisions were mostly aimed at getting a personal benefit out of forest clearing and logging operations. The simplistic assumptions of open access and legal dualism underlying GELOSE led its proponents to overlook these complex local regulatory systems.

A second key assumption was that GELOSE would improve on previous 'participatory' approaches to conservation that, instead of devolving power, actually co-opted community efforts in order to further their own agendas – what Cooke and Kothari (2001) have labelled the 'tyranny of participation'. Participation was already a fashionable concept in Madagascar at that time and was widespread in ICDPs. The proponents of GELOSE, inspired by Ollagnon's (1991) *gestion patrimoniale*,[1] argued that a shift was needed from participation to negotiation (Weber, 1996). Instead of 'participating' in the implementation of an external agenda designed by conservation biologists dedicated to the preservation of a pristine nature, local stakeholders would be recognized as being part of the ecosystem to be managed and invited to negotiate their own management objectives. The negotiation would involve the environmental mediator, whose role would be to give equal weight to the voices of all stakeholders on the negotiation table. This was certainly a laudable intention but it could also prove quite a naïve one. The approach overlooked the fact that the granting of rights to clear forest, not to conserve it, is the main purpose of local forest management institutions (Pollini, 2007; Keller, 2008; Muttenzer, 2010); whereas putting an end to tropical deforestation is a non-negotiable goal among global conservation actors and their donors, upon whom implementation of GELOSE would depend. In the end, GELOSE was not so different in practice from approaches framed within the 'participation' paradigm. Its initial ideal was quickly perverted, as we will see in the next section.

The perversion of the GELOSE ideal

The first perversion of GELOSE concerns its community ideal. Initially, the intention was to transfer management rights to communities, that is, to the *fokonolona* (Figure 8.1), a Malagasy term that designates a group of people who live together and take decisions together regarding the management of their territory. Pollini and Lassoie (2011) show that this did not happen: the GELOSE management contracts are signed by a new institution, which although named the 'grassroots community' (COBA) in the legal text, is in fact an association: that is, a group of individuals who belong to the community and decide to collaborate to achieve certain goals, because they share common interests. If all members of a given *fokonolona* chose

The transfer of natural resource management rights to local communities 177

to form a COBA, then this COBA could be considered a community, or at least an institution that superimposes itself onto an existing community (Figure 8.1). Some NGOs that support management transfers are committed to creating such inclusive COBAs. But the law itself does not require this to happen, which leaves it open to abuse. For example, a COBA can be created by a very small group of local residents, under the supervision of the Forest Service, in order to harvest and sell a particular forest product (as reported by Bertrand *et al.*, 2009). In this case, the COBA is simply a business, legally able to exclude the actual community (the *fokonolona*) from its customary rights over the resource (Figure 8.1). In such a case, GELOSE supports the privatization of natural resources management, rather than its decentralization, which facilitates resource capture by elites.

A second perversion, in synergy with the first, sealed GELOSE's fate. According to the initial design, it was up to local communities to take the lead in implementing GELOSE, by requesting management contracts with the Forest Service when they deemed it worthwhile. However, when the law was passed, very few bottom-up initiatives occurred, due to insufficient information and understanding of GELOSE in rural areas. Conservation NGOs then started to contact local communities and hastily design contracts that matched their own agendas. In order to achieve their objectives more easily, they called for a simplification of the GELOSE legislation, which was deemed to be too complex. This is when GCF came into play. It was proposed as a simpler alternative, as it did not involve the municipal government nor require environmental mediators or the securing of land tenure. According to Bertrand *et al.* (2009), GCF is exemplary of the perversion of the GELOSE ideal by conservation organizations that attempted to transform natural resource management contracts into biodiversity conservation contracts, thereby diverting it from its sustainable development ideals. By 2004, 453 management transfers (GELOSE or GCF) had been established (RESOLVE, 2005), mostly by large international NGOs such as the World Wide Fund for Nature (WWF) and Conservation International (CI). We will now see what effect this instrumentalization of management transfer contracts for conservation purposes actually had on the ground.

The social impacts of management transfers

In this section, we review case studies published in peer reviewed journals dealing with the implementation of the GELOSE and GCF, in order to show the contrast between 'the GELOSE seen from above and the GELOSE experienced at the bottom' (Goedefroit, 2006, p40).[2]

The reshuffling of control and access to resources

Case studies looking at the implementation of GELOSE and GCF contracts on the ground show similar patterns of social impact. The key process revealed through these studies is the reshuffling of power relations and the shift of management authority from one social group to another. It remains unclear, however, whether these effects will be long lasting or are simply transitional phases before more positive outcomes emerge.

Blanc-Pamard and Rakoto Ramiarantsoa (2007) analysed management transfers in Ambendrana and Amindrabe, two villages close to the western edge of Ranomafana National Park. Local resource management associations (COBA) have been created in these sites and GCF contracts have been signed with these associations, with support from the United States Agency for International Development (USAID) funded project LDI (Landscape Development Intervention), in partnership with CI. According to Blanc-Pamard and Rakoto Ramiarantsoa (2007), GCF contracts established a new zoning of the area that did not respect existing land distribution by lineages. The consequence has been a reshuffling of land tenure and forest use rights, with some communities taking advantage of the transfer to take control of access to resources and exclude their competitors. The authors conclude that management transfers lead to the delimitation of new territories and the modification of social relationships between actors.

The modalities of these transformations vary from one site to the other. In some cases, early settlers use management transfers to consolidate their rights over the land and resources and exclude more recent settlers. In other cases, it works the other way around. This was observed in forest management associations created with support of the *Projet de Développement Intégré Forestier Villageois* (PDFIV, funded by the German Government) to eliminate swidden cultivation in Tsinjoarivo, on the western edge of the Malagasy rainforest, close to Ambatolampy. In this area, Merina households, who specialize in paddy rice cultivation in valley bottoms, have migrated to forest land inhabited by Betsimisaraka households, whose livelihood is based on swidden cultivation on hill slopes (Pollini and Lassoie, 2011). The leaders of forest management associations, who were almost all Merina, successfully enforced rules limiting swidden cultivation. Betsimisaraka households, who originally managed the land and lack the resources to create paddy fields, were hostile to these associations, moved away from the region and 'waited for the *vazaha* [foreigners] to leave'[3] with the intention of coming back to reclaim their rights later (Pollini and Lassoie, 2011).

This intended return of Betsimisaraka households reveals another possible outcome. Although the closure of the forest might occur when conservation NGOs maintain a strong presence, access to forest land (following ancient or modified customary rules) might open again when NGOs leave or if they have only a shallow presence. Local people can then either violate and forget management contracts, or use them as a new additional instrument to negotiate access rights between earlier and later settlers (Muttenzer, 2010). Such appropriation of GELOSE institutions by local stakeholders after the departure of GELOSE promoters might be compatible with conserving the customary systems that regulate access to land and resources. The 'modern' institution (the COBA) could remain involved in the establishment and enforcement of these rules though only insofar as the COBA was aligned with those customary institutions. It is too early to assess whether these processes reflect the general tendency but Muttenzer (2010) showed that they do occur in historically settled and ethnically homogeneous areas (such as the eastern forest corridor) as well as in multi-ethnic immigrant areas (such as the north-western lowlands).

The division of communities between COBA members and non-members

Goedefroit's (2006) study of six GELOSE sites explains how the reshuffling of power relations and territorial control are achieved. By creating a new institution (the COBA) whose remit of natural resource management is already performed by customary institutions, GELOSE generally creates divisions within the community. Traditional *fokonolona* leaders with legitimate customary power to control access to land and resources often consider that management transfers are just like any other activity conducted by NGOs and development projects. The COBA is a foreign thing to them and they leave it to young literate people, sometimes outsiders like the village teacher, or recent migrants, to handle its affairs. As a result, migrants with some literacy but little influence on *fokonolona* decisions see taking control of the COBA as an opportunity to claim more rights over access to land and resources (Goedefroit, 2006). Meanwhile, the traditional leaders continue to regulate access to resources just like they did before, until they realize that new territories have been delimited and new rules have been instituted regarding resource use. Conflicts can then arise between *fokonolona* and COBA leaders and their constituencies or clients. However, complaints often decrease after a few years, perhaps because compromises are found between the old and new rules through a genuine negotiation process that involves only local stakeholders. One can wonder, then, whether this social crisis described by Goedefroit (2006) is transitory or will disrupt community institutions in the long term. Muttenzer (2010), whose work will be discussed later, provides support for the first hypothesis.

The limits of environmental mediation

The designers of GELOSE expected such conflicts might occur, which is why they created the role of environmental mediator. However, the mediator is often an urban and literate person hired by the NGO that supports the implementation of GELOSE. He/she is unlikely to be a good mediator if he/she shares the same pro-conservation or 'sustainable development' world view as any other project staff, which unfortunately can hardly be avoided. Goedefroit (2006) notes that when mediators suggest local people should 'change their mentality' regarding resource use, they confuse mediation with environmental education and favour the stakeholders that

180   *Jacques Pollini* et al.

agree, or pretend to agree, with the environmental agenda. As a consequence, the new territories and management rules that are being created reflect the international conservation agenda rather than local people's interest. Goedefroit (2006) calls this process a 'pollution' of the community environment.

GELOSE dina *(rules) as promoters of global conservation goals*

The work of Bérard (2011) provides details about what this 'pollution' consists of. Bérard conducted field research in various sites and reviewed the rules (*dina*) of 32 GELOSE contracts, raising doubts about whether any rights have actually been transferred to communities. She shows that *dina* created through GELOSE are stereotyped and reflect the agenda of the institution (NGO and/or project) that supports the implementation of management transfers, rather than the priorities of the community. They lack the flexibility of traditional rules and are incapable of taking into consideration the specific economic situation of rule breakers. They focus on repression and penalties rather than resource extraction modalities (Bérard, 2011), whereas the GELOSE law explicitly stated that it was intended to improve the economic exploitation of resources by local communities. In sum, unlike traditional *dina*, GELOSE *dina* reflect forest regulations or the agenda of international conservation organizations and fail to capture the complex ethics of village social and economic relationships and the rules and norms that govern behaviour – the so called 'moral economy of the peasant' (Scott, 1976). Moreover, *dina* are usually not applicable to non-COBA members, in spite of the fact that during preparatory GELOSE workshops local communities insisted on the importance of being given the power to control the intrusions of external stakeholders inside their forests (Bérard, 2011).

The dash for standardized contracts

Bérard (2011) reports that under pressure from donors, the many organizations involved in the implementation of management transfers rushed to sign contracts (also noted by Goedefroit (2006), Hockley and Andriamarovololona (2007) and Bertrand *et al.* (2009)). This was not conducive to quality and encouraged stereotyped 'one size fits all' contracts. In some cases, communities were threatened with loss of access to their resources if they did not sign contracts, and promised benefits from future development activities if they did (Hockley and Andriamarovololona, 2007). In this race to sign contracts, local regulatory systems and community management goals were ignored, while stereotyped 'solutions' that reflected external stakeholders' management goals were imposed upon imagined communities reified (that is, made concrete and real) in the form of a COBA.

The reification of imagined communities and land uses

Muttenzer (2010) analyses in further detail the ideologies that lead to the signing of stereotyped contracts by 'communities'. Based on the empirical study of management transfers aimed at i) regulating charcoal production close to Ankarafantsika National Park; and ii) stopping forest clearance on the western edge of the rainforest corridor in the Fianarantsoa province, he shows that rather than protecting forest, community rules in frontier areas are mostly designed to regulate agrarian colonization, that is to organize the conversion of forests into agricultural land.[4] These rules determine, for example, who has rights to clear the land and how, depending on whether one belongs to a group of first settlers, recent migrant, royal lineage, etc. In short, customary systems manage access to land and, once access to forests is granted, land can be cleared for productive purposes. Forest land is thus a reservoir of land for future agricultural expansion. This is the land ethic of rural Malagasy people more generally (Keller, 2008; see also Chapter 5 by Scales).

The recognition and legitimization of these customary rules and land ethic was incompatible with implementing the conservation agenda of the Malagasy NEAP, of which GELOSE was a component. NEAP's key players nevertheless agreed with GELOSE promoters that because the failure of state services had left a political vacuum, customary rules should be restored (Muttenzer, 2010). But this restoration did not concern the existing customary rules used to manage forest clearance. It was, instead, the 'recognition' of imagined local rules more compatible with external notions of sustainability (Muttenzer, 2010). In this imaginative effort, the customary system was artificially segregated into a supposedly open access system that resulted from the collapse of local institutions (*tavy* for clearing),[5] whose eradication would satisfy the international conservation agenda, and a functional system with controlled access (*tavy* for cultivation),[6] compatible with the idealized vision of farmers' communities living in harmony with nature if not disturbed by external forces, which was to be recognized and strengthened (Muttenzer, 2010).

In reality, the two types of *tavy* are two facets of the same process: the extension of cultivated areas to secure existing rights in the case of slow colonization by descendants of first settlers ('*tavy* for cultivation'), or in the case of fast colonization by migrants who become clients of the first settlers ('*tavy* for clearing') (Muttenzer, 2010). The 'customary system' that is being 'recognized' by management transfers thus does not exist. It is the outcome of a contemporary reinterpretation of pre-colonial customs that proponents of management transfer instrumentalize to deny the existence of customary arrangements between first settlers and more recent migrants, and to provide an idealized version of farmer communities living in harmony with nature if undisturbed by the outside world.

The real customary system, which is hybridized with state institutions, and which involves a broader range of stakeholders and management purposes, is thus ignored, which leads to the reshuffling of power relations and the territorial reorganizations described by Kull (2002, 2004), Rabesahala Horning (2004), Goedefroit (2006), Blanc-Pamard and Rakoto Ramiarantsoa (2007) and Pollini and Lassoie (2011).[7] Unexpectedly, some natural and social scientists jointly overlooked these issues and embarked upon GELOSE policy with the same enthusiasm because it flattered their dreams of harmonies – the win–win scenarios of conservation biologists trapped within the NEAP's technocratic structures and social scientists' belief in the existence of 'ecologically noble savages' (Hames, 2007). This led the Malagasy NEAP to reify imagined communities and land uses, through communication efforts that rely more on myths and symbols than facts and empirical observations, and look more like rituals than acts of cognition (Pollini, 2011).

The 'politics of recognition'

Ribot (2011) reviewed case studies worldwide that resonate strongly with findings in Madagascar. He shows that international organizations and NGOs involved in natural resource decentralization policies often choose to create new institutions, instead of supporting existing ones. Drawing on Taylor's (1994) 'politics of choice and recognition' model, he demonstrates that the creation of these new institutions is in itself a political act. It implies identifying, choosing and recognizing (by granting them powers) the institutions to be created or modified and the actors that will be key players (Ribot, 2011). These choices and recognitions are driven by the specific culture, as well as the purposes and interests of the organizations that make them. They are determined by expectations about how these institutions and individuals should behave (they should be 'good' stewards of the land according to western sustainability criteria) rather than acceptance of how they actually behave (they authorize forest clearing). This desired identity, once 'recognized', can be reified through multiple supports (capacity building, training, participation in workshops) that are provided by the intervening agency.

Following this model, COBAs are 'recognized' institutions created by external actors (NGOs and projects). They enter into competition with existing ones (e.g. *fokonolona* assemblies) in the taking of decisions regarding access to land and resources, which fragments the local political and social arena. Their leaders are 'chosen' among villagers more willing to adopt the conservation and sustainable development agenda, or smart and educated enough to fake this willingness. The decisions of COBA leaders thus reflect the interests of the intervening agencies that finance management transfers (conservation and sustainable development organizations and their donors), whose agenda they spread at the local level (Pollini and Lassoie, 2011). Given the huge gaps between the power, knowledge, agenda and culture of local and international stakeholders, negotiations cannot take place, which is conducive to the reification of donors' models, unless the community finds ways to resist or to manipulate the new institutions and merge them with their own (Muttenzer, 2010).

Limits to critiques of GELOSE

In formulating our critique of GELOSE, we are not arguing that local communities should necessarily be given the unfettered right to manage natural resources as they wish. Local communities must surely be involved in managing natural resources but 'traditional' or 'customary' are not synonyms of 'good' or adaptive practices. Appropriate management of public goods (such as biodiversity conservation) may have to be defined at a broader scale than the local landscapes managed by farmer communities, which might lend some legitimacy to external stakeholders pushing their own agenda. However, unless the real purposes of community institutions are clearly recognized, as a prerequisite to genuine negotiation, the divergence of interests is overlooked, negotiation does not occur and the 'tyranny of participation' (Cooke and Kothari, 2001) unfolds instead. The final outcome is the making of an unjust, socially and politically 'violent environment' (Peluso and Watts, 2001) that is conducive to social injustices and conflicts rather than conservation or poverty alleviation.

The impact of management transfers on the environment

While the social and political processes involved in CBNRM are important, so too are their outcomes. In this section, we therefore assess whether and why management transfers have been successful in reducing natural resource degradation. However, perhaps the more important question is whether any impacts will last.

Agricultural intensification and reduction in deforestation after the closure of the forest frontier

Evaluation reports (e.g. RESOLVE, 2005), often suggest that COBAs succeed to a certain extent in reducing deforestation rates, though robust evidence of this remains elusive, as it does for conventional protected areas and CBNRM initiatives around the world (e.g. Bowler *et al.*, 2012). In the seven sites they visited in various parts of Madagascar, Hockley and Andriamarovololona (2007) found that all COBAs had made some effort to honour the law enforcement responsibilities delegated to them by the state. They had attempted to exclude outsiders from logging and mining, prohibited forest clearing and patrolled their forests. But no significant material benefits or more secure access to natural resources resulted. COBAs could thus not be expected to continue to deliver this low-cost conservation without external support, and several appeared to be collapsing.

In the absence of external support, the development of alternative livelihoods by farmers themselves might nevertheless enable long-term conservation success. Toillier *et al.* (2011) studied the impact of management transfer on livelihood strategies in the same sites in the eastern rainforests near Fianarantsoa studied by Blanc-Pamard and Rakoto Ramiarantsoa (2007). They showed that farmers developed various short- and long-term strategies – including agricultural intensification, the development of cash crops, the conversion of marginal land and migration – to adapt to the restrictions on access to land and resources that result from management transfers. Some even took opportunity of the new regulations, land distribution and zoning to justify the establishment of new permanent fields to the detriment of forests. However, Toillier *et al.* (2011) also demonstrate that the poorest farmers bear the costs of this 'success', having a lower food intake and income due to loss of access to land. In the long term, the new strategies might provide benefits to a larger number of households, and the community may be better off overall, as happened as a consequence of mainstream conservation policies in areas studied by Laney (2002) in northern Madagascar. But this would be at a high social and economic cost during the transition period, especially for the most vulnerable groups. Moreover, one could hardly find a difference between this 'success' and the successful adaptation by some farmers living around protected areas that adopt the fortress conservation model.

The case of lake and marine ecosystems

Case studies dealing with management transfers implemented in lake or marine ecosystems provide more plausible success stories. In the Manambolomaty Lakes complex (Tsiribihina watershed in western Madagascar), the Peregrine Fund used the GELOSE law to help local fishermen organize themselves against overfishing by migrants (Watson *et al.*, 2007). It hired a mediator to facilitate negotiations between local fishers, migrants and other stakeholders; provided communities with scientific data about the impact of overfishing; and helped monitor the impact of management rules on resource stocks. It successfully controlled overfishing by migrants, the population of fish and catches increased, and populations of fish eagles – which depend on the same resources as humans – recovered. But this success required continuous support over more than ten years and one can wonder whether the operation could be scaled up given its high transaction costs.

On the coral reefs of south-western Madagascar, the British NGO Blue Ventures supported the establishment of a no-take zone in the fisheries area of the village Andavadoaka in 2004 (Cripps and Harris, 2009). Interestingly, this was not done through GELOSE legislation, but the relative success achieved makes it worth studying this CBNRM case. Octopus catches increased many fold and neighbouring villages adopted the same measures. They constituted the Velondriake Locally Managed Marine Area (LMMA), operational since 2007 and now benefiting about 10,000 people in 25 villages. Temporary closures (two or three months) of small sections of the reef (less than 10 per cent of the reef surface that is controlled by a village), together with fishing gear restrictions in the remaining areas, were implemented in response to coral reef degradation. However, rather than the full-scale enclosure of natural resources through GELOSE and GCF contracts in forest ecosystems, this may in fact represent a 'renaissance of traditional conservation methods' (Johannes, 2002) through a process of 'creeping enclosure', whereby cumulative access restrictions lead to something like full enclosure (Murray *et al.*, 2010). The pre-existing and long-standing common property rules of reef and lagoon tenure remain in place in the remaining 90 per cent of the reef area, little social disruption or resource capture occurs, and the measures can be considered a case of 'adaptive co-management'.

So why is there no social disruption or resource capture as a consequence of progressive enclosure in this case? The answer is because a common property regime is being maintained in spite of ongoing intensification, degradation and access restrictions. Traditionally, reefs were already divided into named sections that are each attributed to a particular village. The harvesting in a given section of a territory is done by residents from this village and immediate neighbours. It is not a case of open access but a convention of overlapping village commons. Everyone knows that access rules could have been different or might be modified in the future. When the associations set up by the project changed the rules (Andriamalala and Gardner, 2010), the project managers found out that the newly created access restrictions are acceptable only if each village decides on its own no-take zone, if contiguous villages close and open at the same time, and if there is no more than one closure per year.

These results show that community conservation with a win–win outcome is possible. However, in the case of both the Peregrine Fund intervention in the Manambolomaty Lakes complex and Blue Venture's experiences in Andavadoaka, the approach benefits from a congruence of interests between local and external stakeholders. In the case of forests ecosystems, there is usually no congruence (Antona *et al.*, 2004) because in many cases the most valuable function of forests for local people is to provide land for agriculture, meaning forests will ultimately be cleared under customary systems, unless the land is not suitable for agriculture (and with the exception of sacred groves whose area is dwindling). Hunting and gathering of forest products exist but only as long as the population is small and does not need to expand its agricultural land. In the case of lake or marine resources, fishing is the only possible resource use and its sustainability, and indeed profitability, depends on a successful conservation of the natural ecosystem.

The impact of management transfers on the local economy

The impact of management transfers on the local economy appears on the whole to be positive in the case of lake and marine resources. In the case of forest resources, GELOSE and GCF contracts tend to disrupt local social systems, but partly succeed in reducing resource degradation, at least temporarily, and possibly trigger or stimulate agricultural intensification. How will these tendencies balance each other? Will local communities concerned by management transfers be better or worse off in the end? Beyond ethical and political considerations, economic analyses might provide a pragmatic answer.

The Fandriana-Vondrozo and Ankeniheny-Zahamena corridors (eastern rainforests)

Hockley and Andriamarovololona (2007) conducted an economic analysis of a stratified sample of seven COBAs in the Fandriana-Vondrozo and Ankeniheny-Zahamena corridors, in the intervention area of USAID-funded project Eco-Regional Initiative (ERI). These COBAs were intended to manage a buffer zone surrounding protected areas, consistent with the commitment of the Malagasy government to develop new joint forest management schemes. The key assumption behind this strategy is that win–win economic scenarios that reconcile the conservation and development agendas can be developed jointly with communities.

We have already seen that such win–win scenarios are unlikely and Hockley and Andriamarovololona (2007, p17) make a stronger case for this. They observe that, given 'increasing populations, a stagnant economy and infertility of existing land, it may often be in the interests of rural people to convert forest to agriculture'. As all the management transfers they studied banned forest clearance, their impact on local communities is basically the same as that of other conservation programmes. This fact is overlooked by conservation and development actors who, in an 'attempt to extract a free lunch ... maximise the value they derive from management transfer contracts, through placing increased restrictions on COBAs' activities' (Hockley and Andriamarovololona, 2007, piii). As a consequence, the main revenue that the COBAs receive is the payment of fees by their members and, as most of this income is spent on meeting their externally defined responsibilities, COBAs are a financial drain on the community, while creating little or no value (Hockley and Andriamarovololona, 2007). This situation leads to unstable COBAs with declining membership, the disruption of existing resource management institutions, 'political disputes and local power struggles' (Hockley and Andriamarovololona, 2007, p29) and 'failure of conservation and a waste of donor investment' (Hockley and Andriamarovololona, 2007, p10). They confirm many of the negative social impacts of the GELOSE already analysed but also provide a material explanation for it: the erosion of local income, at least as long as management transfers fulfil the expectations of their external promoters.

The dry forest of the Mahafaly plateau in south-western Madagascar

Ramamonjisoa and Rabemananjara (2012) compared the economic benefits generated by the conversion of forests into agricultural land to those of community conservation in south-west Madagascar. Regarding the conversion of forests, they found that the local practice of maize cultivation on forested land is highly lucrative. The product can easily be sold on national and international markets and Mahafaly farmers impoverished by successive years of drought are highly attracted by this activity (see Chapter 5 by Scales for more on maize cultivation in western and southwestern Madagascar). Moreover, the trees cut on the cleared land are used to make charcoal that provides additional income. Since the majority of the farmers come from the south, the temporary migration that they undertake to this frontier area permits them to transfer monetary flows of more than US$1.7 million a year to their regions of residence. This cash transfer helps their family to sustain their livelihoods during years of drought (*kere*) and contributes to the development of their home villages. Regarding community conservation, on the other hand, they found that it does not provide significant benefits locally because no forest resources have yet been identified whose sustainable extraction could generate significant income.

These studies, however, do not show the economic long-term impacts of management transfers. As we have seen, restriction of resource access can trigger or accelerate the development of alternative land uses (Toillier *et al.*, 2011), with local positive economic impact in the long term, although with social cost borne by the poor at least during a transition period. But the same would happen in the case of conventional conservation programmes, as shown by Laney (2002). Hence this possible long-term conservation 'success', even if it occurred, would not validate the management transfer approach. Its outcomes, both at the ecological (less forest clearing) and socio-economic (difficult transition especially for the poor) level, would be similar to those achieved by the fortress conservation approach. They may be an anticipation of the same processes, described by Boserup (1965), that would have occurred once the forest was totally cleared, even in the absence of external intervention.

Conclusion

Management transfers in Madagascar create a double bind upon rural households and their institutions. They are aimed at involving local communities in achieving natural resource management goals (conservation and 'sustainable development') that are not their own, while at the same time pretending that if local communities are given the rights to manage natural resources, these same goals will be more easily achieved.

In the short term, and as long as the external actors that support management transfers are present on the ground, the 'best' case scenario (from the conservation perspective) is the division of communities between people who succeed in adopting this external agenda and people who cannot or do not want to adopt it, for example because they need or want to clear forest to expand their economic activities. In the worst case, it results in the transfer of management authority from one social group to another – for example from first settlers to migrants, from migrants to settlers or from one clan to another – without changing local management objectives (e.g. forest clearing for agricultural expansion). In the longer term, the 'best' case scenario is improved natural resources management with economic costs that peak during a transition period (and tend to be borne by the most vulnerable people). In the worst case, the result is long-lasting negative impacts on social organization, local livelihoods and the environment.

In between these scenarios, it is hard to predict what transformations will actually occur and which actors will achieve their goals in the end. But looking at the history of external interventions to which farming communities have already been subjected, one could hypothesize that external actors will be strong only while present on the ground. Their attempts to close the commons may only have temporary effect. In the long term, local actors may be successful in designing and implementing their own institutional arrangements. They are likely either to reopen the forest frontier or to evolve toward more 'sustainable' land uses, in which case their practices will gain more legitimacy.

Could this latter scenario be called a success? If it is, we believe that rather than being the success of management transfer policy, it is the success of local communities of farmers that successfully adapt to the new economic and political constraints generated by the adoption of a new global environmental regime since the Rio Conference in 1992. The adaptive processes realized in the 'best case' scenarios are, in fact, the same that happen in the surroundings of national parks and in other areas where conventional conservation policies are enforced. These processes were described by Laney (2002) and agronomists and economists have long recognized them (Boserup, 1965; Mazoyer and Roudart, 2006). They have happened in most regions of the world where forest resources have been depleted and landscapes reached their carrying capacity.

Management transfer thus simply appears to be a new buzzword; a dream of optimistic social engineers whose main function is to legitimate conservation policies to a broader audience by giving to it a more ethical colouration. It transfers duties rather than rights to local people and when it achieves conservation, it is indeed by the same mechanism as in the case of national parks designed on the fortress model. As Bérard (2011, p110) puts it, 'state authorities rule like before, but at a lower cost, by delegating the tasks of control, punishment and conflict management to villagers or village committees',[8] while conservation NGOs design GELOSE *dina* that enable them to 'interfere with local regulation while defending their own logic and interests' (Bérard, 2011, p110).[9] Management transfers, therefore, have tended to operate in the opposite way to that intended, helping to transfer power from local communities – who managed their resources in ways that matched with their own interests (e.g. agricultural expansion) – to the state and its international partners, thanks to the supervision apparatus of COBAs and *dina* that they put in place. The consequence is that management transfers, as with protected areas, have often resulted in 'the poorest people [in Madagascar]… subsidizing the benefits of conservation felt by the rest of the country and the world' (Hockley and Andriamarovololona, 2007, p42). Marine and other water resources, however, may provide an exception, because there appears to be a greater congruence between local and international objectives in their case.

Notes

1 The word *patrimoine* has no exact equivalent in English but is close to the notion of 'heritage'. According to Ollagnon (1991, in Babin and Bertrand, 1998), *patrimoine* is made out of 'all the material and non-material elements that work together to maintain and develop the identity and autonomy of their holder in time and space through adaptation in a changing environment'.
2 Translation by the authors. Original citation in French: '*le décalage entre la GELOSE "vue d'en haut" et la GELOSE "vécue en bas"*.'
3 Citation from an interview with a local leader. Several persons made similar statements. Two persons also reported cases of murder: one forest association leader was alleged to have been killed with a machete, and another one poisoned, both by Betsimisaraka people hostile to the forest associations. We could not formally verify this information.
4 Muttenzer distinguishes primary rights, which are claimed as soon as a group settles into an area and delimitates a forest territory as being its own and to be later cleared, and secondary rights, which are established when this settler clears forest to cultivate the land and establish pastures, for himself, his clients and his descendants.
5 *Tavy* is the Malagasy word that designates swidden cultivation systems encountered in the east and north, where forests or secondary vegetation is cleared to grow rice. See Chapter 5 by Scales for more discussion of the terminology surrounding swidden cultivation and forest clearance.
6 *Tavy défricheur* and *tavy cultivateur* in the original text in French (Bertrand and Randrianaivo, 2003).
7 This reinterpretation of customary systems is convenient from the perspective of conservation organizations because it justifies limiting forest land conversion by local people. It is also convenient for social scientists who need to 'discover' local alternatives to capitalist expansion and other western hegemonies. This is why both 'conservative' conservationists (e.g. CI) and its critics (e.g. *Centre de Coopération Internationale en Recherche Agronomique pour le Développement* (CIRAD)) put GELOSE at the core of their agenda, although giving it a quite different shape on the ground (citation CIRAD report Didy).

8 Translation by the authors. Original text in French: '*les autorités étatiques font le droit comme avant, mais à moindre coût en déléguant les fonctions de surveillance, de punition et de gestion des conflits à des villageois ou à des comités de villageois*'.
9 Translation by the authors. Original text in French: '*interférer avec les régulations locales en poursuivant leur propre logique et leurs intérêts*'.

References

Andriamalala, G. and Gardner, C. J. (2010) 'L'utilisation du dina comme outil de gouvernance des ressources naturelles: leçons tirés de Velondriake, sud-ouest de Madagascar', *Tropical Conservation Science*, vol 3, pp447–472.

Antona, M., Motte Bienabe, E., Salles, J. M., Pechard, G., Aubert, S. and Ratsimbarison, R. (2004) 'Rights transfers in Madagascar biodiversity policies: achievements and signifiance', *Environment and Development Economics*, vol 9, pp825–847.

Babin, D. and Bertrand, A. (1998) 'Devising strategies to involve multiple partners in sustainable forest management: examples from Africa', *Unasylva*, vol 49, no 194.

Bérard, M. H. (2011) 'Légitimité des normes environnementales dans la gestion locale de la forêt à Madagascar', *Canadian Journal of Law and Society*, vol 26, pp89–111.

Bertrand, A. and Randrianaivo, D. (2003) 'Tavy et déforestation', in S. Aubert, S. Razafiarison and A. Bertrand (eds) *Déforestation et Systèmes Agraires à Madagascar: Les Dynamiques des Tavy sur la Côte Orientale*, CIRAD-CITE-FOFIFA, Antananarivo, Madagascar.

Bertrand, A., Rabesahala Horning, N. and Montagne, P. (2009) 'Gestion communautaire ou préservation des ressources renouvelables: histoire inachevée d'une évolution majeure de la politique environnementale à Madagascar', *VertigO*, vol 9, no 3, pp1–18.

Blanc-Pamard, C. and Rakoto Ramiarantsoa, H. (2007) 'Normes environnementales, transferts de gestion et recompositions territoriales en pays Betsileo: la gestion contractualisée des forêts', *Natures Sciences Sociétés*, vol 15, pp253–268.

Boserup, E. (1965) *The Conditions of Agricultural Growth: The Economics of Agrarian Change Under Population Pressure*, Earthscan, London.

Bowler, D. E., Buyung-Ali, L. M., Healey, J. R., Jones, J. P. G., Knight, T. M. and Pullin, A. S. (2012) 'Does community forest management provide global environmental benefit and improve local welfare?', *Frontiers in Ecology and the Environment*, vol 10, pp29–36.

Brandon, K. E. and Wells, M. (1992) 'Planning for people and parks: design dilemmas', *World Development*, vol 20, pp557–570.

Brockington, D., Duffy, R. and Igoe, J. (2008) *Nature Unbound: Conservation, Capitalism and the Future of Protected Areas*, Earthscan, London.

Chambers R. (1983) *Rural Development: Putting the Last First*, Prentice Hall, London.

Cooke, B. and Kothari, U. (2001) *Participation: The New Tyranny?*, Zed Books, London.

Cripps, G. and Harris, A. (2009) *Community Creation and Management of the Velondriake Marine Protected Area*, Blue Ventures, London.

Dressler, W., Buscher, B. B., Schoon, M., Brockington, D., Hayes, T., Kull, C. A., McCarthy, J. and Shrestha, K. (2010) 'From hope to crisis and back again? A

critical history of the global CBNRM narrative', *Environmental Conservation*, vol 37, pp5–15.
Goedefroit, S. (2006) 'La restitution du droit à la parole', *Etudes Rurales*, vol 178, pp39–64.
Hames, R. (2007) 'The ecologically noble savage debate', *Annual Review of Anthropology*, vol 36, pp177–190.
Hardin, G. (1968) 'The tragedy of the commons', *Science*, vol 162, pp1243–1248.
Hockley, N. J. and Andriamarovololona, M. M. (2007) *The Economics of Community Forest Management in Madagascar: Is There a Free Lunch?*, United States Agency for International Development and Development Alternatives, Antananarivo, Madagascar.
Hockley, N. J. and Razafindralambo, R. (2006) *A Social Cost-benefit Analysis of Conserving the Ranomafana-Andringitra-Pic d'Ivohibe Corridor in Madagascar*, Conservation International and United States Agency for International Development, Antananarivo, Madagascar.
Johannes, R. E. (2002) 'The renaissance of community-based marine resource management in Oceania', *Annual Review of Ecology and Systematics*, vol 33, pp317–340.
Keller, E. (2008) 'The banana plant and the moon: conservation and the Malagasy ethos of life in Masoala', *American Ethnologist*, vol 35, pp650–664.
Kull, C. A. (2002) 'Empowering pyromaniacs in Madagascar: ideology and legitimacy in community-based natural resource management', *Development and Change*, vol 331, pp57–78.
Kull, C. A. (2004) *Isle of Fire: The Political Ecology of Landscape Burning in Madagascar*, University of Chicago Press, Chicago.
Laney, R. M. (2002) 'Disaggregating induced intensification for land-change analysis: a case study from Madagascar', *Annals of the Association of American Geographers*, vol 92, pp702–726.
Mazoyer, M. and Roudart, L. (2006) *A History of World Agriculture: From the Neolithic to the Current Crisis*, Monthly Review Press, New York.
Murray, G., Johnson, T., McCay, B. J., Danko, M., St. Martin, K. and Takahashi, S. (2010) 'Creeping enclosure, cumulative effects and the marine commons of New Jersey', *International Journal of the Commons*, vol 4, pp367–389.
Muttenzer, F. (2010) *Déforestation et Droit Coutumier à Madagascar: Les Perceptions des Acteurs de la Gestion Communautaire des Forêts*, Karthala, Paris.
Ollagnon, H. (1991) 'Vers une gestion patrimoniale de la protection et de la qualité biologique des forêts', *Forest, Trees and People*, vol 3, pp2–35.
Peluso, N. L. and Watts, M. (2001) *Violent Environments*, Cornell University Press, Ithaca.
Pollini, J. (2007) 'Slash-and-burn cultivation and deforestation in the Malagasy rain forests: representations and realities', PhD thesis, Cornell University, Ithaca.
Pollini, J. (2011) 'The difficult reconciliation of conservation and development objectives: the case of the Malagasy Environmental Action Plan', *Human Organizations*, vol 70(1), pp74–87.
Pollini, J. and Lassoie, J. P. (2011) 'Trapping farmer communities within global environmental regimes: the case of the GELOSE legislation in Madagascar', *Society and Natural Resources*, vol 24, pp1–17.

Rabesahala Horning, N. (2004) 'The cost of ignoring rules: forest conservation and rural livelihood outcomes in Madagascar', *Forests, Trees and Livelihoods*, vol 15, pp149–166.

Ramamonjisoa, B. and Rabemananjara, Z. (2012) 'Une évaluation de la foresterie communautaire', *Les Cahiers d'Outre-Mer*, vol 257, pp125–155.

RESOLVE (2005) *Evaluation et Perspectives des Transferts de Gestion des Ressources Naturelles dans le Cadre du Programme Environnemental 3: Rapport Final de Deuxième Phase*, Ministère de l'Environnement, Antananarivo, Madagascar.

Ribot, J. (2011) 'Choice, recognition and the democracy effects of decentralization', *ICLD Working Paper*, no 5, Swedish International Centre for Local Democracy, Visby, Sweden.

Scott, J. (1976) *The Moral Economy of the Peasants: Rebellion and Subsistence in Southeast Asia*, Yale University Press, New Haven.

Taylor, C. (1994) 'The politics of recognition', in A. Guttman (ed.) *Multiculturalism*, Princeton University Press, Princeton.

Toillier, A., Serpantié, G., Hervé, D. and Lardon, S. (2011) 'Livelihood strategies and land use changes in response to conservation: pitfalls of community-based forest management in Madagascar', *Journal of Sustainable Forestry*, vol 30, pp20–56.

Watson, R. T., René de Roland, L. A., Rabearivony, J. and Thorstrom, R. (2007) 'Community-based wetland conservation protects endangered species in Madagascar: lessons from science and conservation', *Banwa*, vol 4, pp8–97.

Weber, J. (1995) 'L'occupation humaine des aires protégées à Madagascar: diagnostic et éléments pour une gestion viable', *Nature Sciences Sociétés*, vol 3, pp157–164.

Weber, J. (1996) *Conservation, Développement et Coordination: Peut-on Gérer Biologiquement le Social?*, colloque panafricain: gestion communautaire des ressources naturelles renouvelables et développement durable, Harare, 24–27 June 1996.

9 Conservation politics in Madagascar

The expansion of protected areas

Catherine Corson

> Earlier versions of this chapter appear in Journal of Peasant Studies
> and Society and Natural Resources.

In 2003, Madagascar's former president, Marc Ravalomanana, announced his intention to triple the country's protected areas in five years[1] to cover a total of six million hectares (ha)—approximately 10 percent of the country's territory. First known as 'the Durban Vision' and later entitled the *Système des Aires Protégées de Madagascar* (SAPM—System of Protected Areas in Madagascar), the initiative aimed to meet the International Union for Conservation of Nature's (IUCN) target of protecting 10 percent of every country's major biomes. The former president underscored that the new protected areas would adhere to IUCN guidelines, which endorse parks ranging from those that prohibit human entry to those that allow sustainable use; encourage consultation with potentially affected local populations; and promote co-management with a variety of public and private entities (IUCN, 2004; Dudley and Phillips, 2006). By December 2010, and despite the 2009 political crisis that ousted Ravalomanana from power, this group had created 125 new protected areas and sustainable forest management sites that, together with pre-existing parks, covered 9.4 million hectares in total (Repoblikan'i Madagasikara, 2010a, b).

The president's proclamation represented a significant international conservation success in one of the world's highest priority biodiversity regions: it was, in the words of Conservation International (CI) President Russell Mittermeier, 'one of the most important announcements in the history of biodiversity conservation' (CI, 2011). It endeared Ravalomanana to conservationists for his efforts to save a country that British Prince Phillip had proclaimed, almost twenty years earlier, was 'committing environmental suicide', and it marked the culmination of an ongoing effort by foreign and Malagasy scientists and policy-makers to prioritize the protection of critical biodiversity ecosystems in Madagascar. To a great extent, it took over the 15-year National Environmental Action Plan (NEAP), co-implemented by foreign aid donors and non-governmental organizations (NGOs)—see also Chapter 7 by Kull.

Advocates publicized the initiative as a ground-breaking way of establishing and managing parks—one that involved communities to a greater extent than previous approaches—and numerous policies underscored the need to consult with potentially affected communities (e.g. World Bank, 2005; Repoblikan'i Madagasikara, 2005a; Borrini-Feyerabend and Dudley, 2005b; Commission SAPM, 2006). Donors and conservation NGOs touted the program's merits: in the words of one U.S. Agency for International Development (USAID) report, the initiative 'represent[ed] a major shift in how... protected areas are understood in Madagascar' (USAID, 2007, p22). Yet, my research revealed limited community engagement in the initial establishment of the protected areas. While the president of Madagascar made the official decision to expand the nation's protected areas, non-state actors from outside Madagascar—including foreign aid donors, international NGOs, consultants, and private commercial interests—shaped the boundaries, rights, and authorities associated with the new protected areas. Via active participation in the design and implementation of Madagascar's protected area governance, these actors legitimated their claims both to forest lands themselves and to the authority to determine forest policy. Ultimately, through the promotion of private and NGO management of Madagascar's new parks, as well as the accommodation of mining interests, the SAPM consolidated non-state and foreign access to and control over the process of determining land and resource rights, and delegitimated local claims to both resources and decision-making authority over conservation policy. I contend that while the high-profile announcement successfully mobilized funding for biodiversity conservation, the resulting political attention in fact undermined the consultation process. In fact, rather than effectively engaging rural communities, the program reinforced non-local decision-making power by creating a mechanism around which foreign conservationists, working with national government agencies, could influence Madagascar's forest policy.

In this chapter, I examine how negotiations among the Madagascar state, multilateral and bilateral donors, private sector organizations, and transnational conservation groups shaped SAPM's implementation. Drawing on interviews representing a range of perspectives about the park expansion program and observations at public meetings, I trace the steps through which the initial protected areas were established. I begin by discussing the history and rationale behind the idea to expand Madagascar's park system and the negotiations around the 2003 announcement. Then, I explore the everyday contestations and compromises that took place among various branches of the state, donors, conservation organizations, mining companies, and community leaders in the process of implementing the announcement—specifically by mapping, classifying, and designating resource uses in the new areas. I focus in particular on how and why only limited consultations

Conservation politics in Madagascar 195

with rural populations took place in the process of designating initial protected areas in the Ankeniheny-Zahamena and Fandriana-Vondrozo biological corridors in Madagascar's eastern rainforest, which were established as temporary parks in 2005 and 2006, respectively. Finally, I illustrate the potential for the initiative to not only produce 'paper parks'—parks that exist only on paper—but to catalyze increased deforestation.[2]

Formulating the parks' announcement

The Fifth World Parks Congress was held in Durban, South Africa, in 2003. A 10-yearly event, it provides the major forum for setting the decadal agenda for international protected areas policy. Numerous scholars have argued that donors and international NGOs have had a powerful influence on environmental policy-making in Madagascar (e.g. Kull, 1996; Duffy, 2006; Horning, 2008; Corson, 2011, 2012). The 'Malagasy' delegation to the Parks Congress, where the former president made his announcement, included government officials from the Madagascar *Ministère de l'Environnement, des Eaux et Forêts* (MinEnvEF—Ministry of Environment, Water and Forests) and *Association Nationale pour la Gestion des Aires Protégées* (ANGAP—National Association for the Management of Protected Areas).[3] Yet, as is common in delegations to international environmental meetings, it also contained NGO and multinational organization delegates. It included foreign organizations such as the British Durrell Wildlife Conservation Trust (DWCT), Madagascar-based conservation organizations Fanamby and *l'Institut pour la Conservation des Ecosystèmes Tropicaux* (Institute for the Conservation of Tropical Ecosystems), and transnational conservation NGOs CI, World Wild Fund for Nature (WWF-Madagascar), and Wildlife Conservation Society (WCS), as well as the representatives from USAID and the World Bank.

These organizations enthusiastically commended the announcement (Brockington et al., 2008; Horning, 2008), and, having made such a public announcement, Ravalomanana faced substantial international pressure to implement it successfully. Through an association first entitled the 'Durban Vision Group' and later called the 'SAPM Commission', an alliance of Madagascar government, foreign aid donors, consultants, and national and international NGOs based primarily in Madagascar's capital city, Antananarivo, scrambled to make the parks' announcement a reality. The official executive steering committee included only Madagascar government officials (Pollini, 2007). However, the environmental officer for USAID-Madagascar and the Director General of the Malagasy National Environment Office led its technical secretariat, which administered the specifics of the program and oversaw several working groups[4] that were responsible for various aspects of SAPM's implementation (Commission SAPM, 2006). Through these working groups, Madagascar government representatives, foreign aid donors, consultants, scientists, and national and transnational conservation NGOs oversaw the development of legal and management guidance, as well as the creation of the SAPM maps. They drafted, shared, marked up, and finalized guidance, policies, and laws that were ultimately issued officially by the Madagascar government. In this manner, they undertook functions that would otherwise have been state responsibilities. One of these was the prioritization of certain areas for conservation (Corson, 2011).

Biodiversity priority setting

Since the end of the nineteenth century, conservationists around the world have been steadfastly committed to the establishment of protected areas and more recently to networks as the most effective means to protect biological diversity and the most important indicator of conservation success. In the 1980s and 1990s, the 'Yellowstone Park models' of exclusionary national parks—where local people were excluded from utilizing resources within park boundaries—gave way to approaches that decentralized management to and/or provided economic benefits for local residents. While rhetoric advocating community involvement continues to characterize contemporary international conservation discourse, new approaches such as ecoregional and transboundary areas, as well as privately managed parks, have recentralized policy-making processes and reduced investments in local communities (Brosius and Russell, 2003; Wolmer, 2003). In some cases, even the rhetoric has been dropped, however, as evidenced by the call for a return to exclusionary parks, or what critiques call 'fortress conservation', in which local people are excluded, by force if necessary, from utilizing resources within park boundaries (for analysis see Brechin et al., 2002; Wilshusen et al., 2002; Adams and Hutton, 2007). Concurrent with these transformations in conservation ideology and practice has been a global effort to expand protected area networks. The IUCN has embraced the recommendations made at the 1992 Fourth World Congress on National Parks and Protected Areas 'that protected areas cover at least ten percent of each biome by the year 2000' (McNeely, 1993; also cited in Brooks et al., 2004, p1081). Similarly, other international conventions and agendas have modified and adopted this aim, including the Millennium Development Goals and the 2010 Biodiversity Convention targets (Secretariat, 2012; UN Statistics Division, 2012). At the 2010 Conference of the Parties to the Convention on Biological Diversity, the target was increased to 17 percent of terrestrial and inland water and 10 percent of coastal and marine areas by 2020 (Corson et al., in press). As protected areas spread across the globe, researchers continue to document detrimental effects—from restricted resource uses to outright displacement—on local populations (e.g. Brockington et al., 2008; Corson, 2011).

In Madagascar specifically, the idea to expand the number of parks was not new: increasing parks was one of the overall objectives of the NEAP's third phase. Years of advocacy by scientists and policy-makers to expand Madagascar's park network to include representation of all the country's major ecosystems preceded the Durban announcement (see also Chapter 7 by Kull). Nicoll and Langrand (1989) had proposed the expansion of Madagascar's protected area network to ensure that Malagasy biodiversity was fully represented in the 1980s. In April 1995, the Global Environment Facility (GEF) funded a major scientific workshop in Madagascar, which concluded that a significant portion of conservation and research priorities were located outside of the existing parks (Hannah et al., 1998), and which led to subsequent efforts to expand the park network. USAID then funded a park expansion planning exercise that culminated in a National Protected Area Management Plan entitled *PlanGrap*, which proposed additional sites for protected status (ANGAP, 2003). An ensuing biodiversity priority setting exercise, entitled *Réseau de la Biodiversité de Madagascar*, or REBIOMA, in turn developed maps to help policy-makers identify conservation priorities by overlaying Geographic Information Systems (GIS) layers of estimated animal and plant species ranges with estimated threat levels (Randrianandianina et al., 2003; Kremen et al., 2008). However, these initiatives had three main limitations.

First, reflecting the interests and expertise of the donors, NGOs, and scientists engaged in these endeavors, they had disproportionately focused on biodiversity prioritization. None had incorporated substantial socioeconomic data, such as where people lived or what environmental resources they used for their livelihoods. Thus, when the SAPM process began, policy-makers had considerable information about forest biodiversity, but very little information about how rural peasants used forest resources. While there were regional and local efforts to map community conservation sites and differentiated land uses, the resulting national protected area maps were based exclusively on biodiversity priorities and, in effect, they erased inhabitants, their livelihoods, and their existing community-managed areas from the targeted landscapes.

Second, the national maps of the new protected areas did not include many pre-existing, community-managed forests for several intertwined reasons. In many cases, while community-based natural resource management (CBNRM) paper maps existed, the spatial data used to create them had been lost. Moreover, *Communautés de Bases* (COBAs—local communities) members frequently did not have copies of their own CBNRM paper maps, and community use patterns often bore little resemblance to the delimitations of these maps (Vokatry Ny Ala, 2006). Similarly, most sacred forests—or forests protected by villagers over decades for cultural or religious reasons— were not included on national SAPM maps, despite the IUCN's emphasis that cultural heritage sites could be protected areas. In fact, sacred forests within CBNRM sites were seldom on CBNRM maps because villagers considered them 'already protected', and thus not necessary to map.[5] At the end of 2006, there were regional attempts to incorporate community management transfers, but they had not yet been incorporated in national SAPM maps.

Finally, the Durban announcement sped up the process of park expansion, increased its extent, and brought international and national political pressure to bear on it. As one senior Madagascar governmental official recounted: 'The original plan was to increase protected areas by 6–7 percent with no limit on the time frame, but then [the conservation] organizations pushed for 10 percent in five years.' This pressure laid the foundation for intense conflicts among scientists, commercial interests, state officials, NGOs, and foreign aid donors to decide how to manage the use of Madagascar's resources (Corson, 2011).

Struggles to control Madagascar's resources

Immediately after Ravalomanana's announcement, and in an effort to protect their own interests, mining and timber extractors rushed to exploit forest lands that might become national parks. They used legal and illegal means, both by working through the formal permitting process and by exploring without permits.[6] In reaction, donors and NGOs urged the government to halt new mining and timber permits in any potential new protected areas until the parks could be established. Scientists who had been working on biodiversity prioritization were suddenly catapulted into the limelight, as policy-makers scrambled to put together a map that would show the locations of new protected areas in which mining and timber permits should be banned. The resulting order, issued by the government in October 2004, suspended all mining and timber permits for two years in any area proposed for protected status, as defined by the accompanying map, which had been created in the biodiversity prioritization exercise (Repoblikan'i Madagasikara, 2004c). This new territory included the vast majority of Madagascar's forests, and the order was to be renewed every two years.

The hastily constructed map frustrated both mining and conservation advocates. Representatives of mining firms complained that the map creators had used inaccurate GIS data that included un-forested, low biodiversity areas. Conservationists retorted that certain high-priority biodiversity areas had been excluded, which left these areas to the mercy of miners.[7] Thus began a series of both informal and official negotiations between international conservation NGOs and mining companies.[8] In some cases, NGO representatives and mining agents made informal joint field visits to contested areas to ground reference the map and negotiate new boundaries.[9] In other areas, large mining companies, such as Dynatec and QIT Madagascar Minerals, agreed to set aside alternative areas as private reserves and/or contribute to the donor-funded Biodiversity Trust Fund in an effort to have a net positive effect on biodiversity (Sarrasin, 2006).

These informal dialogues between conservation and mining interests paralleled the official negotiations that took place under the *Comité Interministériel des Mines et des Forêts* (Inter-ministerial Mining and Forest Commission), which was established in 2004 to coordinate between the Ministry of Energy and MinEnvEF specifically to mediate conflicts where mining permits had been granted in environmentally sensitive areas (Repoblikan'i Madagasikara, 2004a, b). Prior to the 2004 order, the *Bureau du Cadastre Minier* (Office of Mining Registration) had issued mining permits in many of the areas claimed for SAPM, and it had continued to grant mining permits in these areas after 2004.[10] In the Fandriana-Vondrozo corridor, for example, regional SAPM committee members discovered that by 2006, mining permits had already been granted in four-fifths of the areas slated for protected status.[11] The 2006 order that established the Fandriana-Vondrozo corridor as a new protected area attempted to resolve this conflict by stating that permits granted before the October 2004 prohibition of mining and logging in new protected areas would be honored, although owners would have to conduct environmental impact assessments (EIAs) before proceeding (Repoblikan'i Madagasikara, 2006).

In 2006, as the government prepared to renew the 2004 two-year protection order for another two years with a revised and improved map that incorporated the outcome of these negotiations, the *Direction Générale des Eaux et Forêts* (DGEF—Directorate General of Water and Forests) added a new twist. In an attempt to reclaim forest land for production, the DGEF protested that SAPM's overwhelming focus on conservation ignored the need to protect timber and fuel wood supplies. A former DGEF official summarized: 'The forestry administration is terrified that if this expansion of the protected area turns out better, it will be the ANGAP who will have all the forests of Madagascar.' With assistance from USAID Jariala, a forest sector reform project managed by US contractor The International Resources Group, the Director-General of Water and Forests proposed that the SAPM maps also include 'sites of sustainable development', entitled 'KoloAla',[12] that would promote development through sustainable use of timber and non-timber forest resources. The result was the addition of sites of sustainable forest management in the renewal of the 2004 order. Ultimately, despite its claims of recognizing local rights, SAPM legitimated the authority of consultants, private companies, scientists, transnational conservation NGOs, and foreign aid donors to shape Madagascar's future forest policy, including who could use and make money from them and how.

Through these informal and formal means, branches of the state, NGOs, and mining companies negotiated the boundaries and rights associated with the new parks. The efforts ultimately minimized the extent to which the expansion of protected areas interfered with the rapidly expanding mining industry while simultaneously limiting use rights by local residents.

Debating rights to use forest resources

As new borders were being gazetted, government agencies, foreign aid donors, and conservation NGOs based in the capital city Antananarivo engaged in lengthy debates over the extent to which the rural people who lived in and around the new parks should be able to use resources within them. While most SAPM commission members agreed that villagers living in and around the parks should not be allowed to exploit forests to the detriment of the biodiversity contained therein, they also thought that local residents needed 'economic incentives' to preserve forests. The most significant debates focused on the degree to which and where use should be allowed and what exactly the economic incentives should be.

While it is difficult to categorize the complexity of views of different members in a short chapter, they fell generally into two groups. Relatively strict conservation advocates, typified by CI and the WCS, urged that rights be limited to traditional, non-commercial use. They asserted that if communities were permitted to engage in economic activities, they would not be able to stand up to the economic and political power of timber and mining interests and the resulting commercial extraction.[13] Instead, these organizations advocated 'biodiversity valuation' and non-consumptive uses, where sales of medicinal plants, ecotourism, carbon credits, and payments for environmental services would compensate villagers directly for the reduction in their ability to meet their livelihood needs due to park restrictions. They pushed for strict *sites de conservation*, or conservation sites, which would forbid any other form of commercial exploitation within park boundaries.

The countering view was represented by a coalition comprised of German Cooperation, French Cooperation, the Malagasy organization *Service d'Appui à la Gestion de l'Environnement* (Environmental Management Support Service, a privatized state organization), the United Nations Educational, Scientific and Cultural Organization (UNESCO), the United Nations Development Program (UNDP), and the WWF. This group argued that small-scale (as opposed to industrial) commercial extraction would motivate peasants to protect forests and could be managed sustainably under strict management plans that the DGEF, NGOs, or other organizations providing technical assistance could approve. Specifically, the DGEF and French Cooperation put forth a concept called *Le Territoire de Conservation et Développement* (Development and Conservation Territory), which advocated core sites protected for biodiversity and surrounding areas for use and which coincided ideologically with IUCN Category V[14] (Pollini, 2007). This idea circulated among SAPM committee members for several months before it became clear that stricter conservation interests would prevail, at least in the short term, and the French and other pro-commercial advocates temporarily left the Durban Vision negotiations in 2006.[15]

MinEnvEF officials mediated these groups, trying to preserve the financial backing of primarily U.S.-based conservationists, while advocating privately to allow community resource use and protect domestic economic mining and timber interests (Pollini, 2007). Many regional Madagascar government officials, as well as regionally based conservation NGOs and contractors, were more open to the idea of allowing sustainable use of resources in the new parks than their Antananarivo counterparts. They also argued that rural development and agricultural assistance to compensate small farmers for reductions in access to land and resources should accompany SAPM. Government officials across a number of ministries and quasi government departments expressed frustration at the overwhelming emphasis by the donor community, in particular the Americans, on conservation issues.

Most importantly, despite the rhetoric promoting sustainable use of park resources, SAPM introduced the possibility that the authority granted in some CBNRM projects to raise funds through commercial exploitation could be rescinded. From 2004 to 2006, Antananarivo-based government agencies, foreign aid donors, and conservation NGOs engaged in lengthy debates over the extent to which people who lived in and around the new parks should be able to use resources within them. The existing park legislation, the *Code des Aires Protégées* (COAP—Code for Managing Protected Areas), and associated decrees prohibited most human uses within park boundaries, except traditional non-commercial use, and forbade commercial exploitation unless explicitly authorized as an exception by a management plan. They also laid out strict penalties for violations (Repoblikan'i Madagasikara, 2001, 2005a, 2005b). Once the CBNRM sites were included in national parks, they fell under the COAP and, thus, were legally subject to these penalties, which directly contradicted prior rights granted in some *Gestion Locale Sécurisée* (GELOSE—Locally Secured Resource Management) and *Gestion Contractualisée des Forêts* (GCF—Contracted Forest Management) contracts to generate economic benefits through forest product use, such as small-scale commercial timber extraction (see Chapter 8 by Pollini et al. on GELOSE and Corson, 2011).

The rush to implement the Durban Vision

During these national and regional discussions, interviewed rural people living in the Ankeniheny-Zahamena and Fandriana-Vondrozo corridors, where I focused my study, expressed very little, if any, knowledge about SAPM. While limited community engagement took place at the mayoral level, the effort to meet the five-year deadline superseded any consultation efforts with villagers themselves, and some consultation leaders and/or Antananarivo-based SAPM leaders even suppressed mayoral dissent (Corson, 2012).

When I first discussed the program with SAPM commission members in October 2005, they were rushing to establish a million hectares by the end of the year, and SAPM had become the primary focus of Madagascar's foreign and domestic environmental communities. Between the November 2003 announcement and December 2004, commission members devoted attention to deciding overall biodiversity priorities, and no new parks were established. By January 2005, with only four years left to meet the president's target, the Minister for Environment, Water and Forests decided to establish one million ha of protected areas per year. This target formed the proximate cause of the hurry.

However, by October 2005, only about 100,000 ha of new protected area had been created across Madagascar. It was clear that one million hectares would not be established by the end of 2005 if extensive rural consultations about SAPM took place. In particular, it was impossible to consult with all potentially affected rural villages in the two eastern rainforest corridors, which comprised the two largest proposed protected areas. A massive debate ensued in Antananarivo over how much consultation was really necessary, and it split groups who might otherwise have been allies. A number of national and regional Madagascar government staff stressed the importance of thorough consultation so as not to cause resistance and burning of forests as a result, and organizations such as the WWF and regionally based U.S.-funded contractors advocated a slow down in the process. Others argued that, given the importance of Madagascar's biodiversity, it was critical that the program be implemented rapidly, and CI, in particular, pushed to meet the deadline.

Ultimately, in order to meet the minister's deadline, but to allow for future consultation, the SAPM Commission decided to establish new parks under *arrêtés* (orders), which granted temporary protected status to the new protected areas, which would be renewable after two years, and which delineated park boundaries and allowable resource uses based on limited consultations. These orders left specific details about park management and uses within sub-zones of larger parks to be decided upon in future management plans. Moreover, as the final policy guidance and legislation were not issued until 2008 and 2009, these temporary parks were established without clear guidance on how to develop management plans or conduct community consultations.

The rural consultation process

The organizations involved in the community consultation process that did occur in the two corridors reflected the make-up of the Antananarivo-based SAPM Commission to a great degree. Regional representatives of the DGEF and ANGAP, as well as primarily USAID-funded conservation and agricultural development NGOs and contractors, formed committees to design and implement the SAPM initiative. The DGEF and CI representatives based in the regional cities of Toamasina, Moramanga, and Fianarantsoa coordinated the initial consultations in the two eastern rainforest corridors in 2005. The USAID-sponsored Miaro conservation program, administered by CI in collaboration with ANGAP, the WWF and WCS, provided much of the funding and technical guidance. In 2006, the DGEF took on a larger coordination role, although without funding or practical implementation capacity, as bureaucrats often worked without phones, faxes, computers, or vehicles. Moreover, because U.S. grantees and contractors could more easily communicate with their Antananarivo-based counterparts, they often received information about SAPM Commission decisions before their Malagasy government colleagues. Thus, they continued to play a behind-the-scenes role, cemented by their human resources, vehicles, and GIS skills.

In both corridors, pressure to reach the minister's annual deadline, combined with insufficient staff and funding in both government and NGOs to reach villages located in remote areas meant that consultation efforts only reached the district level, relying on mayors as the spokespeople of the tens of thousands of people living in the communes. One member commented that, because it would take more than two days to walk to reach certain sites, it would require enormous human resources, time, and material to conduct a thorough consultation. In the Fandriana–Vondrozo corridor, a Fianarantsoa regionally based international NGO representative noted: 'We tried to gather people because it was impossible for us to visit 66 municipalities with 10 districts and five regions.'

In theory, villagers could later contest the park boundaries agreed upon during these discussions. However, in order to grant the temporary protected status establishing the new parks, key political figures, including ministers, regional political leaders, and mayors, had to agree to these initial limits. Thus, by the time temporary protection status was granted, there was already a substantial buy-in at various political levels. Furthermore, such an approach empowered mayors as the key liaisons between national and regional governments and 'the people', and as the primary 'local' decision-makers. It assumed that mayors would consult with affected populations and completely ignored intra- and inter-village tensions; gender and class differences; and mayors' tendencies to favor their own villages (e.g. Kull, 2002). Several interviewees reported that mayors did not consult with villagers before agreeing to particular limits.[16]

Sustaining this assertion, a related study conducted in 2006 with the Malagasy NGO Vokatry ny Ala specifically to assess village awareness of SAPM among pre-existing CBNRM sites in the communes of Ikongo, Ambatofotsy, Miarinarivo, Andranomiditra, and Ialamarina in the Fandriana-Vondrozo corridor found that, other than mayors and presidents of COBAs, very few individuals had heard of SAPM. The mayors and COBA presidents had all attended SAPM consultations, but, in most cases, they had not reported back to villagers nor solicited their input. One COBA president justified this by saying that he felt it would be better to report to the population after the final policy had been decided. Of the 130 people interviewed in this study, 75 percent had never heard of SAPM, and 85 percent of the 32 people who had heard of it were mayors, COBA presidents, or COBA members who had been invited to SAPM sessions (one commune held a workshop to which it invited some COBA members). The other five had learned about it from the radio. Villagers who knew about it thought it was a form of CBNRM, and even COBA presidents and mayors saw it as a mechanism that would help secure local access and control over resources by officially recognizing, albeit not officially transferring, land rights (Vokatry Ny Ala, 2006).

Intriguingly, when talking about the consultation process, interviewees often conflated consultation with *sensibilisation*, a French term that might be translated into English as 'persuasive education', 'outreach', or 'awareness-raising'. Development organizations often use the term to refer to their field training activities, such as health education or agricultural improvement techniques. For example, a Fianarantsoa-based government official said: 'In the districts, we do the *sensibilisation*-consultation. We say that to achieve the president's declaration, you should give a part [of your land] to be categorized according to IUCN guidance.' Similarly, a Toamasina-based official noted: 'We tried to inform the participants about the necessity of conservation and the important role played by forest resources.' Another Antananarivo-based official commented: 'People should understand what [SAPM] is, and that it is for their good.' Thus, the endeavor's goal was not to consult the population about whether or not they wanted the parks, but to instill in them the notion that parks were 'for their own good'. 'Consultation' began with informing the participants what they *should* think, specifically that the conservation areas were necessary. Effectively, the processes in these two corridors served to convince the population to agree to pre-decided areas, which were based on biodiversity priorities.

I contend that regional consultation leaders conflated 'consultation' with *sensibilisation* precisely because their mandate from Antananarivo was to persuade, not to consult. While various workshops designed to raise awareness about the initiative were held throughout the country, there was no clear guidance on how to conduct local consultation. The December 2005 decree simply said that affected populations should be consulted and their interests taken into account, but did not say how to do that (Repoblikan'i Madagasikara, 2005a). Similarly, IUCN consultants stressed the need to 'engage at least some community representatives in the preliminary identification of the protected areas' followed by more extensive consultation later (Borrini-Feyerabend and Dudley, 2005b, p14). However, their guidance on how to conduct consultation simply stated that it should occur through 'proven and coherent methods and tools, in particular with regard to social communication and participatory governance approaches' (Borrini-Feyerabend and Dudley, 2005a, p11). Finally, draft policy guidance was said to 'begin consultations with communal councils and/or mayors, regional authorities, technical services and development programmes to ensure their commitment towards the creation of new areas protected' (Commission SAPM, 2006) and only to consult with communes, villages, and hamlets before final park creation. While it did not define consultation or how the interests of stakeholders should be taken into account, it did underscore that the goal was 'to ensure' local commitment to park creation (Commission SAPM, 2006).

Regional representatives of the Madagascar government, NGOs, and contractors had to mediate between conservation visions emanating from Antananarivo and local realities. While the Antananarivo-based commission members, isolated from field realities, debated theoretical problems, such as how much consultation would be necessary, regional staff had to carry out their mandates. Regional agents complained about disorganization, incoherence, and political pressure from Antananarivo-based policy-makers who seemed unaware of the program's implementation challenges. One regional government official said that: 'CI or other donors, they all have different concepts ... so we are lost. Yet at the same time they tell us to do it all before 2005.' In particular, regional contractors, government officials, NGO staff, and mayors repeatedly emphasized that SAPM should be accompanied by rural development assistance to compensate villagers for reductions in access to land and resources. Yet, very limited funds materialized to this end.

Finally, some interviewees reported that mayoral resistance was stifled. In the Ankeniheny-Zahamena corridor, where CI was avidly pushing for the rapid establishment of a protected area, a regional contractor commented: 'The tendency was that commune mayors didn't want to make definitive decisions. But that got lost in the shuffle, perhaps due to the bias of the people getting pressured to hurry.' In the Fandriana-Vondrozo corridor, during the initial mayoral discussions, the government, NGOs, and contractors who led the consultations did not use the biodiversity prioritization map because it did not have communes marked on it and was therefore meaningless to mayors who needed to see political, as well as ecological, boundaries. Instead, the leaders asked mayors to propose areas within their communes for protected status. The resulting map contained a series of unconnected parks, violating the fundamental conservation goal, which was to preserve the forest corridor. The SAPM technical secretariat instructed the consultation team to go back and *persuade* the mayors to create a connected corridor, which they did.[17] The delimitations from this amended mayoral 'consultation' were then included in the 2006 one million hectares of protected areas.

At the end of 2006, DGEF agents in both regions stressed that they planned to do village-level consultation in the future, but that they did not have the funds to pay staff to moderate the discussions or to cover the vehicle expenses to get to remote locations. Nor did they have the equipment or staff to develop maps. So they continued to rely on NGOs and consultants.[18] However, the CI regional office did not have a budget for such consultations, and the disbursement of World Bank funding for consultations from the national DGEF office to the regional offices was delayed for various bureaucratic reasons.

Many regional governmental and non-governmental actors anticipated future protests; a DGEF regional agent expressed concern for his/her own safety: 'If we set limits ... fast without consultation, it will be us—field agents—who will become victims of the people.' In 2008, the government renewed the 2006 order with an updated map of protected areas (Figure 9.1); introduced a revised COAP law; and issued a series of policy documents that both allowed limited commercial resource extraction and reinforced non-state powers. A USAID consultant report, written in 2010, summarized upon reflection about the entire process: 'Among many rural people SAPM gained an early reputation for being top-down and largely engineered by outsiders. This has bred skepticism and hostility that will be difficult to overcome' (Freudenberger, 2010, p47). I contend that it was their marginalization from decision-making processes—from

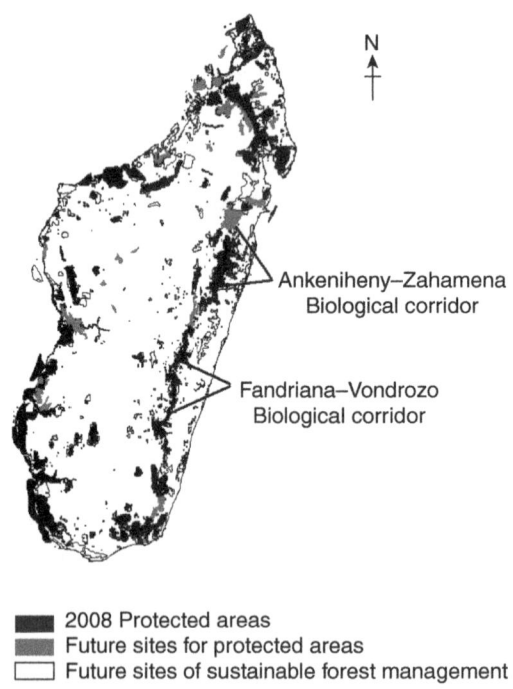

Figure 9.1 Map showing the expansion of protected areas and location of case studies discussed in the chapter

the authority to determine boundaries, resource rights, and acceptable resource uses on their customary land—that represented the ultimate dispossession (Corson, 2012).

Conclusion

In December 2010, the government passed two new orders, which renewed the 2008 temporary protection order, streamlined some of the regulations, included more explicit restrictions on the use of resources in marine and coastal protected areas, created an intergovernmental SAPM Commission, explicitly separated the sites of sustainable forest management from 'protected areas', and codified 171 new protected areas and sustainable forest management sites covering 9.4 million hectares in total (Repoblikan'i Madagasikara, 2010a, 2010b). More ethnographic research is needed to analyze the politics behind the 2008 and 2010 changes and, therefore, detailed analysis of these orders is beyond the scope of this chapter. Nonetheless, it is important to highlight the dynamic nature of the negotiations over the authority to determine who can access and benefit from Madagascar's natural wealth. While the 2010 official creation of an intergovernmental commission may represent a move to reinstate state control over the process of expanding protected areas, a recent World Bank document summarizes the continuing strong influence by transnational NGOs, working across international, national, and regional scales, on Madagascar forest policy. Using numbers that exclude 2.5 million hectares of sustainable forest management sites, the report summarizes:

> The protected area network in Madagascar [includes] 2.4 million hectares of protected areas managed by Madagascar National Parks and 4.5 million hectares of new protected areas that are being developed predominantly by NGOs (including CI, WCS and WWF) ... Triggered by the Convention on Biodiversity Conference of Parties (CBD COP) in Nagoya in October 2010, informal discussions have recently commenced amongst the Government and NGOs as to the feasibility of increasing coverage of the network to cover 16–18 percent of the country's surface.
>
> (World Bank, 2011, p31)[19]

The Madagascar case reveals not just the strong influence of transnational organizations on domestic forest policy, but also how marginalizing state processes can be produced through unelected and unaccountable entities, dispersing both claims to and authority over natural resources and subjects across local, national, and international boundaries. Despite SAPM advocates' insistence that the creation of new protected areas would engage communities that live in or utilize resources in the parks, minimal village-level consultation took place in the two eastern rainforest corridors studied. Notwithstanding the millions of dollars supporting conservation in Madagascar, Antananarivo-based donors, government agencies, and NGOs provided inadequate financial support for and guidance on how to conduct local consultations as they debated what kinds of resource uses they thought should be allowed and what the economic incentives for conservation should be in the absence of information from villagers about what they needed and wanted from the forest. National and international political pressure to implement the program rapidly compelled the day-to-day attention of regional staff toward creating biodiversity prioritization maps and meeting numerical targets and away from consultation with affected villagers. This pressure led to a consultation process that only reached the mayoral level, where mayors were empowered as liaisons between national government and villagers and often failed to disseminate information to villagers, who then remained unaware of the program. Additionally, the consultations were processes of persuasive education, which began with maps that outlined park boundaries based on biodiversity priorities rather than local resource use patterns, and in which mayoral resistance was stifled in order to reach annual targets. I assert that regional consultation leaders conflated 'consultation' with *sensibilisation* because their mandate from Antananarivo was to persuade, as evidenced by the push to establish parks quickly, the request to redo the mayors' map of unconnected sites, and the guidance to 'ensure' local commitment to conservation. Ultimately, the initiative superseded, rather than supported, previous community conservation initiatives by introducing the possibility of rescinding previously granted community commercial use rights (Corson, 2012).

The case of protected area expansion in Madagascar illustrates how conservation policy is negotiated through transnational alliances among state and non-state organizations. Through activities ranging from membership on the Malagasy delegation to the Parks Congress and the SAPM implementing committees to informal negotiations, foreign and non-state actors also cemented their power and authority to recognize the validity of particular claims to forest land. In this manner, the 'state' gave way to the consortium of state officials, foreign donors, transnational NGOs, contractors, and private sector agents who in turn served in its place. In short, SAPM not only recognized the claims of the forest service, foreign conservation organizations, and mining interests over those of rural peasants, it also legitimated the power and authority of these agents to determine the fate of Madagascar's forests and, in doing so, delegitimized the power and authority of rural peasants to control the source of their livelihoods and ultimately garner wealth from forest resources. It was rural people's marginalization from decision-making processes about these resources that represented the ultimate dispossession (Corson, 2011).

Policy recommendations

In the context of an expanding mining, oil, and gas industry in Madagascar, there are enormous stakes in how the expansion of protected areas is implemented. Ironically, the country's main challenge is not the creation of more protected areas, but the effective management of those already existing. The initiative to triple Madagascar's protected areas took as a starting premise that Madagascar had 1.7 million hectares of existing parks, which included only the national parks, strict nature reserves, and special reserves that ANGAP managed strictly for conservation. However, I contend, based on the IUCN's progressive protected area categories allowing sustainable use, that Madagascar had already met the 10 percent goal before the SAPM announcement. As Simsik (2003) noted, if biosphere reserves, hunting reserves, forest stations, and reforested areas—as well as the four million hectares of classified forests that the forest service managed for wood supplies—are added, a total of 6.6 million hectares were already protected in Madagascar by the early twenty-first century. However, neither the forest nor parks service could manage these areas due to mutually reinforcing lack of human, material, and financial resources. In fact, much of Madagascar's increasing illegal wood exploitation and mining occurs in protected areas. The 2009 exposure of considerable illegal logging in national parks in Madagascar (Global Witness and EIA, 2009) revealed that the creation of parks on paper is insufficient to protect the country's biodiversity. The lack of extensive consultation, before the initial park boundaries were established, with the only people with the proximate power to manage the forest in the face of growing timber and mining was a major misstep.

In sum, rather than expanding potentially ineffective conservation territories, conservationists should have focused instead on making the existing parks system effective for three key reasons. First, the new protected areas will likely have limited effect on biodiversity conservation. Second, regardless of their conservation outcome, by restricting resource uses and by limiting their ability to profit from resources, the new parks will alienate many rural people from their livelihood sources and impede their ability to improve their standard of living. Third, and most important, the policy process of developing the parks has legitimated transnational and Antananarivo-based power and authority to control access both to forests and to the wealth generated from them. In direct contradiction of the community-based conservation initiatives that emerged in Madagascar in the 1990s and early twenty-first century and the rhetoric surrounding SAPM, it marginalized rural people's control over decision-making processes about forest rights, as well as access to and the ability to benefit from Madagascar's forests.

While more recent ideas, such as the proposed federation of CBNRM leaders (Freudenberger, 2008) could help to institutionalize a village voice in SAPM negotiations, these organizations would be subject to existing power dynamics both within and among villages and within the network of government agencies, NGOs, donors, private companies, and others who have a stake in the expansion of Madagascar's protected areas. An effective future consultation process would begin, rather than end, with funding and resources devoted to long-term discussions with villagers and the development of maps that recognize village resource needs, current use patterns, and community protected zones. Discussions would involve, as Richard and Dewar (2001) propose, two-way conversation and negotiation, rather than persuasion. This would necessitate letting go of the aim to educate people about the importance of conservation and taking instead a receptive stance. To this end, discussions could include trained community participation specialists on consultation teams (e.g. Campbell and Vainio-Matilla, 2003); support villagers in leading their own unhurried discussions, participatory mapping, management plan development, and land use documentation (e.g. Chambers, 1994); and take measures to address gender, class, and other power dynamics that transpire in public discussions (Mosse, 1994). Ultimately, successful forest management in Madagascar will entail empowering people affected by conservation policies to participate in *decisions* about how to structure the management of natural resources, as well as the actual *management* of resources (Corson, 2011, 2012).

Acknowledgement

I would like to thank Taylor & Francis for permission to reproduce material from two papers: 'Territorialization, enclosure and neoliberalism: non-state influence in struggles over Madagascar's forests', by C. Corson, *Journal of Peasant Studies*, vol 38:4, 2011, pp703–726, and 'From rhetoric to practice: how high profile politics impeded community consultation in Madagascar's new protected areas', by C. Corson, *Society and Natural Resources*, vol 25:4, 2012, pp336–335, available at www.tandfonline.com. The first article was reprinted in its entirety in: N. Peluso and C. Lund. eds. New Frontiers of Land Control, Routledge. Research support came from the National Science Foundation, Andrew W. Mellon/American Council of Learned Societies, Rural Sociological Society, University of California Berkeley Center for African Studies, the Foreign Language and Area Studies Program and the Beahrs Environmental Leadership Program. An especial thanks to Vokatry ny Ala, and Rebioma provided the shape files for the map.

Notes

1 The deadline was later extended to 2012.
2 The material presented herein draws on document analysis, participant observation, and 144 semi-structured interviews conducted with former and current representatives of national, regional, and local branches of the Madagascar government; foreign aid donors; international conservation and development NGOs; Malagasy NGOs; private sector companies; consultant groups; and

scientific organizations based in Antananarivo, the regional cities of Toamasina and Fianarantsoa, and selected villages in the Ankeniheny-Zahamena and Fandriana-Vondrozo eastern rainforest corridors. Because of the sensitive nature of some of the interviews, I agreed to protect confidentiality for all interviewees, and thus all information is reported anonymously in that sources are identified only by general position; village identities are kept confidential; and interview dates are not revealed.
3 Now 'Madagascar National Parks'.
4 The Durban Vision Group oversaw five technical groups: 1) management and categorization, which was responsible for setting up a management system coherent with the IUCN guidelines; 2) biodiversity prioritization, which continued the biodiversity prioritization work; 3) communication, which coordinated communication with regional and central authorities and the general public; 4) legal framework, which developed legislation relative to the program; and 5) funding.
5 Informal communication with Vokatry ny Ala staff.
6 Interviews with a conservation NGO representative, a bilateral donor official, and a foreign scientist.
7 Interviews with mining and international conservation NGO representatives.
8 The mining industry in Madagascar comprises two sub-sectors: the large-scale mining sector, which focuses primarily on various industrial ores, and the small-scale informal, unregulated gemstone mining sector. The vast majority of mining takes place in the small-scale, informal, unregulated mining sector in Madagascar. Both illegal and legal mining are expanding rapidly in Madagascar (Bilger, 2006). The negotiations referred to in this section are between conservation NGOs and large-scale mining companies.
9 Interviews with mining and international conservation NGO representatives.
10 Interviews with contractors and a regional Madagascar government official.
11 Interviews with regional international conservation NGO representatives.
12 In Malagasy, this can be translated as 'tending the forest'.
13 Interviews with international conservation NGO representatives.
14 The most recent version of the World Conservation Union's system of categories for protected areas was endorsed in 1994. It includes seven categories of protected areas: Category Ia: Strict Nature Reserve/Wilderness Protection Area managed mainly for science or wilderness protection; Category Ib: Wilderness Area: protected area managed mainly for wilderness protection; Category II: National Park: protected area managed mainly for ecosystem protection and recreation; Category III: Natural Monument: protected area managed mainly for conservation of specific natural features; Category IV: Habitat/Species Management Area: protected area managed mainly for conservation through management intervention; Category V: Protected Landscape/Seascape: protected area managed mainly for landscape/seascape conservation or recreation; Category VI: Managed Resource Protected Area: protected area managed mainly for the sustainable use of natural resources (Phillips et al., 2004; Dudley and Phillips, 2006).
15 Interviews with bilateral donor officials.
16 Interviews with villagers and contractors.
17 Interview with regional contractors and government officials.
18 Interviews with a regional government official and a contractor.
19 This would meet newly negotiated CBD targets to protect, by 2020, at least 17 percent of terrestrial and inland water, and 10 percent of coastal and marine areas.

References

Adams, W. M. and Hutton, J. (2007) 'People, parks and poverty: political ecology and biodiversity conservation', *Conservation and Society*, vol 5, pp147–183.

ANGAP (2003) *Madagascar Protected Area System Management Plan: Revised*, Association Nationale pour la Gestion des Aires Protégées, Antananarivo, Madagascar.

Bilger, B. (2006) 'The path of stones: the race for Madagascar's jewels', *The New Yorker*, October 2, pp66–79.

Borrini-Feyerabend, G. and Dudley, N. (2005a) *Elan Durban... Nouvelles Perspectives pour les Aires Protégées à Madagascar*, report of the first IUCN mission to Madagascar: World Commission on Protected Areas and Committee on Environmental, Economic and Social Policy, World Conservation Union and MIARO, May 2005.

Borrini-Feyerabend, G. and Dudley, N. (2005b) *Les Aires Protégées à Madagascar: Bâtir le Système à Partir de la Base: Rapport de la Seconde Mission UICN (Version Finale)*, World Conservation Union Commission on Environmental, Economic and Social Policy and World Commission on Protected Areas, September.

Brechin, S. R., Wilshusen, P. R., Fortwangler, C. L. and West, P. C. (2002) 'Beyond the squarewheel: toward a more comprehensive understanding of biodiversity conservation as social and political process', *Society and Natural Resources*, vol 15, pp41–64.

Brockington, D., Duffy, R. and Igoe, J. (2008) *Nature Unbound: The Past, Present and Future of Protected Areas*, Earthscan, London.

Brooks, T. M., Bakarr, M. I., Boucher, T., Da Fonseca, G. A. B., Hiltontaylor, C., Hoekstra, J. M., Moritz, T., Olivieri, S., Parrish, J., Pressey, R. L., Rodrigues, A. S. L., Sechrest, W., Stattersfield, A., Strahm, W. and Stuart, S. N. (2004) 'Coverage provided by the global protected-area system: is it enough?', *BioScience*, vol 54, pp1081–1091.

Brosius, J. P. and Russell, D. (2003) 'Conservation from above: an anthropological perspective on transboundary protected areas and ecoregional planning', *Journal of Sustainable Forestry*, vol 17, pp39–65.

Campbell, L. M. and Vainio-Matilla, A. (2003) 'Participatory development and community-based conservation: opportunities missed for lessons learned', *Human Ecology: An Interdisciplinary Journal*, vol 31, pp417–438.

Chambers, R. (1994) 'The origins and practice of participatory rural appraisal', *World Development*, vol 22, pp953–969.

CI (2011) 'Madagascar to triple areas under protection: plan calls for the creation of a 6-million-hectare network of terrestrial and marine reserves', www.conservation.org/newsroom/pressreleases/Pages/091603_mad.aspx (accessed January 14, 2011).

Commission SAPM (2006) *Procédure de Création des Aires Protégées du Système d'Aires Protégées de Madagascar (SAPM)*, Commission SAPM, Antananarivo, Draft 8 June.

Corson, C. (2011) 'Territorialization, enclosure and neoliberalism: non-state influence in struggles over Madagascar's forests', *Journal of Peasant Studies*, vol 38, pp703–726.

Corson, C. (2012) 'From rhetoric to practice: how high profile politics impeded community consultation in Madagascar's new protected areas', *Society and Natural Resources*, vol 25, pp336–351.

Corson, C., Gruby, R., Witter, R., Hagermann, S., Suarez, D., Greenburg, S., Bourque, M., Gray, N. and Campbell, L. M. (in press) 'Everyone's solution?

Defining and re-defining protected areas in the Convention on Biological Diversity', *Conservation and Society*.
Dudley, N. and Phillips, A. (2006) 'Forests and protected areas guidance on the use of the IUCN protected area management categories', in A. Phillips (ed.) *World Commission on Protected Areas (WCPA) Best Practice Protected Area Guidelines Series No.12*, IUCN, Cambridge.
Duffy, R. (2006) 'Non-governmental organisations and governance states: the impact of transnational environmental management networks in Madagascar', *Environmental Politics*, vol 15, pp731–749.
Freudenberger, K. (2010) *Paradise Lost? Lessons from 25 Years of USAID Environment Programs in Madagascar*, USAID, Washington, DC.
Freudenberger, M. S. (2008) *Ecoregional conservation and the Ranomafana – Andringitra Forest Corridor: A Retrospective Interpretation of Achievements, Missed Opportunities, and Challenges for the Future*, July 2, 2008, version 2, unpublished manuscript.
Global Witness and EIA (2009) *Investigation into the Illegal Felling, Transport and Export of Precious Wood in Sava Region Madagascar*, in cooperation with Madagascar National Parks, the National Environment and Forest Observatory and the Forest Administration of Madagascar, December.
Hannah, L., Rakotosamimanana, B., Ganzhorn, J. U., Mittermeier, R. A., Olivieri, S., Iyer, L., Rajaobelina, S., Hough, J., Andriamialisoa, F., Bowles, I. and Tilkin, G. (1998) 'Participatory planning, scientific priorities, and landscape conservation in Madagascar', *Environmental Conservation*, vol 25, pp30–36.
Horning, N. R. (2008) 'Strong support for weak performance: donor competition in Madagascar', *African Affairs*, vol 107, pp405–431.
IUCN (2004) *Governance of Natural Resources: The Key to a Just World that Values and Conserves Nature?* Briefing Note 7, Commission on Environmental, Economic and Social Policy, World Commission on Protected Areas and the World Conservation Union.
Kremen, C., Cameron, A., Moilanen, A., Phillips, S. J., Thomas, C. D., Beentje, H., Dransfield, J., Fisher, B. L., Glaw, F., Good, T. C., Harper, G. J., Hijmans, R. J., Lees, D. C., Louis Jr. E., Nussbaum, R. A., Raxworthy, C. J., Razafimpahanana, A., Schatz, G. E., Vences, M., Vieites, D. R., Wright, P. C. and Zjhra, M. L. (2008) 'Aligning conservation priorities across taxa in Madagascar with high-resolution planning tools', *Science*, vol 320, pp222–226.
Kull, C. (1996) 'The evolution of conservation efforts in Madagascar', *International Environmental Affairs*, vol 8, pp50–86.
Kull, C. (2002) 'Empowering pyromaniacs in Madagascar: ideology and legitimacy in community-based natural resource management', *Development and Change*, vol 33, pp57–78.
McNeely, J. A. (ed.) (1993) *Parks for Life: Report of the IVth World Congress on National Parks and Protected Areas*, IUCN Communications Division, Gland, Switzerland.
Mosse, D. (1994) 'Authority, gender and knowledge: theoretical reflections on the practice of participatory rural appraisal', *Development and Change*, vol 25, pp497–526.
Nicoll, M. E. and Langrand, O. (1989) *Madagascar: Revue de la Conservation et des Aires Protégées*, World Wide Fund for Nature, Gland, Switzerland.
Phillips, A., Stolton, S., Dudley, N. and Bishop, K. (2004) *Speaking a Common Language: An Investigation into the Uses and Performance of the IUCN System of*

Management Categories for Protected Areas, final draft report, World Conservation Union and Cardiff University.

Pollini, J. (2007) *Slash and Burn Cultivation and Deforestation in the Malagasy Rain Forests: Representations and Realities'*, PhD thesis, Department of Natural Resources, Cornell University, Ithaca.

Randrianandianina, B. N., Andriamahaly, L. R., Harisoa, F. M. and Nicoll, M. E. (2003) 'The role of protected areas in the management of the island's biodiversity', in S. Goodman and J. Benstead (eds) *The Natural History of Madagascar*, University of Chicago Press, Chicago.

Repoblikan'i Madagasikara (2001) *Code de Gestion des Aires Protégées*, loi no. 2001/05.

Repoblikan'i Madagasikara (2004a) *Arrêté Interministériel Complétant les Dispositions de l'Arrêté n°7340/2004 Portant Création d'un Comité Interministériel des Mines et des Forêts*, le Ministère de l'Energie et des Mines and le Ministère de l'Environnement des Eaux et Forêts, no. 12720/2004, 8 July.

Repoblikan'i Madagasikara (2004b) *Arrêté Interministériel Portant Création d'un Comité Interministériel des Mines et des Forêts*, Ministère de l'Energie et des Mines and Ministère de l'Environnement des Eaux et Forêts, no. 7340/2004.

Repoblikan'i Madagasikara (2004c) *Arrêté Interministériel Portant Suspension de l'Octroi de Permis Minier et de Permis Forestier dans les Zones Réservées comme 'Sites de Conservation'*, Ministère de l'Environnement des Eaux et Forêts and Ministère de l'Energie et des Mines, no.19560/2004, 18 October.

Repoblikan'i Madagasikara (2005a) *Appliquant l'Article 2 Alinéa 2 de Loi no 2001/15 Portant Code des Aires Protégées*, Ministère de l'Environnement des Eaux et Forêts, décret 2005–848.

Repoblikan'i Madagasikara (2005b) *Organisant l'Application de la Loi no 2001-005 du 11 Février 2003 Portant Code de Gestion des Aires Protégées*, Ministère de l'Environnement des Eaux et Forêts, décret 2005-013.

Repoblikan'i Madagasikara (2006) *Arrêté Interministériel Portant Protection Temporaire de l'Aire Protégée en Création Dénommée 'Corridor Forestier Fandriana-Vondrozo'*, Ministère de l'Energie et des Mines and Ministère de l'Environnement des Eaux et Forêts, no. 16 071/2006.

Repoblikan'i Madagasikara (2010a) *Arrêté Interministériel Modifiant l'Arrêté Interministériel Mine-Forêts no 18633 du 17 Octobre 2008 Portant Mise en Protection Temporaire Globale des Sites Visés par l'Arrêté Interministériel no. 17914 du 18 Octobre 2006 et Levant la Suspension de l'Octroi des Permis Miniers et Forestiers pour Certains Sites*, Ministère de l'Environnement des Forêts, 52005/2010.

Repoblikan'i Madagasikara (2010b) *Arrêté Interministériel Portant Création, Organisation et Fonctionnement de la Commission du Système des Aires Protégées de Madagascar*, Ministère de l'Environnement des Forêts, 52004/2010.

Richard, A. F. and Dewar, R. E. (2001) 'Politics, negotiation and conservation: a view from Madagascar', in W. Weber, L. J. T. White, A. Vedder and L. Naughton-Treves (eds) *African Rain Forest Ecology and Conservation: An Interdisciplinary Perspective*, Yale University Press, New Haven.

Sarrasin, B. (2006) 'The mining industry and the regulatory framework in Madagascar: some developmental and environmental issues', *Journal of Cleaner Production*, vol 14, pp388–396.

Secretariat (2012) *The Convention on Biological Diversity: 2010 Biodiversity Target: 1) Goals and Subtargets and 2) Indicators*, www.cbd.int/2010-target/goals-targets.aspx, accessed November 30, 2012.

Simsik, M. (2003) *Priorities in conflict: Livelihood practices, environmental threats, and the conservation of biodiversity in Madagascar*, thesis (Ed.D), Education, University of Massachusetts, Amherst.

UN Statistics Division (2012) Progress towards the Millennium Development Goals, 1990–2005 Goal 7 – Ensure environmental sustainability, http://unstats.un.org/unsd/mi/goals_2005/Goal_7_2005.pdf, accessed November 30, 2012.

USAID (2007) *USAID's Biodiversity Conservation and Forestry Programs, FY 2005*, USAID, Washington, DC.

Vokatry Ny Ala (2006) *Evaluation et Analyse du Dynamique de Transferts de Gestion dans le Corridor Ranomafana-Andringitra: Cas d'Ampatsy et d'Andranomiditra*, a collaboration between EcoRegional Initiatives–Fianarantsoa and the University of California at Berkeley Beahrs Leadership Program, Vokatry ny Ala, Fianarantsoa, Madagascar.

Wilshusen, P. R., Brechin, S. R., Fortwangler, C. L. and West, P. C. (2002) 'Reinventing a square wheel: critique of a 'resurgent protection paradigm' in international biodiversity conservation', *Society and Natural Resources*, vol 15, pp17–40.

Wolmer, W. (2003) *Transboundary Protected Area Governance: Tensions and Paradoxes*, paper presented at the Transboundary Protected Areas in the Governance Stream of the 5th World Parks Congress in Durban, South Africa, September 12–13.

World Bank (2005) *Environmental and Social Safeguard Policies: Policy Objectives and Operational Principles*, World Bank operational manual: operational policies section 4.12, World Bank, Washington DC.

World Bank (2011) *Project Paper on a Proposed Additional IDA Credit in the Amount of SDR26 Million (US$42 Million Equivalent) and a Proposed Additional Grant from the Global Environment Facility Trust Fund in the Amount of US$10.0 Million to the Republic of Madagascar for the Third Environmental Program Support Project (EP3)*, report no. 61964-MG.

10 The Durban Vision in practice
Experiences in the participatory governance of Madagascar's new protected areas

Malika Virah-Sawmy, Charlie J. Gardner and Anitry N. Ratsifandrihamanana

Introduction

The International Union for the Conservation of Nature's (IUCN) fifth World Parks Congress, held in Durban, South Africa, in 2003, marked the dawn of a new paradigm for protected areas (Phillips, 2003). In response to an increasing realisation that the establishment of protected areas could have high social costs that were borne disproportionately by local people (Ghimire and Pimbert, 1997; Adams *et al.*, 2004; West *et al.*, 2006), the theme of the congress was 'Benefits Beyond Boundaries', and for the first time debates about conservation's impact on wider society became mainstream (Roe, 2008). A key topic of the Congress concerned the role of protected areas in poverty alleviation, and it was agreed that in pursuing their goal of biodiversity conservation, conservation agencies should not increase poverty or undermine the livelihoods of the poor (Adams *et al.*, 2004). The Durban Action Plan that emerged from the meeting contained targets to ensure that protected areas were established in full compliance with the rights of indigenous peoples and local communities (IUCN, 2003a), while the Durban Accord highlighted the need to involve those living near protected areas in their governance (IUCN, 2003b).

To promote the greater integration of local communities and other stakeholders into protected area governance, a set of good governance principles were proposed for adoption by protected area management agencies (Graham *et al.*, 2003; Borrini-Feyerabend, 2004), and the IUCN published guidelines promoting a transition to more participatory forms of governance. These included a matrix of four protected area governance models – governance by government; shared governance; private governance; and governance by indigenous peoples and local communities – that was incorporated into revised IUCN guidelines for applying protected area management categories (Borrini-Feyerabend, 2007; Dudley, 2008). Despite the publication of such guidelines, the conservation literature globally remains sparse when it comes to illustrations of how good governance principles and new governance models are implemented in real-world conservation, and we know little about the challenges faced by conservationists in applying these principles in the management of new protected areas.

During the same congress, the former President of Madagascar, Marc Ravalomanana, made his famous announcement to triple Madagascar's protected area coverage in order to reach the IUCN recommendation of each nation protecting 10 per cent of its territory (Freudenberger, 2010; see Chapter 9 by Corson). Implementation of the Durban Vision, as it became known, entailed radical changes to the way that protected areas were conceived and managed in the country, changes that paralleled the global trends that became apparent in Durban with the Durban Accord. In Madagascar, national-level priority setting exercises were carried out using distribution models for over 800 species of vertebrates, invertebrates and plants to determine where the new protected areas should be created (Kremen *et al.*, 2008; Rasoavahiny *et al.*, 2008), but most priority areas were in landscapes with large human populations that depend to varying extents on the use of natural resources for their livelihoods. As a result, the steering committees (hereafter referred to as the 'Durban Vision Community') established to advise on the implementation of the Durban Vision realised that the country's existing model of strict protected areas (IUCN Category Ia, II and IV, Table 10.1) would not be appropriate, and recommended that new protected areas should instead be created as multiple-use protected areas (IUCN Category III, V and VI, Table 10.1) in which sustainable natural resource use by local communities would be permitted (Freudenberger, 2010). The state's objectives for the protected area system also changed. While the established network of 46 sites were managed for biodiversity conservation as well as scientific research and recreation (Randrianandianina *et al.*, 2003), the objectives of the expanded system also included the conservation of Madagascar's cultural heritage and the sustainable use of natural resources for conservation and development (Commission SAPM, 2006). To accommodate the new vision and provide appropriate governance models for the new multiple-use protected areas, the IUCN's governance matrix was adopted, for the first time allowing community, private or shared governance of protected areas and promoting the adoption of good governance principles (Graham *et al.*, 2003; Table 10.2). Previously, all the country's protected areas had been managed by the state via the parastatal Madagascar National Parks (MNP), formerly *Association Nationale pour la Gestion des Aires Protégées* (ANGAP).

In order to oversee the management of the expanded protected area system (*Système des Aires Protégées de Madagascar*, SAPM), a Directorate (*Direction du Système des Aires Protégées*, DSAP) was established in 2003. The national legislation regulating protected areas – the *Code des Aires Protégées* (COAP) – was revised to permit new protected area categories and governance types and was submitted to the Senate in 2008. Although the legislation has not been ratified due to the political crisis of 2009, DSAP has continued its efforts to ensure that novel approaches are adopted to promote the integration of

Table 10.1 IUCN protected area categories and how they are being applied in Madagascar

IUCN categories	IUCN definition	Application in Madagascar
I (Strict Nature Reserve or Wilderness Area) Known in Madagascar as *Réserves Naturelles Intégrales* (RNI) and managed by MNP	Set aside to protect biodiversity and also possibly geological/geomorphological features, where human visitation, use and impacts are strictly controlled and limited to ensure protection of the conservation values.	Many Strict Nature Reserves have been re-categorised as National Parks (Category II) in order to allow for ecotourism for revenue generation. Others that still exist as Strict Nature Reserves include: Bemaraha (divided into a park and a reserve), Betampona, Lokobe, Tsaratanana and Zahamena (divided into a park and a reserve).
II (National Park) Known in Madagascar as *Parc National* (PN) and managed by MNP	Large natural or near-natural areas set aside to protect large-scale ecological processes, along with the complement of species and ecosystems characteristic of the area, which also provide a foundation for environmentally and culturally compatible spiritual, scientific, educational, recreational and visitor opportunities.	The 19 National Parks have specially focussed on the conservation of biodiversity with the promotion of recreation and tourism. These parks include: Andohahela, Andringitra, Ankarafantsika, Ankarana, Baie de Baly, Bemaraha, Isalo, Kirindy Mitea, Mananara Nord, Andasibe-Mantadia, Marojejy, Masoala, Midongy Befotaka, Montagne d'Ambre, Ranomafana, Tsimanampesotse, Tsingy de Namoroka, Zahamena and Zombitse Vohibasia.

IUCN categories	IUCN definition	Application in Madagascar
III (Natural Monument) Known in Madagascar as *Monument Naturel* Part of Durban Vision new protected area categories	Set aside to protect a specific natural monument, which can be a landform, seamount, submarine cavern, geological feature such as a cave or even a living feature such as an ancient grove. They are generally small protected areas and often have high visitor value.	The *Monument Naturel* is a new category as part of the Durban Vision and can be managed by a range of actors. In Madagascar, this category will be applied particularly to sacred forests (see later on Ankodida, a new protected area as part of Durban Vision).
IV (Habitat/species management area) Known in Madagascar as *Réserves Spéciales* (RS) and managed by MNP	Aim to protect particular species or habitats and management reflects this priority. Many Category IV protected areas will need regular, active interventions to address the requirements of particular species or to maintain habitats but this is not a requirement of the category.	The *Réserves Spéciales* were created specifically for conserving their unique biodiversity. *Réserves Spéciales* include: Ambatovaky, Ambohijanahary, Ambohitantely, Analamerana, Andasibe Analamazaotra, Andranomena, Anjanaharibe Sud, Bemarivo, Beza Mahafaly, Bora, Cap Sainte Marie, Ivohibe, Kalambatritra, Kasijy, Mangerivola, Maningoza, Manombo, Manongarivo, Marotandrano, Forêt d'Ambre and Tampoketsa Analamaintso.

Continued

Table 10.1 IUCN protected area categories and how they are being applied in Madagascar, *continued*

IUCN categories	IUCN definition	Application in Madagascar
V (Protected landscape/seascape) Known in Madagascar as *Paysage Harmonieux Protégée* Part of Durban Vision new protected area categories	Where the interaction of people and nature over time has produced an area of distinct character, with significant ecological, biological, cultural and scenic value, and where safeguarding the integrity of this interaction is vital to protecting and sustaining the area and its associated nature conservation and other values.	*Paysage Harmonieux Protégée* is a new Durban Vision Category and can be managed by a range of actors. According to Gardner (2011), possible examples of sustainable interactions in Madagascar are not common but could include the sclerophyllous scrub and alti-montane prairies of Andringitra, which are at least partly maintained by cattle grazing and fires, the fire-maintained Tapia woodlands of the central highlands, spiny forest on the Mahafaly Plateau managed by Mahafaly pastoralists. In addition, sacred forests, found in all three forest ecoregions (humid forest, dry forest and spiny forest) in Madagascar can also be included in this category. Furthermore, Virah-Sawmy (2009) demonstrated that even non-sacred forests, such as the littoral forest natural fragments, were relatively well-managed by Tanosy communities until recently when mining and migration changed the local dynamics. The sacred forest of Ankodida is an example of a new protected area under Category V Most proposed protected areas are likely to be of this category because the following Category VI, the other option, are for landscapes that are relatively intact, whereas the Durban Vision protected areas tend to be in areas with significant human interactions.
VI (Sustainable use area) Known in Madagascar as *Réserve de Ressources Naturelles*	Conserve ecosystems and habitats, together with associated cultural values and traditional natural resource management systems. They are generally large, with most of the area in a natural condition, where a proportion is under sustainable natural resource management and where low-level, non-industrial use of natural resources compatible with nature conservation is seen as one of the main aims of the area.	Known in Madagascar as *Réserve de Ressources Naturelles* and can be managed by a range of actors. Corridor Forestier Ambositra Vondrozo (COFAV) is the best example of this category in Madagascar. Most of the peripheral zones have community-based natural resource management (CBNRM) and a large part of the corridor is relatively intact. This category is particularly relevant for new marine reserves, including those managed by Madagascar National Parks (see Table 10.3).

Source: definitions from Dudley (2008).

Table 10.2 IUCN protected area governance types and how they are being applied in Madagascar

Governance types	IUCN definition	Application in Madagascar
Type A: Government Managed Protected Areas (state governance)	A government body (national or regional) holds the authority, responsibility and accountability for managing the protected area, determines its conservation objectives (such as the ones that distinguish the IUCN categories), develops and enforces its management plan and often also owns the protected area's land, water and related resources.	This is the case for all *Parcs Nationaux*, *Réserves Naturelles Intégrales* and *Réserves Spéciales* that are managed by MNP. A special decree allows the delegation of parks and above-mentioned reserves from the state to MNP to manage.
Type B: Co-Managed Protected Areas (shared governance) Part of Durban Vision new protected area governance type	Complex institutional mechanisms and processes are employed to share management authority and responsibility among a plurality of (formally and informally) entitled governmental and non-governmental actors.	Most of the proposed Category V and VI new protected areas will most likely fall under this governance type and will involve a multitude of actors such as the Ministry of Environment, the National Forestry Administration, Regional Forestry Administration and local associations with delegators (non-governmental organisations (NGOs) mainly and local stakeholders).

Continued

Table 10.2 IUCN protected area governance types and how they are being applied in Madagascar, *continued*

Governance types	IUCN definition	Application in Madagascar
Type C: Private Protected Areas (private governance) Part of Durban Vision new protected area governance type	Private governance comprises protected areas under individual, cooperative, NGO or corporate ownership. The setting of the area for conservation may be not-for-profit or for-profit. Typical examples are lands and resources acquired by NGOs explicitly for conservation purposes. Many individual landowners also pursue conservation objectives out of respect for the land and desire to maintain its beauty and ecological value.	Qit Madagascar Minerals (QMM) co-manages the new Protected Areas in the littoral forest with local communities as part of their biodiversity offset strategy.
Type D: Community Conserved Areas (community governance) Part of Durban Vision new protected area governance type	Natural and modified ecosystems including significant biodiversity, ecological services and cultural values voluntarily conserved by indigenous, mobile and local communities through customary laws or other effective means. Here authority and responsibility rest with communities through a variety of forms of ethnic governance or locally agreed organisations and rules. Rules generally intertwine with cultural or religious values and practices.	Distinctions between Community Conserved Areas and Co-Managed Protected Areas are not very easy because all protected area governance, other than those managed by Madagascar National Parks, has had to involve government authorities. However, in this chapter, we point out that some protected areas such as Ankodida with its network of sacred sites and a variety of clan governance and locally agreed organisations and rules falls more under this category of Community Conserved Areas.

Source: definitions from Borrini-Feyerabend (2007).

local communities into protected area governance. DSAP delegates the mandate of establishing new protected areas to delegators, usually NGOs, who follow SAPM guidelines to establish protected areas in a two-step process: temporary protected status was first granted to 29 sites (though more sites were prioritised for new protected areas but do not have temporary status), conferring protection against mining interests and timber exploitation; following this definitive protected area status is granted if legislative requirements are fulfilled. In order to acquire definitive status, delegators are expected to work closely with local communities to i) define or redefine protected area boundaries; ii) define and implement zoning of the protected area; iii) establish participatory governance structures integrating relevant stakeholders; and iv) develop a social safeguards policy to mitigate the social impacts of the protected area on affected communities. The delegators have mostly included the large environmental NGOs operating in Madagascar, including the WWF, CI and WCS, as well as a number of smaller Malagasy and international conservation organisations (see Table 10.3 for new Durban Vision protected areas with temporary status). The mining company Qit Madagascar Minerals, a subsidiary of Rio Tinto, is the only private company involved in the process of protected area creation as part of their biodiversity offset strategy. Although promoted by NGOs, most new protected areas have established governance structures that integrate local communities to various extents (Raik, 2007; Gardner, 2011): in the long term, SAPM aims for delegators to withdraw into more advisory roles once local management structures are functional and able to take on management responsibility.

Despite DSAP's efforts to ensure the greater integration of local community interests through the promotion of new governance types and multiple-use protected area categories, the expansion of the protected area

Table 10.3 New protected areas within the Durban Vision and the delegated institutions supporting them

Protected areas	Delegated institutions	Area (hectares)
Allée des Baobabs	Fanamby	320.42
Ambalabe	Missouri Botanical Garden (MBG)	3,118.16
Ambato Atsinanana (Sainte Luce)	Qit Madagascar Minerals (QMM)	1,310.83
Ambatotsirongorongo	Wildlife Conservation Society (WCS)	1,053.85
Amoron'i Onilahy	World Wide Fund for Nature (WWF)	158,194.66
Analalava	Missouri Botanical Garden (MBG)	224.85

Continued

Table 10.3 New protected areas within the Durban Vision and the delegated institutions supporting them, *continued*

Protected areas	Delegated institutions	Area (hectares)
Andreba	Wildlife Conservation Society (WCS)	30.00
Anjozorobe Angavo	Fanamby	52,298.07
Ankodida	World Wide Fund for Nature (WWF)	10,550.94
Antrema	Projet Pilote Bioculturel d'Antrema	20,646.63
Complexe de zones humides de la Mangoky	Asity Madagascar (with Birdlife)	221,907.22
Complexe Mahavavy Kinkony	Asity Madagascar (with Birdlife)	301,700.72
Corridor Forestier Ankeniheny Zahamena	Conservation International (CI)	369,909.87
Corridor Forestier Bongolava	Conservation International (CI)	113,097.99
Corridor Forestier Ambositra-Vondrozo	Conservation International (CI)	291,054.21
Forêt de Tsitongambarika	Asity Madagascar (with Birdlife)	60,335.33
Ibity	Missouri Botanical Garden (MBG)	5,960.56
Lac Alaotra	Durell Wildlife Conservation Trust (DWCT)	46,827.37
Loky-Manambato	Fanamby	248,425.18
Makira	Wildlife Conservation Society (WCS)	372,179.41
Mandena	Qit Madagascar Minerals (QMM)	230.51
Menabe Antimena	Fanamby	211,147.08
Mikea	Madagascar National Parks (MNP)	184,639.56
Montagne des Français	SAGE Madagascar	6,113.44
Nord-Ifotaky	World Wide Fund for Nature (WWF)	22,280.59
Nosy Hara (Marine Reserve)	Madagascar National Parks (MNP)	125,522.62
Nosy Tanikely (Marine Reserve)	Madagascar National Parks (MNP)	178.84
Ranobe PK 32	World Wide Fund for Nature (WWF)	168,500.24
Tampolo	ESSA Forêt (University of Antananarivo)	674.61

Note: all these protected areas are still in the establishment phase and have only temporary protected status.

system, like the wave of CBNRM that preceded it (see Chapter 8 by Pollini *et al.*), has been criticised as lacking in true participation (see Chapter 9 by Corson). Corson (Chapter 9) also provides evidence that the Durban Vision itself represents an international biodiversity conservation agenda rather than a domestic one (see also Duffy, 2006), and this dichotomy of interests characterises both the implementation of CBNRM and the establishment of new protected areas: in both cases it is argued that these new forms of land management are implemented by outside interests who have different objectives to the rural communities whose customary land is affected, and the imposition of new governance structures can have major impacts on the social dynamics of the communities involved (see Chapter 8 by Pollini *et al.* and Chapter 9 by Corson).

In this chapter, we explore the establishment of participatory governance structures for new protected areas from a practitioner's perspective. Using two case studies of new protected areas established as part of the Durban Vision in southern Madagascar, we explore the application of good governance using a synthetic framework adapted from Graham *et al.* (2003) and Lockwood (2010), incorporating five good governance principles: i) legitimacy; ii) inclusiveness; iii) fairness; iv) accountability and transparency; and v) direction and effectiveness. As conservation practitioners closely involved with the establishment of the case study protected areas since the launch of the projects in 2005, we offer a practitioner's perspective on the challenges of implementing participatory governance for new protected areas as a complement to academic critiques and a contribution to the development of best practice. We first provide a brief overview of the history of protected areas in Madagascar, illustrating a transition from strictly protected, centrally governed protected areas to the multiple-use, participatory governed sites that characterise the Durban Vision. We then present the two case studies of Ankodida and Ranobe PK32; two sites selected because of our personal experience of their establishment and management, and because their radically different social contexts help illustrate the different challenges faced by protected area delegators. Finally, we explore the case studies in greater detail through the lens of the five good governance principles.

Protected areas in Madagascar

Protected areas (Box 10.1) in Madagascar, as in many parts of the world, represent the cornerstone of conservation efforts as they aim to achieve the long-term conservation of nature with associated ecosystem services and cultural values (Dudley, 2008). In Madagascar, protected area coverage has increased steadily since their first establishment in 1927 in order to fully represent the diversity of species and ecosystems on the island (see Kremen *et al.*, 2008) with associated ecosystem services. This expansion can be viewed in three phases: i) the creation of reserves during the colonial period (1896–1960) – see also Chapter 6 by Scales; ii) the

Box 10.1 Definition of terms concerning protected areas

- Protected areas are defined as 'clear geographical spaces, recognised, dedicated and managed, through legal or other effective means, to achieve the long term conservation of nature with associated ecosystem services and cultural values' (Dudley, 2008, pp8–9).
- Protected area categories refer to the management objectives of the protected areas (Table 10.1). The categories system is international but national labels for protected areas may vary and a gradation of human intervention is implied in the six categories (Table 10.1).
- Protected area delegation in the context of the Durban Vision refers to the act of giving another party (private and NGOs) the responsibility of carrying out the process of protected area establishment agreed to in a contract with the government (Ministry of Environment). The delegator is the party who assumes the responsibility of performing this duty.
- *Dina* are traditionally sets of local rules or social norms that govern behaviour and resource use in rural communities. Under CBNRM contracts, they may be legalised and gain the status of by-laws, and thus are often used to regulate natural resource use in management transfers and protected areas.
- Protected area governance refers to the degree to which protected area decision-making practices and structures follow fair, legitimate, inclusive and other principles across an array of different protected area management types and categories (Graham et al., 2003). The governance types is based on 'who holds de facto decision making authority and responsibility for the area and resources in question or, more simply, who decides how the area and resources are to be managed towards what specific purpose/management category' (Borrini-Feyerabend, 2007, p8) (see Table 10.3).
- Protected area zoning divides the protected area into units (e.g. buffer zone, recreation zone, sustainable use zone, core conservation zone, etc). These units represent a range of restrictions and possibilities, according to the ecological and social specificities, in order to represent the management objectives (the category) and prevent compromises to the protected area conservation objectives. In Madagascar, the buffer zones of the new protected areas are often under CBNRM contracts.

expansion of national parks in 1990–2003 during the implementation of the National Environmental Action Plan (NEAP) – see also Chapter 7 by Kull; and iii) the expansion of protected areas with different governance models and categories from 2004 as part of the Durban Vision. These three phases saw Madagascar's protected area coverage expand from 450,000 hectares in the 1980s (Nicoll and Langrand, 1989) to almost six million hectares (10 per cent of land area) in 2010, of which 2.65 million hectares will be managed by MNP while 3.25 million hectares would be designated for community governance or shared governance with local communities (Freudenberg, 2010).

The first phase of protected area establishment in Madagascar had been inherited from the colonial park system in France (the protected areas were either *Réserves Naturelles Intégrales* or *Réserves Spéciales*; Table 10.1). The first 10 protected areas were established in 1927, before the signing in 1933 of The Convention Relative to the Preservation of Fauna and Flora in the Natural State. These included Betampona, Masoala, Zahamena, Tsaratanana, Andringitra, Lokobe, Ankarafantsika, Namoroka, Bemaraha and Tsimanampetsotsa, and were followed by the creation of two more reserves, Andohahela in 1939 and Marojejy in 1952, and the first national park of Montagne d'Ambre in 1958 (Nicoll and Langrand, 1989). The network of protected areas in Madagascar established during the French colonial period was not intended to conserve biodiversity specifically, but rather reserves were established in remote areas of outstanding beauty. They were essentially spectacular landscapes but also 'paper parks', in the sense that they lacked vision and goals, as well as basic management requirements to operate effectively.

After independence, protected areas were not on the agenda as Madagascar began to replace its colonial rules with more broad-reaching policies to increase and modernise agricultural production in tandem with less restrictive environmental protection policies – see also Chapter 7 by Kull. Few protected areas were created (e.g. Isalo National Park in 1967 and the Special Reserve of Beza-Mahafaly in 1975). During this period, however, intense negotiations between the government, the IUCN, academic institutions and NGOs, especially in the 1970s and the 1980s, revived the need for expanding the network of protected areas in Madagascar and making existing protected areas more effective. The second phase of protected area expansion (1990–2003) therefore focused on the establishment of a more effective and extensive park system (Freudenberger, 2010), with the creation of ANGAP and partnerships with international organisations. ANGAP was created as a non-profit parastatal association managing protected areas on behalf of the Malagasy people; a decree delegated the management of Madagascar's protected area network to ANGAP. During this phase, new national parks continued to be established with the objective of conservation and revenue generation through tourism, often through the expansion of existing *Réserves Naturelles Intégrales* or *Réserves Spéciales* and their re-categorisation as national parks. These included Andasibe-Mantadia (1989), Ranomafana (1991), Andohahela (1997), Masoala

(1997), Zahamena (1997), Tsingy de Bemaraha (1997), Andringitra (1999), an expanded Isalo (1999), Masoala (1999) and Marojejy (1999).

As a newly created agency, ANGAP coordinated the management of 44 protected areas that included ten National Parks and 34 reserves using funds from USAID (Freudenberger, 2010). With USAID support, Integrated Conservation and Development Projects (ICDPs) were introduced with the aim of compensating local communities on the periphery of National Parks for the restricted use of natural resources. These included programmes to develop sustainable livelihoods, as well as education and health programmes. National parks that benefited from ICDPs included: Andasibe-Mantadia, Ranomafana, Montagne d'Ambre, Masoala, Andohahela, Zahamena, and Isalo (Freudenberger, 2010). In addition to livelihood support, these projects helped the parks have an effective field presence and significant facilities aimed at managing the park and encouraging ecotourism for revenue generation in order for the parks to become more effective at achieving their management goals.

There are currently 19 National Parks, six *Réserves Naturelles Intégrales* and 23 *Réserves Spéciales* in Madagascar. Six of these (Isalo, Andasibe-Mantadia, Ranomafana, Ankarana, Montagne d'Ambre and Tsingy de Bemaraha) attract more than 30,000 visitors annually. Overall, visitation increased nearly 30-fold since the start of the NEAP, from 12,000 people in 1984 to 345,000 in 2008 (Freudenberger, 2010). Tsingy de Bemaraha National Park was further listed as a World Heritage Site in 1990, followed in 2007 by a cluster of six eastern rainforest sites known as the Rainforests of Antsinanana World Heritage Site (Marojejy, Masoala, Zahamena, Ranomafana, Andringitra and Andohahela).

In establishing one of the most extensive networks of protected areas in Africa, Madagascar has experienced issues with social welfare. From a human rights perspective, these parks were established with little regard to the resource requirements of adjacent communities (Durbin and Ralambo, 1994) as their primary management objective was the conservation of biodiversity and generating revenues from ecotourism. The approach during the second phase of protected area expansion was to implement ICDPs in order to compensate for restrictions imposed by the establishment of the protected areas. These policies were seen as a 'dramatic shift away from the State's role in environmental management primarily in terms of exclusion and policing, as it had been since colonial times' (Freudenberger, 2010, p7) and, in many ways, these policies were indeed a radical step from contemporary global conservation practices when also compared to practices in Africa at the time (see West *et al.*, 2006). ICDP approaches at the time represented a more balanced approach to achieving social goals and the conservation of biodiversity. Unfortunately, the design of ICDPs was flawed as it did not acknowledge the mutually incompatible interventions between conservation and development, and the approach has largely been abandoned globally. Instead, many now advocate for either conservation projects with development (the goal here is to conserve the ecosystem and to reduce resource-dependency) and development projects with conservation (the goal is to contribute to social wellbeing that is compatible with conservation). As USAID developed different approaches, MNP became increasingly reliant on funds from the World Bank and the *Kreditanstalt für Wiederaufbau* (KfW), as well as the newly set up Madagascar Foundation for Protected Areas and Biodiversity, which was created in 2005 and is widely recognised as a model and anchor of sustainable finance for Madagascar's protected area network.

The third phase of protected area expansion – the 'Durban Vision' – is based on the recognition that for conservation to be successful over the long term, local uses must be integrated within protected area management. Furthermore, in order to enhance protected area ownership, local stakeholders must be integrated into governance structures. In the next section, we explore the establishment of participatory governance structures using a synthetic framework adapted from Graham *et al.* (2003) and Lockwood (2010) incorporating five good governance principles through our personal experiences of establishing two new protected areas as part of the Durban Vision in southern Madagascar.

Case studies of protected area expansion: are the governance structures of new protected areas within the Durban Vision legitimate, inclusive, fair, accountable, transparent and effective?

Ankodida and Ranobe PK32 are two new protected areas in southern Madagascar that were granted temporary protected status in 2008. Both were established by the WWF to conserve priority areas of the spiny forest, a global priority ecoregion (Olson and Dinerstein, 1998). Both are designated as multiple-use protected areas (Ankodida is proposed as Category V, Ranobe PK32 is proposed as Category VI) and are primarily zoned for sustainable natural resource use by local communities. In both landscapes, WWF has followed SAPM guidelines in the development of governance structures designed to integrate the interests of local communities, but the resulting governance models differ as a result of the radically different social contexts and histories of the two landscapes. The two case studies thus represent two different models of new protected areas within the Durban Vision, one in which governance structures are built upon CBNRM management transfer contracts, and one in which they are not.

Ankodida is a recently established, community-managed protected area that is proposed as Category V because of the human–nature interactions that have succeeded in conserving the forest over recent history (Figures 10.1 and 10.2). The landscape is dominated by a sacred forest, the former home of a pre-colonial Tandroy king that also shelters spirits that play an important role in the spiritual life of the Tandroy people (Gardner *et al.*, 2008). The protected area builds on the strong customary institutions and willingness of the local people to conserve the sacred forest. It is a relatively small protected area (10,744 hectares), with a strict conservation

Figure 10.1 A view of the sacred forest on a hill in Ankodida protected area

Source: photograph by Johannes Ebeling.

Figure 10.2 A local meeting with staff from the WWF for the establishment of Ankodida protected area

Source: photograph by Johannes Ebeling.

zone comprising the sacred forest at its centre (designated as a Category III Natural Monument within the wider Category V landscape), buffered by 8,000 hectares under CBNRM (Figure 10.3). Within the CBNRM areas,

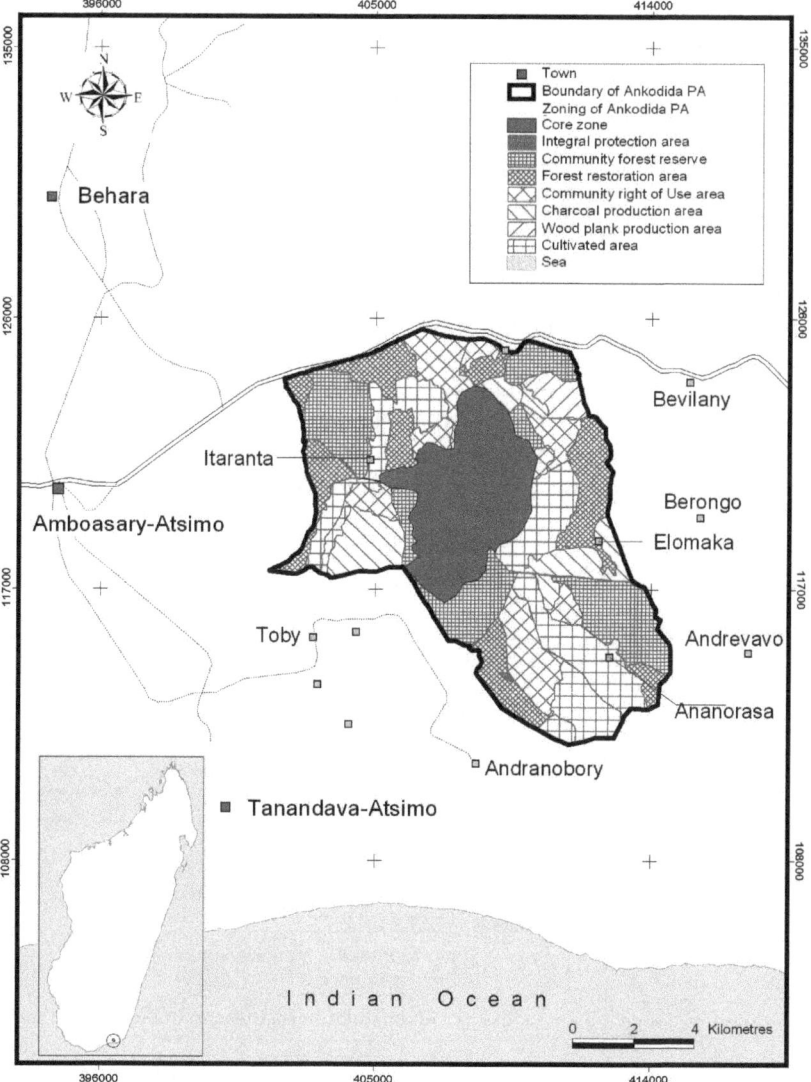

Figure 10.3 Protected area zoning of Ankodida.

Source: created by Anjara Andriamanalina, WWF.

which comprise six CBNRM contracts, there are also smaller sacred forests of various types (where ancestral tombs are located and ceremonies of different kinds are held) that have been designated as the conservation zone of each of the six CBNRM zones (Figure 10.3). Each management committee of the six CBNRM associations is responsible for management of their territory, while the core conservation area of the protected area as a whole is governed by a union, which is made up of the elected representatives of the six CBNRM associations and respected figures from local communities in charge of conflict resolution – this includes local notables and clan leaders in the community (Figure 10.4). The six CBNRM contracts were established prior to creation of the protected area, and their establishment did not supersede existing rules on resource use, in contrast to the case studies reported by Corson (Chapter 9). WWF's efforts since 2006 have focused on strengthening institutions and building capacity of the CBNRM associations and the union, with the objective of the union becoming the legal manager of Ankodida in the long term.

Ranobe PK32 comprises a large (163,000 hectare) landscape of limestone plateaux and coastal sands between the Fiherenana and Manombo rivers in southwest Madagascar (Figure 10.5). The richest area for faunal biodiversity in the region, it was largely uninhabited before 2000, with populations concentrated along the coast and in the two river valleys. As a result, the

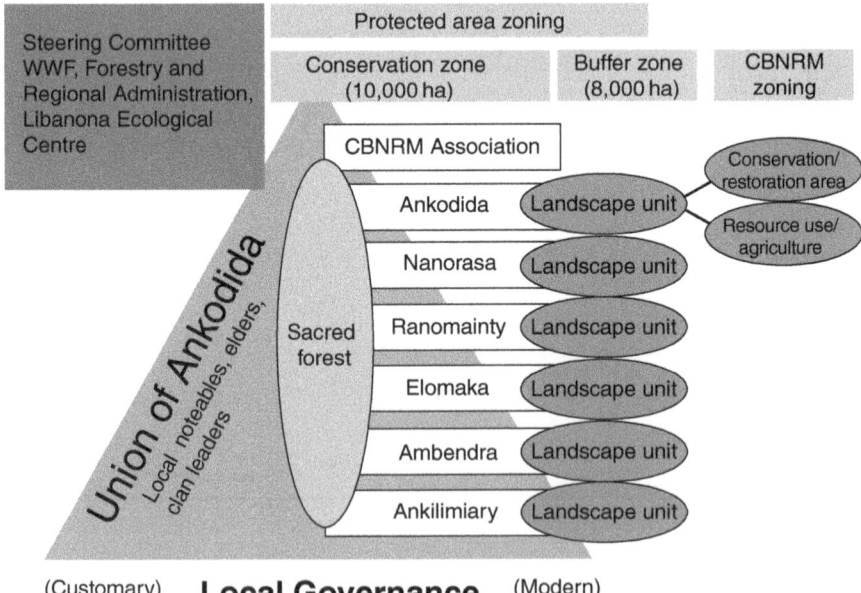

Figure 10.4 Governance structure of Ankodida protected area as a hybrid between traditional and modern forms of governance

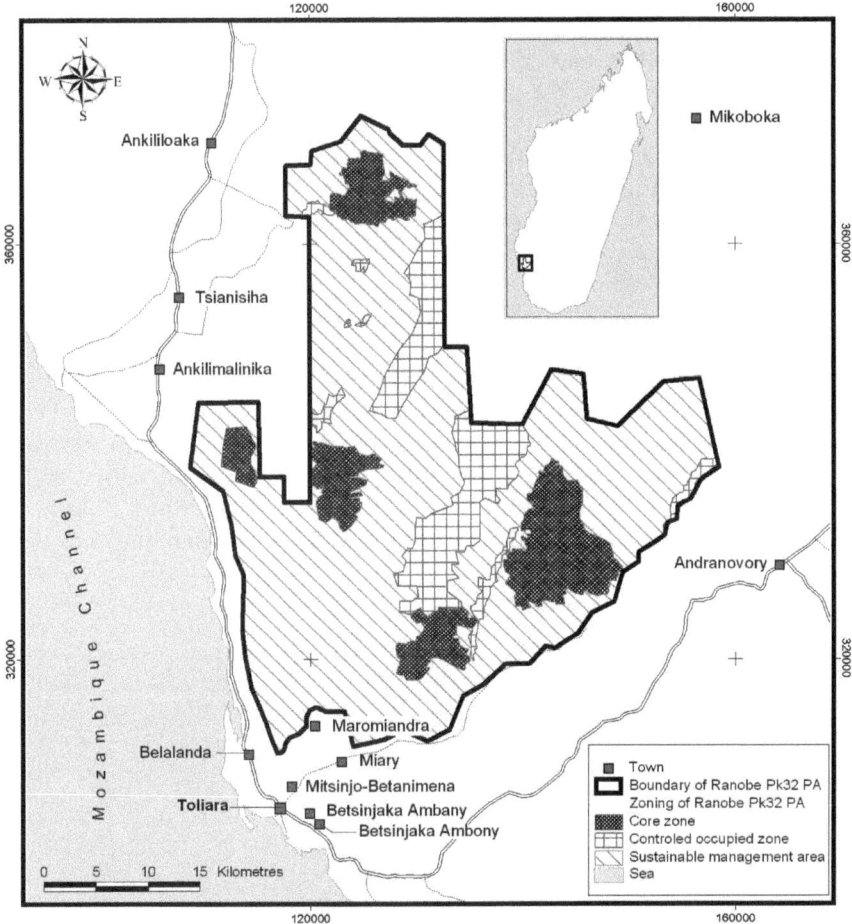

Figure 10.5 Protected area zoning of Ranobe PK32 showing the conservation zone of the protected area, covering only 13.5 per cent of the protected area, with the rest dedicated to sustainable use and commercial exploitation of forest resource

Source: created by Anjara Andriamanalina, WWF.

majority of the landscape had no recent settlement history and no strong customary institutions, in strong contrast to Ankodida. The predominantly Masikoro population who settled in the river valleys over the past centuries have welcomed migrants in more recent times, especially Tandroy migrants who played an important role as agricultural labourers during the cotton boom along the irrigated plains of the Manombo River in the 1970s and early 1980s. When the cotton boom ended, many of these migrants then turned to charcoal production as a result of increasing demand from the city of Toliara. More recently, strong demand for maize as an export crop (Casse *et al.*, 2004) and cyclical droughts encouraging emigration from the far south (Fenn and Rebara, 2003), large numbers of Mahafaly and Tandroy settlers have migrated to the area to practice swidden agriculture (*hatsake*) in unsettled areas of the limestone plateaux (Figures 10.6 and 10.7). Some members of resident communities from within the landscape have also taken up swidden agriculture because the degradation of regional irrigation infrastructure has reduced the productivity of their permanent fields (see also Chapter 5 by Scales for more on the drivers of deforestation and the socio-economic dimensions of swidden cultivation).

When the protected area project was launched in 2006, an inter-communal association (known as MITOIMAFI) was created to group the mayors and councilors of the eight rural communes of the protected area (Figure 10.8). These elected local authorities were chosen as representatives of local communities because, unlike in Ankodida, the landscape was not managed through CBNRM management transfer contracts and no other formal institutions existed apart from those of the State. Over time, however, it became clear that this structure was ineffective in controlling *hatsake* or

Figure 10.6 A view of the extent of *hatsake* (swidden agriculture for maize cultivation) in the forest frontier of Ranobe PK32 from the air

Source: photograph by Xavier Vincke.

Figure 10.7 View of on-the-ground deforestation for maize cultivation, also showing migrant shelter

Source: photograph by Xavier Vincke.

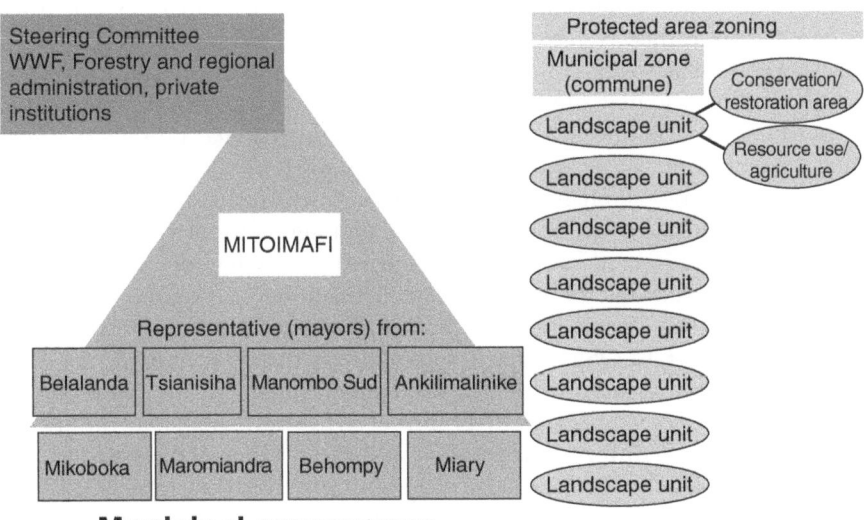

Figure 10.8 Governance structure of Ranobe PK32 in 2006 based mainly on municipal governance

charcoal production and the governance structure of the protected area was revised following a series of village-level consultations. As of 2012, each commune within the protected area is represented by a number of local units (Figure 10.9).

Principle 1: legitimacy

Legitimacy is 'the acceptance and justification of shared rule by a community ... [and] concerns who is entitled to make rules and how authority itself is generated' (Bernstein, 2005, pp142–143). For protected area governance 'legitimacy is a key factor in the ethical acceptability of governance arrangements' (Lockwood, 2010, p758). The legitimacy of governing bodies can be earned in different ways: first, it can be earned 'through their efforts at leadership, through effectiveness at producing outcomes or by generating consensus around a vision' (Lockwood, 2010, p758); this is termed earned or output legitimacy. Furthermore, many indigenous and local communities also earn governance legitimacy through their long-standing connection to particular places. On the other hand, government agencies are legitimised because they have authority indirectly conferred on them through legislation enacted by higher tiers of government (Lockwood, 2010).

Legitimacy is perhaps the most problematic and controversial area of concern for the Durban Vision protected areas. This is not a matter deliberately ignored by institutions establishing new types of protected areas in Madagascar, but rather there is a degree of confusion over legitimacy. Prior to the Durban Vision, no other formal options for conservation existed other than either establishing national parks or through CBNRM. Most areas of importance for biodiversity were placed under CBNRM, which was viewed as less costly to implement than protected area expansion. As a result, many of the new sites for protected areas, such as Ankodida, overlap with zones already under CBNRM, and the governance models of these sites are composed of networks of CBNRM management structures.

By integrating established CBNRM actors into protected area governance and building consensus about the management of these sites, the legitimacy of protected area governance structures has been considered by protected area delegators and state institutions as genuinely earned. However, the legislation for CBNRM in Madagascar, known as *Gestion Locale Sécurisée* (GELOSE), was designed for extractive natural resource management rather than protected area governance and the conservation of biodiversity (see Chapter 8 by Pollini *et al.*). There is a clear difference between these two different precepts that is often not well understood and distinguished by the Durban Vision community. Furthermore, GELOSE has suffered its own problems with legitimacy, in particular with regards to the new local resource management institutions that it created (see Chapter 8).

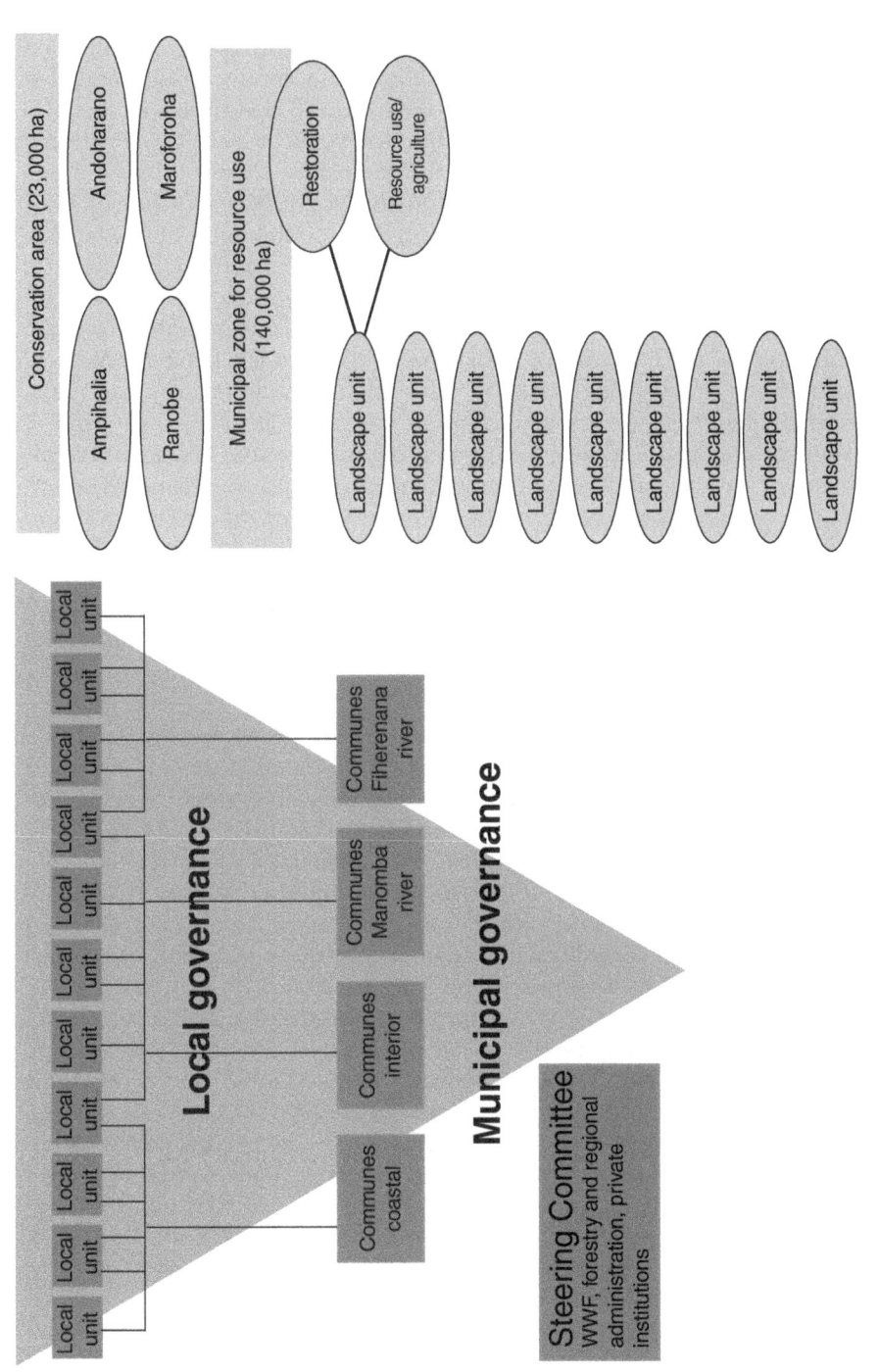

Figure 10.9 A new governance structure for Ranobe PK32 in 2012 to ensure more representation of local actors as a hybrid between local and municipal governance

In Ankodida, locally agreed rules over sacred forests have existed for many years thanks to a multitude of actors that have maintained and enforced those cultural traditions. This includes local notables and traditional clan leaders in the community who enforce the rules and traditional healers who use the sacred forest for rituals and therefore play a central role within a society whose belief systems centre on the respect for ancestors and the prescription and respect of taboos (*fady*) (Gardner *et al.*, 2008; see Chapter 14 by Kaufmann for more on taboos and resource use). The union of Ankodida – the governing body of the protected area – was therefore designed to build on the existing traditional structures and include not only the CBNRM associations but also traditional clan leaders and local notables from the community to form a type of hybrid local governance between modern and traditional rules (Figure 10.3). This synergy between protected area governance and customary management was possible because the role of customary management in the conservation of the sacred forests was clear, the actors favoured protected area management and therefore protected area governance could easily build on these existing rules. Furthermore, many of the local notables and clan leaders also formed part of CBNRM associations, in contrast to situations described by Pollini *et al.* (Chapter 8) where associations bypassed traditional institutions. In this way, designation of the sacred forests as a protected area helped enforce the traditional management of the area (Fritz-Vietta *et al.*, 2011; see also later 'direction and effectiveness').

The degree of congruence between the goals of protected area establishment and customary management of Ankodida's sacred forests is, however, an exception rather than a general rule. In most non-sacred forests, customary landowners and new settlers see large extensive forest as a land reserve for agriculture or a legitimate way to make a claim to land (see also Chapter 5 by Scales). Swidden agriculture practised by customary landowners and settlers can take different forms, ranging from relatively small changes in forest composition and structure to the total removal of forest cover over large areas. Customary land management matters in Madagascar because 80 per cent of land is acquired by the traditional system according to lineage or family relations (Casse *et al.*, 2004) or based on the chronology of clearing in newly settled forested areas (Muttenzer, 2006); only 20 per cent obtain a legal land title (Casse *et al.*, 2004).

Reconciling customary land management practices and conservation poses a significant dilemma for protected area governance, given that rules are constrained by government legislation that has outlawed pasture burning and swidden agriculture since before the colonial era (Kull, 2003; Raik, 2007; see Chapter 6 by Scales). How, therefore, should protected area governance position itself with regards to practices that are not accepted by the state but that are seen as legitimate forms of land use by local communities, particularly when the rules of the state are themselves not accepted as legitimate (Horning, 2003)? As argued by Muttenzer (2006), devolving management of forests to village associations whose members are often also the customary owners of the land raises as many new questions as it can possibly answer. This is even more the case with legitimacy in protected area governance.

Given Lockwood's (2010) assertion that rural communities can earn governance legitimacy through 'their long-standing connection to particular places', the legitimacy of land claims by recent settlers and migrants, with no historical attachment or claims to the land they earmark for deforestation, is debatable. Such is the situation at forest frontiers but, regardless of their legitimacy, the legitimacy of other local actors such as customary landowners is often not well appreciated. In the case of the forest frontier in Ranobe PK32, its recent settlement history, combined with its position as a centre for migration, meant that the development of strongly enforced traditional rules over forest resources was not possible. WWF opted for the creation of a political entity – an inter-communal association called MITOIMAFI – made up of elected representatives (mayors and councillors) from eight administrative communes to help with its establishment. Six years into the process of establishing the governing structures, it has become apparent that the inter-communal association created more discord over natural resource governance than consensus. Here, perhaps ensuring that customary landowners, notables and village elders and representatives of different user groups come together to decide on new rules, and manage them, will ensure greater input into governance. This may perhaps lead to a more successful blend of the strength of the state and on-the-ground institutions. Although there may be no guarantee that policy negotiations result in win–win scenarios, it may nevertheless smooth the path toward consensus in situations where there are large incompatibilities in the interests, values or priorities of different stakeholders.

Principle 2: inclusiveness

Inclusiveness refers to 'the opportunities available for all stakeholders to participate in and influence decision-making processes and actions' (Lockwood, 2010, p760). This concept stems from the 'ethical understanding that each person has an equal right to have a say in matters that affect her or his life' (Lockwood, 2010, p760). Inclusiveness is critical as it provides a platform to clarify and mediate conflict over diverse interests and values, with enhancing public ownership and commitment to solutions and access to many different perspectives and kinds of knowledge (Pimbert and Pretty, 1997).

Inclusive of *whom* is a pertinent question for each protected area, since communities are made up of diverse sets of interests. The two case studies have involved public consultation with all villages within the boundaries of the protected area. The aim of public consultations in villages has been to ensure a clearer understanding of the range of interests and values in order to re-define the boundaries, management rules and zones of the protected area in an open, participatory way. Public consultation has sadly not been the best medium to foster inclusiveness as the latter remains highly problematic in the cultural context of Madagascar where traditional patterns of authority, royalty, reciprocity and social bonds prevent open debates over sensitive issues that affect land management. A good social science background and understanding of social dynamics is needed to clarify diverse interests and values (see Chapter 14 by Kaufmann). Often, conservation agencies are not the best placed to study these social dynamics.

Faced with this challenge, improved processes for fostering inclusiveness are being trialled by WWF. In Ankodida and Ranobe PK32, this includes, for example, having a range of small focus groups (e.g. different user groups, women's and migrant groups) to debate issues over protected area management in smaller land-management units with social scientists invited to be part of these focus groups to avoid conservation agenda biases. The use of focus groups has led to different zoning of Ranobe PK32 compared to sessions during public consultations, with 13.5 per cent of the area dedicated to conservation and the rest to sustainable use. Similarly, in Ankodida, though the zoning was generally considered adapted to the local context by all of the smaller focus groups, it became clear that a set of different rules applicable to each zone of the protected area was needed, as previously only one rule had been formulated in the social contract (*dina*) between the communities and the union concerning the protected area.

Lastly, although new settlers and migrants may not represent legitimate groups as part of the governing structures, nevertheless they should be included during negotiations in the phase of protected area establishment. For example, in Ranobe PK32 consultation with migrants a few years following the project to establish the protected area revealed that they were often employed by wealthy customary landowners to practise swidden agriculture. Large-scale deforestation and a lack of on-the-ground consensus was therefore not just a migrant problem as previously thought. This pre-conception would have been avoided had migrants been consulted at the initial state on an individual basis given the sensitivity of this type of information.

Principle 3: fairness

In the context of protected area governance, fairness is a multi-faceted principle. It includes the respect and attention given to the views of stakeholders, and respect between higher- and lower-level authorities, as well as the recognition of human and indigenous rights and the intrinsic value of nature (Lockwood, 2010). Lockwood (2010, p760) suggests that it is particularly those charged with advancing protected area governance that need to be fair in the exercise of the authority conferred on them, particularly in relation to the distribution of power, the treatment of participants, recognition of diverse values, consideration of current and future generations, and the development of mechanisms to share costs, benefits and responsibilities of decision making and action.

We focus here on the development of mechanisms that delegators are putting into place to share the cost and benefits of establishing a protected area.

Sustainable extraction is permitted in the buffer zones of Madagascar's new IUCN Category V and VI multiple-use protected areas, which make up the bulk of the sites and are often managed through CBNRM mechanisms. This ensures the integration of local uses into protected area management, and has therefore been a key contributing factor to fairness in the provision of natural resources for current and future generations. The forests in our two case studies provide a broad range of resources that contribute to subsistence and household income, including food (fruit, honey, bushmeat and wild tubers such as *ovy* (*Dioscorea* spp.) which are especially important during years of drought or in normal years towards the end of the dry season when household food reserves are running low (Cheban *et al.*, 2009; Scales, 2012)); supplies of firewood; building materials; grazing and shelter for zebu cattle in the dry season (Kaufmann and Tsirahamba, 2006); and traditional medicines. In Ankodida, the CBNRM areas have dedicated zones for resource extraction for local uses, as well as areas set aside for the commercial production of charcoal and wood planks (from *Alluaudia procera*, Didiereaceae) extracted and sold at a nearby urban market (Amboasary Sud).[1] Similarly, in Ranobe PK32 the new zoning will ensure sustainable extraction is permitted in 86.5 per cent of the protected area, the entire surface except for five strict conservation zones (Figure 10.5). This is in contrast to Madagascar's first generation of state-managed protected areas where conservation zones form the largest part of the protected area.

In addition, DSAP has developed a strong legal framework for protected area governance, including a social safeguard policy for each protected area. The safeguard policy is intended to ensure that the costs of protected area establishment, in terms of access restrictions, are compensated through direct livelihood projects or in monetary form, although attempts to calculate the financial costs of restrictions at the level of individual affected households have not been made due to the methodological and logistical complexity of the task. Rather, the social safeguard policy serves as a good framework to help delegators understand the vulnerability of different household types to restrictions as a result of protected area establishment, and identify potential projects that would directly alleviate the impacts of restrictions (for example, the provision of improved poultry types to compensate for restrictions on bushmeat harvesting). However, delegators can rarely support all identified projects, neither can they compensate all individual households across the protected area landscape for income losses incurred as a result of restrictions (in monetary form or through development support), because the real cost may amount to millions of dollars. Unfortunately, funding on this scale is rarely available to delegators of new protected areas, particularly since ICDP approaches have fallen out of fashion, though we believe that better designed ICDPs are critical for these protected areas. It is noteworthy, nonetheless, that 40 per cent of WWF funds in Madagascar is dedicated to livelihood programmes, although these investments remain insufficient given the scale of the issue and the landscape (WWF, 2010). Direct conservation payments also offer a potential solution to compensating local communities for loss of access, although their application in politically unstable Madagascar may be problematic, compounded with high levels of corruption. Furthermore, lessons learnt from Brazilian Amazonia show that schemes that closely align payments with opportunity costs are not necessarily more equitable in outcomes, due to the fact that those who traditionally account for high deforestation (e.g. wealth customary landowners) reap the greatest benefits from payments (Börnor et al., 2010).

The large-scale deforestation that occurs in Madagascar (Harper et al., 2007) rarely lifts rural communities out of poverty; despite hundreds of years of continuous forest clearance, Madagascar remains one of the poorest countries on earth. The profits from large-scale forest clearance have often gone to elites, both Malagasy and foreign (see Chapter 5 by Scales). In many cases, for example at the forest frontier of Ranobe PK32, deforestation is encouraged and financed by wealthy customary landowners (Virah-Sawmy and Vincke, in review). In the long term, however, the landscape becomes depleted of natural resources, ecosystem services dry up and the rest of the rural population are left without the critical safety net that natural resources represent, causing them to slip further into destitution or migrate elsewhere, re-creating similar scenarios in relatively undisturbed forest landscapes in other regions. As such, the problem is displaced in time and space, but never solved. To break this cycle, multiple-use protected areas are designed to provide the safety net of continued, sustainable resource extraction, while catalysing ideally the transition from extensive swidden to more intensive, productive forms of agriculture through the provision of financial and technical support.

Principle 4: accountability and transparency

Accountability and transparency concerns the way decisions are taken. Specifically, accountability concerns the extent to which a governing body is answerable for those decisions to its constituency and also answerable to higher-level authorities (Lockwood, 2010). In accountability, the allocation and acceptance of responsibility for decisions and actions is critical and needs to be appropriate to the institutional levels that best match the scale of issues and values being addressed, especially within co-management structures (Lockwood, 2010). Transparency is the degree to which decision-making processes are made visible, the clarity with which the reasoning behind decisions is communicated and the ready availability of relevant information about a governance authority's performance (Lockwood, 2010).

With new protected areas, some mechanisms exist for upward accountability and transparency to higher authorities in the form of reporting, steering committees, visits and evaluation. For example, delegation contracts are renewed every two years subject to the evaluation of delegators according to their compliance with new national legislation. In reality, however, assessing whether delegators are doing a good job of establishing these protected areas in terms of the management and governance put into place, according to social, ecological and economic criteria, is extremely complex and rarely carried out. Here, steering committees established for each protected area can help delegators deliver better governance outcomes. For example, the steering committee of Ankodida has included the Libanona Ecology Centre, together with regional authorities and state agencies.

Parallel accountability is also important for shared governance and it involves the sharing of power and responsibilities between the delegators, the government and the local governing structures being put into place (Berkes, 2004). For example, how do delegators themselves ensure that they do not 'interfere with local regulations while defending their own logic and interest' (Bérard, 2011, p110; see also Chapter 8 by Pollini *et al.*)? The participation of independent experts in the steering committees offers a powerful way to improve practices.

Accountability to rural households through established governance structures is perhaps the most critical issue, but is so far limited in scope, even in the Ankodida community protected area. Some downward accountability exists for CBNRM associations. This is based on established rules, on which they are evaluated by the Forestry Administration every three or ten years, and requires these associations to convene annual meetings of members during which conflicts and grievances are discussed and resolved. In addition, they also have to hold an annual election of their management committee, although it is known that the associations are not completely democratic, and oversight and support is needed to achieve this aim. More effective downward accountability will be facilitated with improving more democratic processes within CBNRM associations, together with ensuring the representation of elected members of other user groups (pastoralists, for example, are not represented) within networks of small landscape units within the governance structure. These elections would not be based on who holds power or who can write (a common critique of CBNRM), but who best represents the views of the user groups in small landscape units. This strategy would be novel and is being trialled in Ranobe PK32 and requires delegators to view protected area governance beyond representations of only CBNRM actors or local government, though both remain important actors.

Principle 5: direction and effectiveness

The effectiveness of protected areas in achieving both conservation and social goals is a source of intense debate globally. This is largely because it is methodologically complex to measure the effectiveness of protected areas since the assessment needs to control for the inherent biases that arise from biophysical and socio-economic factors, given that protected areas are generally situated on marginal lands with lower pressure for deforestation than non-protected areas (Andam *et al.*, 2008, 2010); therefore direct comparisons between protected and non-protected areas are flawed, as the results would show lower deforestation in protected areas than densely populated areas. Worldwide, conservation biologists have focused on the effectiveness of protected areas in maintaining forest cover, with increasing evidence (even when inherent biases were controlled for) suggesting that protected areas do reduce deforestation and the incidence of fire (see Andam *et al.* (2008) for Costa Rica; Nelson and Chomnitz (2011) for Latin America, Africa and Asia). No studies have controlled for such biases in Madagascar. The evidence that is available suggests that protected areas have suffered from lower deforestation rates than their adjacent buffer and neighbouring areas, as a result of a combination of various factors such as edge effects (less deforestation in the centre of forest blocks, where parks are located, than at the periphery where villages are located), displacement (leakage) of deforestation to non-managed zones and the presence of park agents (see Whitehurst *et al.* (2009) for Kirindy Mitea National Park; Allnutt *et al.* (2013) for a counter example from Masoala National Park).

The assessment of forest cover over time in our two case studies also shows that deforestation has decreased, with the establishment of management interventions, including the implementation and application of robust rules and control mechanisms associated with the protected areas (Virah-Sawmy and Vincke, in review). These data, based on the mapping of individual cropland cleared between 2009 and 2012 within these two protected areas, also show that assessing the effectiveness of protected areas using only rates of deforestation is flawed. Of greater importance is knowledge of how many farmers are involved in forest clearance and, critically, how they clear the forest. The study showed that farmers cleared very small plots in the buffer zones of Ankodida, and generally always outside the sacred forest (median value of 0.38 to 0.12 hectares between 2009 and 2012 respectively) and, together with their use of more diverse cropping, are ensuring that soil fertility is maintained for a longer period (Virah-Sawmy and Vincke, in review). By contrast, the croplands cleared at the forest frontier in Ranobe PK32 were relatively large (median value of 6.27 to 1.58 hectares between 2009 and 2012) and consisted of maize monocultures, which generally impoverishes the soil faster. On these grounds, we can say to a certain extent that these new protected areas have been effective, especially in the case of Ankodida, with its traditional governance and responsible swidden practices.

Conservationists tend to focus on the effectiveness of management, which is the extent to which management is protecting values and achieving goals and objectives. A global tool has been set up, known as the Management Effectiveness Tracking Tool (METT), developed by WWF, IUCN and the World Bank and is used in Madagascar to help measure changes in protected area management effectiveness over time. The term management effectiveness reflects three main 'themes' in protected area management: design issues relating to both individual sites and protected area systems; adequacy and appropriateness of management systems and processes; and delivery of protected area objectives (Hocking *et al.*, 2000). Ankodida and Ranobe PK32 are being auto-evaluated by WWF, in collaboration with local stakeholders, every two years using METT. However, the management goals of these new protected areas remain technocratic (e.g. viability of target species, sustainable funding, implementation of control) with little subtlety in terms of human–nature interactions. Rather, another way to define the goals and values would be from the lens of different resource user groups. We contend that this has been insufficiently done and that addressing lessons learnt through the experiences of practitioners are essential to building best practice.

Anthropologists and other social scientists, on the other hand, have analysed the social and political impacts of protected areas but have not attempted to balance these in terms of their overall achievements. For example, the social impacts of Madagascar's new protected areas, and in particular the impacts of livelihood and development projects on poverty alleviation, remain largely unknown (Gardner *et al.*, 2013). Developing joint reflective practices, based on the three above-mentioned criteria of effectiveness (in protecting biodiversity, in its management effectiveness and in reducing social impacts), together with assessing the governance principles as we did in this chapter, can help the ultimate challenge faced by conservationists – that of shaping human interactions with nature in landscapes of which people are a part (Brockington *et al.*, 2006). This remains critical but is lacking so far.

Conclusions

The celebrated work of Nobel Laureate Elinor Ostrom emphasised that the success of local governance depends on the nature of the institutions to which power is devolved (Ostrom *et al.*, 1999). Similarly, in Madagascar the success of the Durban Vision will be largely contingent on the nature of the local institutions as well as broader institutional arrangements in shared governance. When viewed in comparison to the pre-2003 protected area network, which was managed by the state and forbade the extractive use of natural resources by adjacent communities, the Durban

Vision has clearly made great strides in integrating the interests of local communities into protected area governance and management. Major changes include: i) a shift in protected area management categories from the strictly protected Categories Ia, II and IV to the multiple-use Categories III, V and VI in which human interactions with the environment are explicitly recognised as an integral part of protected area management (although we believe nonetheless that more should be done to expand local uses in new protected areas, as WWF did in the re-zoning of Ranobe PK32); and ii) a shift in protected area governance types from state governance ('governance by government') to community governance or shared governance by NGOs and communities. For example, protected area governance in Ankodida has built on traditional rules and institutions and responsible practices there demonstrate the validity and effectiveness of the approach. As the management capacity of the union of Ankodida grows, regional authorities and the protected area delegators will progressively withdraw, although the latter will be required to provide financial and technical support for, for example, agricultural development that helps farmers reduce their dependence on swidden agriculture. However, from our experience in the establishment of new protected areas in southern Madagascar, there remains much scope for improvement in order to make the Durban Vision transformative in the way that local actors' voices are heard, debated and represented.

The use of the synthetic framework adapted from Graham *et al.* (2003) and Lockwood (2010) incorporating five good governance principles helped us to better understand some of the strengths and the weaknesses of protected area governance. We encourage practitioners and academics to also use these principles when assessing a protected area as a whole, rather than concentrating only on one aspect of the principle. We believe that the weakest aspect of the Durban Vision has been the question of legitimacy of other resource users such as pastoralists, customary landowners, other CBNRM or state actors. For example, pastoralists have played a very important role in ensuring forest management in spiny forest ecosystems of southern Madagascar because the forest is important for grazing and hiding cattle from thieves (Kaufman and Tsirahamba, 2006). But because of the current central role given mainly to CBNRM actors in protected area governance, there is an increasing risk that the governance of protected areas within the Durban Vision simply becomes synonymous with delegated management following the failed approaches of GELOSE policy, instead of being based on the understanding that protected area governance is about ensuring decisions are taken in a legitimate, fair and inclusive way. The Durban Vision community therefore needs to reflect on how best to build on existing legitimate institutions that mediate decision-making at the rural level, as well as aim for more democratic processes within those institutions and representation of other resource users. Ideally, building on a hybrid of traditional customary management including local notables and elders who mediate conflict resolution between different elective representatives from defined groups relevant to each protected area (pastoralists, customary land users and so on) would best combine aspects of legitimacy, fairness, inclusiveness and accountability in order to manage various conflicts with the establishment and management of the protected area. The use of elected representatives in CBNRM (a process known as democratic decentralisation) has already been undertaken in Zambia (Gibson and Marks, 1995) and Burkino Faso (Ribot, 2002). The type of institution proposed, combining the strength of customary management with representation, is being instigated with the union of Ankodida, although in this case the union is involved with mediating conflicts between different CBNRM associations but not between the elected representatives of other user groups. This type of independent local mediation takes a long time to build up for protected area governance and requires a good degree of mutual confidence between local communities and protected area delegators and also requires that the conditions of the protected area are well understood at the outset.

Finally, while the implementation of the Durban Vision has involved great advances in the participation of local communities in protected area governance, it has been a process of learning-by-doing for protected area delegators lacking detailed guidelines or academic literature to provide an evidence base for their decision-making. This is illustrated by the changes in zoning and governance structure over six years of management of Ranobe PK32, brought about partly through the greater voice given to local communities in decision-making. Many weaknesses and imbalances of power remain and true participation continues to be elusive. In this respect, the Durban Vision community needs greater constructive collaboration with social scientists, and to learn to systematically use their approaches and tools in achieving good governance outcomes. Unfortunately, we know of very few academics working in Madagascar in the social sciences that are directly helping agencies on the ground to deliver better outcomes through reflective practices or action-research methods. Whilst we understand that independent agendas are important for social scientists, our experiences suggest that publishing does little to actually change on-the-ground implementation. Rather, platforms such as steering committees, as well as joint action-research projects, should be proactively used by experts and academics who wish to contribute to improved practices. In parallel, the Durban Vision Community should welcome those inputs in order to ensure multiple-use protected areas provide the safety net of continued, sustainable resource extraction, while catalysing the transition from extensive swidden to more intensive, productive forms of agriculture through the provision of financial and technical support. Such collaboration will be essential if the good governance of Madagascar's new protected areas is to be achieved.

Acknowledgements

We would like to thank the on-the-ground team for numerous discussions on ways to improve on governance issues in the new protected areas and implementing them. In particular, many thanks to: Bernardin Rasolonandrasana, Flavien Rebara, Xavier Vincke and their teams. We are also very grateful to Ivan Scales for giving us the opportunity to present a practitioner's perspective and his numerous comments that helped us to significantly improve the chapter.

Note

1 In Madagascar, most of the multi-use protected areas have integrated human use of forest resources for domestic purposes. Experiences abroad and in Madagascar indeed show that the safest approach for protected areas lies in working with local communities to protect and secure traditionally owned resources from unrestrained intrusions of the market and immigrant settlers even if the resources only provide local communities with non-economic benefits. More risky are strategies that offer economic incentives for protected area support, especially if these economic incentives come from international markets where the high-value resources sought distort the local market, creating more conflicts and inequality within communities who are already marginal to regional or national economies (Angelsen, 2010). For example, in Madagascar, efforts to integrate the exploitation of high-valued woods in community forests have failed because the lucrative nature of high-valued woods led to organised trafficking and corruption at every level. On the other hand, community forests aimed at commercialisation of low-valued products such as charcoal for the local and regional markets have worked relatively well (Montagne *et al.*, 2010).

References

Adams, W.M., Aveling, R., Brockington, D., Dickson, B., Elliott, J., Hutton, J., Roe, D., Vira, B. and Wolmer, W. (2004) 'Biodiversity conservation and the eradication of poverty', *Science*, vol 306, pp1146–1149.

Allnutt, T.F., Asner, G.P., Golden, C.D. and Powell, G.V.N. (2013) 'Mapping recent deforestation and forest disturbance in northeastern Madagascar', *Tropical Conservation Science*, vol 6, pp1–15.

Andam, K.S., Ferraro, P.J., Pfaff, A., Sanchez-Azofeifa, G.A. and Robalino, J.A. (2008) 'Measuring the effectiveness of protected area networks in reducing deforestation', *Proceedings of the National Academy of Sciences of the United States of America*, vol 105, pp16089–16094.

Andam, K.S., Ferraro, P.J., Sims, K.R.E., Healy, A. and Holland, M.B. (2010) 'Protected areas reduced poverty in Costa Rica and Thailand', *Proceedings of the National Academy of Sciences of the United States of America*, vol 107, pp9996–10001.

Angelsen, A. (2010) 'Policies for reduced deforestation and their impact on agricultural production', *Proceedings of the National Academy of Sciences of the United States of America*, vol 107, pp19639–19644.

Bérard, M.H. (2011) 'Légitimité des normes environnementales dans la gestion locale de la forêt à Madagascar', *Canadian Journal of Law and Society*, vol 26, no 1, pp89–111.

Berkes, F. (2004) 'Community-based conservation in a globalized world', *Proceedings of the National Academy of Sciences of the United States of America*, vol 104, pp15188–15193.

Bernstein, S. (2005) 'Legitimacy in global environmental governance', *Journal of International Law and International Relations*, vol 1, pp139–166.

Börner, J., Wunder, S., Wertz-Kanounnikoff, S., Tito, M.S., Pereira, L. and Nascimento, N. (2010) 'Direct conservation payments in the Brazilian Amazon: scope and equity implications', *Ecological Economics*, vol 69, pp1272–1282.

Borrini-Feyerabend, G. (2004) 'Governance of protected areas, participation and equity', in *Biodiversity Issues for Consideration in the Planning, Establishment and Management of Protected Areas Sites and Networks*, Secretariat of the Convention on Biological Diversity, Montreal.

Borrini-Feyerabend, G. (2007) 'The "IUCN protected area matrix": a tool towards effective protected area systems', *IUCN World Commission on Protected Areas Task Force: IUCN Protected Area Categories*, paper presented at summit on the IUCN categories in Andalusia, Spain, 7–11 May 2007.

Brockington, D., Igoe, J. and Schmidt-Soltau, K. (2006) 'Conservation, human rights, and poverty reduction', *Conservation Biology*, vol 20, pp250–252.

Casse, T., Milhøj, A., Ranaivoson, S. and Randriamanarivo, J.R. (2004) 'Causes of deforestation in southwestern Madagascar: what do we know?', *Forest Policy and Economics*, vol 6, pp33–48.

Cheban, S.A., Rejo-Fienana, F. and Tostain, S. (2009) 'Etude ethnobotanique des ignames (Dioscorea spp.) dans la forêt Mikea et le couloir d'Antseva (sud-ouest de Madagascar)', *Malagasy Nature*, vol 2, pp111–126.

Commission SAPM (2006) *Procédure de Création des Aires Protégées du Système d'Aires Protégées de Madagascar (SAPM)*, Commission SAPM, Draft 8 June, Antananarivo.

Dudley, N. (2008) *Guidelines for Applying Protected Area Management Categories*, IUCN, Gland, Switzerland.

Duffy, R. (2006) 'Non-governmental organizations and governance states: the impact of transnational environmental management networks in Madagascar', *Environmental Politics*, vol 15, pp731–749.

Durbin, J.C. and Ralambo, J.A. (1994) 'The role of local people in the successful maintenance of protected areas in Madagascar', *Environmental Conservation*, vol 21, pp115–120.

Fenn, M. and Rebara, F. (2003) 'Present migration tendencies and their impacts in Madagascar's spiny forest ecoregion', *Nomadic Peoples*, vol 7, pp123–137.

Freudenberger, K. (2010) *Paradise lost? Lessons Learnt from 25 Years of USAID Environment Programs in Madagascar*, a United States Agency for International Development (USAID) Commissioned Report, Washington, DC.

Fritz-Vietta, N.V.M., Ferguson, H.B., Stoll-Kleemann, S. and Ganzhorn, J.U. (2011) 'Conservation in a biodiversity hotspot: insights from cultural and community perspectives in Madagascar', *Biodiversity Hotspots*, vol 3, pp209–233.

Gardner, C.J. (2011) 'IUCN management categories fail to represent new, multiple-use protected areas in Madagascar', *Oryx*, vol 45, pp336–346.

Gardner, C.J., Ferguson, B., Rebara, F. and Ratsifandrihamanana, A.N. (2008) 'Integrating traditional values and management regimes into Madagascar's expanded protected area system: the case of Ankodida', in J.M. Mallarach (ed.) *Protected Landscapes and Cultural and Spiritual Values*, IUCN, GTZ and Obra Social de Caixa Catalunya. Kasparek Verlag, Heidelberg.

Gardner, C.J., Nicoll, M.E., Mbohoahy, T., Olesen, K.L.L., Ratsifandrihamanana, A.N., Ratsirarson, J., René de Roland, L.-A., Virah-Sawmy, M., Zafindrasilivononona, B. and Davies, Z.G. (2013) 'Protected areas for conservation and poverty alleviation: experiences from Madagascar', *Journal of Applied Ecology*, vol 50, pp1289–1294.

Ghimire, K.B. and Pimbert, M.P. (1997) *Social Change and Conservation: Environmental Politics and Impacts of National Parks and Protected Areas*, Earthscan, London.

Gibson C.C. and Marks, S.A. (1995) 'Transforming rural hunters into conservationists: an assessment of community-based wildlife management programmes in Africa', *World Development*, vol 23, pp941–956.

Graham, J., Amos, B. and Plumptre, T. (2003) *Governance Principles for Protected Areas in the 21st Century*, prepared for the 5th IUCN World Parks Congress held in Durban, South Africa.

Harper, G.J., Steininger, M.K., Tucker, C.J., Juhn, D. and Hawkins, F. (2007) 'Fifty years of deforestation and forest fragmentation in Madagascar', *Environmental Conservation*, vol 34, pp325–333.

Hocking, M., Stolton, S. and Dudley, N. (2000) *Evaluating Effectiveness: A Framework for Assessing the Management of Protected Areas*, IUCN, Gland, Switzerland.

Horning, N.R. (2003) 'How rules affect conservation outcomes', in S.M. Goodman and J.P. Benstead (eds) *The Natural History of Madagascar*, University of Chicago Press, Chicago.

IUCN (2003a) *The Durban Action Plan*. One of the outputs to the 2003, 5th IUCN World Parks Congress held in Durban, South Africa.

IUCN (2003b) *The Durban Accord*. One of the outputs to the 2003, 5th IUCN World Parks Congress held in Durban, South Africa.

Kaufmann, J.C. and Tsirahamba, S. (2006) 'Forests and thorns: conditions of change affecting Mahafale pastoralists in southwestern Madagascar', *Conservation and Society*, vol 4, pp231–261.

Kremen, C., Cameron, A., Moilanen, A., Phillips, S. J., Thomas, C. D., Beentje, H., Dransfield, J., Fisher B.L., Glaw, F., Good, T.C., Harper, G.J., Hijmans, R.J., Lees, D.C., Louis, E. Jr., Nussbaum, R.A., Raxworthy, C.J., Razafimpahanana, A., Schatz, G.E., Vences, M., Vieites, D.R., Wright, P.C. and Zjhra, M.L. (2008) 'Aligning conservation priorities across taxa in Madagascar with high-resolution planning tools', *Science*, vol 320, pp222–226.

Kull, C.A. (2003) 'Deforestation, erosion, and fire: degradation myths in the environmental history of Madagascar', *Environment and History*, vol 6, pp423–450.

Lockwood, M. (2010) 'Good governance for terrestrial protected areas: a framework, principles and performance outcome', *Journal of Environmental Management*, vol 91, pp754–766.

Montagne, P., Razafimahatratra, S., Rasamindisa, A. and Crehay, R. (2010) *ARINA, le Charbon de Bois à Madagascar: Entre Demande Urbaine et Gestion Durable*, Edition CITE, Antananarivo.

Muttenzer, F. (2006) 'Déforestation et droit coutumier à Madagascar: l'historicité d'une politique foncière', PhD thesis, Université de Genève, Geneva.

Nelson, A. and Chomitz, K.M. (2011) 'Effectiveness of strict vs multiple use protected areas in reducing tropical forest fires: a global analysis using matching methods', *PLOS ONE*, vol 6(8), e22722.

Nicoll, M.E. and Langrand, O. (1989) *Madagascar: Revue de la Conservation et des Aires Protégées*, World Wide Fund for Nature, Gland, Switzerland.

Olson, D. and Dinerstein, E. (1998) 'The Global 200: a representation approach to conserving the world's most biologically valuable ecoregions', *Conservation Biology*, vol 12, pp502–515.

Ostrom, E., Burger, J., Field, C.B., Norgaard, R.B. and Policansky, D. (1999) 'Revisiting the commons: local lessons, global challenges', *Science*, vol 284 (5412), pp278–282.

Phillips, A. (2003) 'A modern paradigm', *World Conservation Bulletin*, vol 2, pp6–7.

Pimbert, M.P. and Pretty, J.N. (1997) 'Parks, people and professionals: putting "participation" into protected area management', in K.B. Ghimire and M.P. Pimbert (eds) *Social Change and Conservation: Environmental Politics and Impacts of National Parks and Protected Areas*, Earthscan, London.

Raik, D. (2007) 'Forest management in Madagascar: an historical overview', *Madagascar Conservation & Development*, vol 2, pp5–10.

Randrianandianina, B.N., Andriamahaly, L.R., Harisoa, F.M. and Nicoll, M.E. (2003) 'The role of the protected areas in the management of the island's biodiversity', in S.M. Goodman and J.P. Benstead (eds) *The Natural History of Madagascar*, University of Chicago Press, Chicago.

Rasoavahiny, L., Andrianarisata, M., Razafimpahanana, A. and Ratsifandrihamanana, A.N. (2008) 'Conducting an ecological gap analysis for the new Madagascar protected area system', *Parks*, vol 17, pp12–21.

Ribot, J.C. (2002) *Democratic Decentralisation of Natural Resources: Institutionalizing Popular Participation*, World Resources Institute, Washington, DC.

Roe, D. (2008) 'The origins and evolution of the conservation-poverty debate: a review of key literature, events and policy processes', *Oryx*, vol 42, pp491–503.

Scales, I. R. (2012) 'Lost in translation: conflicting views of deforestation, land use and identity in western Madagascar', *The Geographical Journal*, vol 178, pp67–79.

Virah-Sawmy, M. (2009) 'Ecosystem management in Madagascar during global change', *Conservation Letters*, vol 2, pp163–177.

Virah-Sawmy, M. and Vincke, X. (in review) 'Reducing deforestation using innovative aerial photography to nudge farmers' behaviours in Madagascar', *Conservation Biology*.

West, P., Igoe, J. and Brockington, D. (2006) 'Parks and peoples: the social impact of protected areas', *Annual Review of Anthropology*, vol 35, pp251–277.

Whitehurst, A.S., Sexton, J.O. and Dollar, L. (2009) 'Land cover change in western Madagascar's dry deciduous forests: a comparison of forest changes in and around Kirindy Mite National Park', *Oryx*, vol 43, pp275–283.

WWF (2010) 'Africa poverty and conservation policy: the MWIOPO diagnostic, Antananarivo', World Wide Fund for Nature, Antananarivo.

Part 4

Making conservation pay?

Incentive-based conservation, the commodification of Madagascar's nature and conflicting views of landscape and nature

11 Tourism, conservation and development in Madagascar
Moving beyond panaceas?

Ivan R. Scales

This chapter draws on Scales, I. R. (in press) 'Trees, tourists and trade-offs: the political ecology of rainforest tourism, forest clearance and biodiversity conservation in Madagascar', in Prideaux, B. (ed.) *Rainforest Tourism, Conservation and Management: Challenges for Sustainable Development*, Earthscan, London.

Introduction

Madagascar presents researchers and policy makers with a classic conservation conundrum: how to protect biodiversity at the same time as delivering economic growth and creating alternative livelihoods that place less pressure on ecosystems and biological diversity. Since the 1980s, Madagascar has been a hotbed of conservation activity (see Chapter 7 by Kull). Efforts to protect Madagascar's flora and fauna have centred primarily on the designation of protected areas (see Chapter 10 by Virah-Sawmy *et al.*). However, there has been a growing recognition amongst policy makers of the need to move beyond simple strategies of 'fortress conservation'. Not only have protected areas often failed to stop impacts such as logging and forest clearance but the creation of protected areas has often imposed significant costs on rural households due to loss of access to natural resources (Ferraro, 2002).

In an attempt to improve the performance of protected areas and reduce conflicts with rural households, conservation organizations and government ministries have experimented with a wide range of schemes to involve or create benefits for communities living around parks. Over the past ten years, policy has increasingly turned to nature tourism to try and solve the challenges of both biodiversity conservation and poverty alleviation. In this chapter, I discuss the reasons why most forms of nature tourism in Madagascar have had a limited impact as tools for integrated conservation and development. I start by providing a brief overview of Madagascar's tourism sector. Having set the scene, I use a political ecology approach to argue that tourism's lack of success to date has been due to two main factors:

i) conflicting perceptions and priorities between different stakeholders; and ii) an uneven distribution of the costs and benefits of managing nature for tourism. As a result, expectations that tourism will deliver a 'win–win' solution are often overly ambitious. There are important questions about the power relations in such schemes, with nature tourism helping to reinforce the 'top-down' politics of forest management (see also Chapter 7 by Kull and Chapter 9 by Corson). I argue that, rather than seeking cure-all solutions, policy makers must be more willing to engage with alternative perceptions and priorities when it comes to natural resource management, and should be ready to accept trade-offs and compromise if nature tourism is to play a more effective role as a tool for conservation and development in Madagascar.

An overview of tourism in Madagascar

Tourists mostly come to Madagascar for one reason – its wildlife (Figure 11.1). A survey carried out in 2000 found that more than half of visitors came to the island for nature tourism (Christie and Crompton, 2003). Accurate data on tourism in sub-Saharan Africa are hard to come by and Madagascar is no exception. The following statistics must therefore be treated with caution. It is safe to say that in comparison to other African destinations, Madagascar's tourism sector is small. According to the Malagasy Ministry for Tourism, in 2011 there were 225,005 foreign visitors to Madagascar (Ministère du Tourisme, 2012). The number of visitors peaked in 2008 with 375,010 visitors, but dropped in 2009 following a political crisis.[1] To put this in context, it is worth looking at the visitor numbers of nearby African countries with significant nature-based tourism sectors. In 2007, Kenya received approximately 1,644,000 visitors and earned over US$1.5 billion in tourist receipts; Tanzania received 692,000 visitors and earned over US$1 billion; and the small island of Mauritius received 906,971 visitors and earned more than US$1.6 billion (Twinning-Ward, 2009). Meanwhile, in the same year Madagascar received 344,348 visitors and earned approximately US$506 million, which equates to roughly three per cent of Gross Domestic Product (GDP) (Christie and Crompton, 2003; Twinning-Ward, 2009).

The majority of visitors to Madagascar are European. French tourists dominate (60 per cent), followed by Italians (12 per cent), Americans (4.2 per cent), Swiss (2.9 per cent), Germans (2.8 per cent) and British (2.2 per cent) (Christie and Crompton 2003). The prevalence of French tourists is primarily down to Madagascar's history – the island was a French colony between 1896 and 1960 and French remains an official language (see Chapter 6 by Scales for more on the island's colonial history). Furthermore, the large majority of direct scheduled flights from Europe are from Paris, making travel more complicated and expensive for visitors outside France.

Figure 11.1 Ring-tailed lemurs (*Lemur catta*) in Berenty Nature Reserve, one of Madagascar's biggest tourist draws

Source: photograph by Helen Scales.

The small number of visitors to Madagascar is related to a range of factors. First, it is expensive to get to due to the small number of airlines operating scheduled flights to the island (Christie and Crompton, 2003; Mercer *et al.*, 1995). Madagascar is also difficult to travel around, with long distances, often rudimentary lodging, roads in poor condition, and internal flights that are both costly and unreliable (Christie and Crompton, 2003; Durbin and Ratrimoarisaona, 1996; Mercer *et al.*, 1995). The combination of poor infrastructure and high cost means that Madagascar is unable to cater to either luxury or budget travellers and is left as a niche destination for more adventurous nature lovers.

The political ecology of tourism in Madagascar

Madagascar's ecosystems involve a large and diverse set of stakeholders, from rural households to government ministries and international conservation organizations. Nature-based tourism adds to these complex politics of environmental management by involving tour companies and tourists, who have their own perspectives and priorities. It therefore should come as no surprise that the politics of resource management and biodiversity conservation are fraught with tension. Research carried out under the banner of political ecology has provided useful insights into the causes of conflicts over natural resources and the factors that have limited the success of nature-based tourism as a tool for integrated conservation and development.

Political ecology seeks to understand the political and economic drivers of environmental change and the conflicts that occur over natural resources. It starts from the premise that natural resources are used and contested by multiple stakeholders, who perceive nature in different ways and often have contrasting and conflicting priorities (Stott and Sullivan, 2000). Not only that, but the costs and benefits of resource use are unevenly distributed and actors differ in the power they have over access to and control of resources. Political ecology can thus be defined as 'research based explorations to explain linkages in the condition and change of social-environmental systems, with explicit consideration of relations of power' (Robbins, 2004, p12).

Conflicting visions of nature

With Madagascar's unique and charismatic biodiversity, combining nature tourism with conservation and poverty alleviation seems an obvious solution. The idea has certainly caught the imagination of conservationists and government ministers. An influential World Bank-funded study of the island's tourism sector concluded that: 'In Madagascar, where rural poverty is widespread and where the poor put stress on the natural resource base, tourism could generate positive externalities' (Christie and Crompton, 2003, p1). Norris (2006, p264) believes that: 'Ultimately, nature-based tourism may offer the greatest promise as a sustainable driver of biodiversity protection', while Frank Hawkins, the former head of Conservation International (CI) in Madagascar, stated that: 'If you could replace *tavy* [swidden agriculture] with tourism, the lemurs would have nothing to worry about' (McGrath, 2005). Meanwhile, the government's Madagascar Action Plan boldly proclaimed that:

> We will become a 'green island' again ... The world looks to us to manage our biodiversity wisely and responsibly – and we will. Local communities will be active participants in environmental conservation under the guidance of bold national policies ... we will develop industries around the environment such as eco-tourism.
>
> (MAP, 2007, p97)

The current network of national parks that nature tourism depends on largely took shape in the 1990s under the National Environmental Action Plan (NEAP; see Chapter 7 by Kull). Tourism has played a central role in conservation policy from an early stage. In 1990, the country's first Environmental Charter identified tourism as a key mechanism for generating money to fund national parks and help to integrate conservation and development (MEP, 1990). Phase 1 of the NEAP (1991–1996) involved the creation of a national association for the management of the island's protected areas – the *Association Nationale pour la Gestion des Aires Protégées* (ANGAP, now called Madagascar National Parks (MNP))[2] – which was charged with managing the island's protected areas with particular emphasis on tourism as a potential source of revenue. The perceived benefits are such that some strict nature reserves, where all human visitation is highly restricted, have been re-categorized as 'National Parks' in order to allow tourist visitors (see Chapter 10 by Virah-Sawmy *et al.*).

As Chapters 5 (Scales) and 7 (Kull) have highlighted, the logic that drives such a focus on tourism is clear. The 'problem' is that Madagascar's ecosystems (and therefore its endemic flora and fauna) are threatened by the practices of rural households, especially swidden cultivation but also grassland burning and over-harvesting of certain species of animals and plants. The 'cause' of such practices is poverty and a lack of livelihood alternatives. The 'solution' is therefore equally clear. If households can be provided with alternative sources of income – ones that depend on maintaining ecosystem functions and resource stocks – they will be motivated to protect wildlife. It is no wonder then that tourism seems so attractive. It offers the prospect of a non-consumptive source of income from the island's ecosystems, with tourists 'taking only pictures' while 'leaving only footprints' and more importantly money.

However, this received wisdom is problematic for a number of reasons (see Chapter 5 by Scales for an in-depth discussion of the following factors). First, there is a growing body of research pointing to the fact that the received wisdom ignores the role of other land uses in environmental degradation, particularly large-scale cash crop production. Second, it is based on a poor understanding of the factors that drive the resource use decisions of rural households. Looking at forest clearance, for example, it ignores the role of wealthy elites, who have been heavily involved in forest clearance by using their control of access to land and their ability to hire additional labour to clear large areas of forest for the purposes of growing exportable cash crops (Minten and Méral, 2006; Scales, 2011). As research has shown, there is no simple relationship between population growth, poverty and natural resource use (Geist and Lambin, 2001; Reardon and Vosti, 1995; Steneck, 2009). This means that raising the incomes of poor rural households through nature tourism has the potential to mitigate only one of the many drivers of environmental degradation and, rather than being seen as a panacea, must be part of a broader set of solutions.

There is an even deeper problem at the heart of the received wisdom on nature tourism and conservation in Madagascar, which boils down to radically different conceptions between different stakeholders of what constitutes a 'rational' use of natural resources. International conservation non-governmental organizations (NGOs) and government ministries tend to see the island's ecosystems primarily in terms of global conservation, where the priority is the maintenance of maximum biological diversity through the creation of protected areas. Furthermore, environmental discourse and policy tend to appeal to ideas of 'pristine nature', 'wilderness' and 'global heritage' (Kull, 1996; Scales, 2012; see Chapter 7 by Kull). This is epitomized by the island's status as the world's 'hottest biodiversity hotspot'[3] (Ganzhorn et al., 2001). Accordingly, the only people allowed to enter national parks are tourists and scientists and all other uses of natural resources are banned. Over the past twenty years, policy has seen a move away from schemes based on consumptive uses (for example attempts to improve the sustainability of timber harvesting) to non-consumptive uses of wildlife and landscapes (Scales, 2012; see Chapter 7 by Kull).

Through this lens, practices such as swidden agriculture are seen as inherently destructive, irrational and contradictory: 'It is paradoxical that the same people who depend heavily on the natural forest are the ones most heavily involved in its destruction' (Favre, 1996, p39). The problem with this assumption is that it is based on a fundamental misunderstanding of the ethnoecology (indigenous environmental knowledge) of rural communities. Rather than being seen as something to be protected from human use, resources such as forests are considered to be *tany fivelomana* – land where one can create a livelihood – and practices such as swidden agriculture are seen as ways of making land productive (Keller, 2008; Scales, 2012). From this perspective, the logic of forest clearance is clear – cutting and burning vegetation provides increased light and a nutrient rich ash for crops and swidden agriculture is a risk averse, low labour and low capital system for making otherwise nutrient poor soils productive (Scales, 2012; see Chapter 5 by Scales).

Ormsby and Mannie (2006, p283) suggest that conservation programmes based on nature tourism may 'serve to mediate conflict by explaining to local residents the purpose of conservation'. However, the difference in perceptions between different actors means that rather than mediating conflict, such programmes have often increased tensions. Not only do they restrict access to natural resources while allowing foreign tourists to enter parks, they attempt to impose a radically different vision of what constitutes a valid use of nature. They prioritize the interests of global conservation and non-consumptive nature tourism over local livelihoods and consumptive uses of natural resources.

A powerful example of this can be found in western Madagascar where the iconic 'Baobab Alley' (Figure 11.2), near the town of Morondava, has recently caused considerable tension between conservation organizations

Figure 11.2 Baobab Alley: rows of Grandidier's baobab (*Adansonia grandidieri*) near Morondava. Tourist curiosity or conservation icon?

Soure: photograph by Ivan Scales.

and rural households. The site is one of the region's biggest tourist attractions. The rows of baobabs (*Adansonia grandidieri*) are certainly striking. However, in terms of biodiversity conservation, the site is somewhat of a curiosity (Marie *et al.*, 2009). It has a long history of human occupation and modification and had until recently not been identified as having significant biological value by conservation organizations. While *A. grandidieri* is classified by the International Union for Conservation of Nature (IUCN) as endangered, the 'Baobab Alley' is a long way away from the dry-deciduous forest that these baobabs are normally found in (Baum, 1996). The trees are most likely the last remnants of prior forest cover that has been cleared over a period of more than a century (Baum, 1996; Scales, 2011). It is their very isolation amongst strikingly verdant rice paddies that makes the scene so photogenic and attractive to tourists. So another way of looking at the 'Baobab Alley' is that it is the site's human history that makes it such a draw. If tourists simply want to see baobabs, they can do so in the villages or dry-deciduous forests nearby.

However, it is the same human actions that created the site that have now led to significant conservation attention. It has been suggested that regular flooding of the soil around the trees due to agriculture is causing the roots of the baobabs to rot and leading to an increase in mortality (Baum, 1996; Marie *et al.*, 2009).[4] Following the Durban Vision announcement (see Chapter 9 by Corson and Chapter 10 by Virah-Sawmy *et al.*), Fanamby – a Malagasy NGO – proposed a new *Site de Conservation* in the Central Menabe region, which expanded the geographical scope of the region's existing protected areas (Fanamby, 2002, 2005; Raharinjanahary, 2004; see Chapter 10 by Virah-Sawmy *et al.*). The *Site de Conservation* has become the *Aire Protégée Menabe-Antimena*, placing constraints on resource use in areas that were previously outside protected areas, for example banning certain agricultural practices. This includes the 'Baobab Alley'. In 2007, the area was classified as an IUCN Category III 'Natural Monument' (see Chapter 10 by Virah-Sawmy *et al.* for details on different IUCN categories).

Unsurprisingly, this process has been strongly contested. On the one hand, conservation organizations and government ministries see the site as a vital part of both the local and national economy and by extension crucial as a tourist draw and potential source of funding for conservation and development. On the other hand, locals see the interests of foreign tourists trumping their own. Not only that, but they receive few benefits from passing tourists (Marie *et al.*, 2009). To them, the baobabs seem less of a blessing and more of a curse:

> They [conservation organizations] want us to plant baobabs trees. But why should we plant more baobabs when the ones already here are causing us so many problems? We're not allowed to clear land and we're not allowed to grow rice. What are we supposed to do?[5]

This example shows how important nature tourism is as a force for redefining how landscapes are perceived and reconfiguring what is deemed to be of conservation importance. This has happened elsewhere in Madagascar. For example, a few kilometres up the road from 'Baobab Alley' lies the 10,000-hectare 'Kirindy Forest', a reserve created in 1978 with financial and expert help from Coopération Suisse (Swiss Aid) to carry out research on improved forestry techniques and to teach these new techniques to local loggers (Tonganiriko et al., 2002). However, the reserve now hosts a tourist camp and the forestry trails cut in the 1980s have become the perfect way for visitors to move through the forest and see lemurs. This change reflects the fact that conservation policy in the region has increasingly seen a move away from schemes to improve logging and agriculture to projects based on non-consumptive uses such as nature tourism (Scales, 2012). Similarly, large parts of the area that went on to become Ranomafana National Park, in eastern Madagascar, saw heavy logging for decades (Bohlen, 1993). However, in 1991, new legislation banned all forest exploitation, allowing only scientific research and tourism (Peters, 1999). This is not to say that the ideas that underpin environmental management cannot change or that protected areas cannot evolve over time but to stress that a narrow focus on nature tourism at the expense of other possible uses reflects a certain set of external values and priorities.

The uneven costs, benefits and politics of tourism

As well as dealing with the different perceptions that stakeholders have with regards to natural resources, political ecology also pays attention to how the costs and benefits of natural resource use are distributed, and who has the power to control this process.

There is a large literature pointing out how uneven the costs and benefits of conservation projects can be. The establishment of protected areas has involved the eviction of communities and restrictions over the use of natural resources within their boundaries (Brockington and Igoe, 2006; West et al., 2006). The national parks visited by tourists in Madagascar are no different, having led to the displacement of thousands of rural households and severe restrictions on natural resource use (Ghimire, 1994; Pollini, 2011). As a result, Madagascar's national parks have led to the disruption of household economies, property systems, traditional skills and cultural values (Peters, 1999). One study estimated the cost imposed on communities living around Ranomafana National Park, due to lost access to natural resources, to be $39/year per household, equivalent to as much as 25 per cent of household income (Ferraro, 2002).

The argument made by policy makers is that tourism might help to compensate for such losses. However, at present, tourism is concentrated in a few geographical regions and a small number of protected areas. Four national parks (Andasibe-Mantadia, Isalo, Ranomafana, Montagne d'Ambre) and one special reserve (Ankarana) attracted over 88 per cent of the visitors between 1992 and 2000 (Christie and Crompton, 2003). These parks are relatively easy to access and well catered for in terms of accommodation and park facilities. There seems to be only so many protected areas that tourists are willing to visit and only so much wildlife that even passionate nature lovers want to see. This is also the case with marine areas:

> [E]cotourism is far from being a panacea for Madagascar's coastal challenges – given the enormous scale of this continental island there are simply not enough tourists to bring sustainable revenue to manage 30 per cent of the country's 5,500 km coastline. Beyond a few model sites blessed with adequate communications infrastructure, tourism services and reliable visitor numbers, sustainable marine conservation finance from tourism is an unrealistic expectation for Madagascar.
> (Harris, 2011, pp11–12)

While tourism is clearly not a solution for managing all of Madagascar's forests or coastal environments, perhaps it might succeed for the national parks that are well visited by tourists? Unfortunately, even for parks popular with tourists, there are serious constraints to the effectiveness of tourism as means of creating incentives for conservation. For nature tourism to work as a tool for conservation and development, it must be capable of creating benefits for rural households equal to the costs incurred from the existence of protected areas and the loss of access to resources. In theory, tourism could create a range of benefits: revenue from entrance fees; employment opportunities (for example, as park guards, guides or in hotels); as well as broader benefits to local economies through tourist spending.

Dealing first with general economic benefits, tourism's impact as a tool for incentivizing conservation is limited in Madagascar by the sector's small size. The lack of data on tourist spending and the tourism value chain makes it impossible to ascertain how much or how little tourist spending stays in local economies around the parks, but there is likely to be considerable 'leakage' (the failure of tourist spending to remain in the destination economy). Although Madagascar lacks international hotel chains and tourism mainly involves small tour companies, the sector is dominated by foreign operators and wealthy Malagasy (Christie and Crompton, 2003; Peters, 1998). The problem is exacerbated by the fact that air fares account for over half the money that tourists spend on holidays to Madagascar, reducing the amount of money left to spend in country (Christie and Crompton, 2003).

Not only are benefits small in comparison to other livelihoods activities, they are also unevenly distributed, both between and within communities living around protected areas (Chaboud et al., 2004). Peters (1998), in a study of communities living around Ranomafana National Park, found that tourism directly employed just over 100 people (with less than half coming from the local population of 27,000); indirectly benefited fewer than 100 people; and led to infrastructural improvements in fewer than a dozen of the 160 villages surrounding the park.

The reality is that, to date, tourism in Madagascar has created few employment opportunities. Those that are available tend to favour more educated individuals, who have the necessary language skills to deal with tourists (Durbin and Ratrimoarisaona, 1996; Walsh, 2005). For example, research on the tourism-related benefits of Masoala National Park found that tourism 'particularly benefits residents who own a hotel or café ... or are employed as park guides' (Ormsby and Mannie, 2006, p281). Similarly, tourism in the Mikea Forest has largely failed to create incentives for conservation as economic benefits accrue to a minority of hotel owners and staff, most of whom come from outside the region (Seddon *et al.*, 2000). Walsh (2005) reminds us that the majority of people living around protected areas in Madagascar do not have the skills, inclinations or connections necessary to profit from nature tourism.

With regards to shares of park revenues, there is a different sort of problem. While these benefits might be more evenly distributed to communities around the park, they do not create a direct link between the existence of a park and rural livelihoods. The standard model is that a share of entrance fees is given to a management committee to be used for development projects, such as healthcare and education facilities (Durbin and Ratrimoarisaona, 1996). The benefits thus accrue at the community level. In Masoala National Park, for example, visitor fees have been used for road improvements, the construction of wells and sanitation projects (Ormsby and Mannie, 2006). While these have been welcomed, they do not create alternative sources of income and therefore do not reduce dependence on the forest. According to Durbin and Ratrimoarisaona (1996, p351), 'it is hard to see how these community-level benefits will change the behaviour of individual households that rely for most of their livelihood on exploiting resources within the parks'. Ultimately, the fundamental economic challenge to nature tourism in Madagascar is that it is currently incapable of generating sufficient benefits to outweigh the costs incurred from the loss of access to natural resources imposed by protected areas.

Given the large gap between costs and benefits, it is not surprising that national parks and tourism have led to considerable tension and conflict with rural households (Keller, 2008; Peters, 1999; Walsh, 2005). Many Malagasy living around national parks have compared government ministries, conservation organizations and tourists to the previous colonial masters, who claimed the island's natural resources as property of the state and set aside large areas of forest for commercial exploitation and the benefit of outsiders (Keller, 2008; Walsh, 2005; see Chapter 5 by Scales).

Finally, when weighing up the pros and cons, it is also important to note that tourism can itself have significant environmental impacts. For example, Stephenson (1993) found that in the *Reserve Speciale d'Analamazoatra*, in the eastern rainforests, visitors often walked off established pathways to improve their view of lemurs and in doing so trampled vegetation. This potentially led to increases in herbaceous plants, especially exotic floral species, and produced microhabitats unsuitable for small endemic mammals but favourable to introduced species such as rats (*Rattus rattus*). Belle *et al.* (2009, p32), who looked at marine reserve management in southwestern Madagascar, found that

> as in many places, tourism in the Toliara region is both an asset and a threat. Some of the pressures directly or indirectly caused by tourism include: destructive anchorage due to the lack of appropriate mooring lines, physical contact with the reef, demand from the hotels for seafood, pollution from hotels due to the lack of sewage facilities, and the curio trade involving the collection and sale to tourists of marine organisms such as shells and starfish.

Seddon *et al.* (2000) argue that in the Mikea Forest (in southwestern Madagascar), tourism has had more negative than positive social and environmental impacts – coastal scrub land has been purchased and cleared to build hotels; the growth in hotels and restaurants has led to increases in demand for charcoal and building materials as well as placing increased pressure on limited water resources; and cultural tourism has deepened rifts between local ethnic groups in the Mikea Forest due to the uneven distribution of benefits.

Conclusions

Faced with the twin challenges of protecting biodiversity and reducing the poverty of rural households, conservation organizations and government ministries in Madagascar have increasingly turned to nature tourism. It offers the tantalizing possibility of paying for conservation, creating alternative livelihoods and reducing poverty through the non-consumptive use of nature. Little wonder that it tends to be seen as a panacea for integrated conservation and development.

However, there is currently little evidence that nature tourism in Madagascar is capable of providing genuine viable alternatives for most households living around Madagascar's protected areas. To date, benefits have been small and unevenly distributed. As well as these socio-economic realities, there is a deeper issue that political ecology encourages us to look at, namely the question of power. Although there have been increased efforts to involve communities in decision-making, it is clear that the politics of conservation in Madagascar are still very much 'top-down' (Dressler *et al.*, 2010; Duffy, 2008; Pollini, 2011). Despite the central role Madagascar's forests and marine resources play in the lives of some of the poorest people on the planet, they are increasingly framed in terms of global biodiversity and global natural heritage (see Chapter 7 by Kull). Nature tourism only adds to this process, by tying the management of many of the island's ecosystems to the income generated by international tourists. These outsiders are allowed to visit large areas that rural households are barred from. As a result, rather than healing the rift between conservation and development, nature tourism has often led to further tensions – between consumptive and non-consumptive uses, and between global conservation priorities and local natural resource use necessities. As long as this is the case, many rural households will continue to see biodiversity conservation and nature tourism as new forms of domination reminiscent of the colonial period, with the perceptions and priorities of outside actors once again trumping their own.

Contrary to the exuberance shown by policy makers and conservationists, I have argued that nature tourism by itself cannot be expected to solve Madagascar's environmental problems. Research suggests that in order to be more effective, tourism needs to do more to increase local participation, both in the political and economic sense. Rather than aiming for an impossible 'win–win', policy makers must be willing to accept compromise and trade-offs and to work *with* rather than *against* local perceptions and priorities.

Notes

1 In 2009, president Marc Ravalomanana was unconstitutionally ousted by a political movement led by the major of Antananarivo, Andry Rajoelina. At the time of writing this chapter, Rajoelina was still in charge of the 'High Transitional Authority', with a timetable for new presidential elections yet to be decided. Political instability and an increase in crime and social unrest has impacted on tourist numbers.
2 ANGAP was replaced in 2006 by the *Système des Aires Protégées de Madagascar* (SAPM) when Madagascar's protected area network was expanded under the 'Durban Vision'. It is now referred to as *Parc Nationaux Madagascar* (Madagascar National Parks (MNP)). See Chapter 7 by Kull for more on the evolution of Madagascar's environmental institutions.
3 Biodiversity hotspots are defined as 'areas featuring exceptional concentrations of endemic species and experiencing exceptional loss of habitat' (Myers *et al.*, 2000).
4 It must be noted that this interpretation of what is happening to the baobab trees at 'Baobab Alley' has been contested (Marie *et al.*, 2009).
5 Interview carried out by author, Bekonazy village, 14 July 2006.

References

Baum, D. A. (1996) 'The ecology and conservation of the Baobabs of Madagascar', in J. U. Ganzhorn and J. P. Sorg (eds) *Primate Report 46-1: Ecology and Economy of a Tropical Dry Forest in Madagascar*, Deutsches Primatenzentrum, Gottingen.
Belle, E. M. S., Stewart, G. W., De Ridder, B., Komeno, R. J. L., Ramahatratra, F., Remy-Zephir, B. and Stein-Rostaing, R. D. (2009) 'Establishment of a community managed marine reserve in the Bay of Ranobe, southwest Madagascar', *Madagascar Conservation and Development*, vol 4, pp31–37.

Bohlen, J. T. (1993) *For the Wild Places: Profiles in Conservation*, Island Press, Washington, DC.
Brockington, D. and Igoe, J. (2006) 'Eviction for conservation: a global overview', *Conservation and Society*, vol 4, pp424–470.
Chaboud, C., Méral, P. and Andrianambinimina, D. (2004) 'Le modele verteux de l'écotourisme: mythe ou realité? L'exemple d'Anakao et Ifaty-Mangily à Madagascar', *Mondes en Developpment*, vol 32, pp11–32.
Christie, I. T. and Crompton, D. E. (2003) *Africa Region Working Paper Series No. 63: Republic of Madagascar: Tourism sector study*, World Bank, Washington, DC.
Dressler, W., Buscher, B., Schoon, M., Brockington, D., Hayes, T., Kull, C. A., McCarthy, J. and Shrestha, K. (2010) 'From hope to crisis and back again? A critical history of the global CBNRM narrative', *Environmental Conservation*, vol 37, pp5–15.
Duffy, R. (2008) 'Neoliberalising nature: global networks and ecotourism development in Madagascar', *Journal of Sustainable Tourism*, vol 16, pp327–344.
Durbin, J. and Ratrimoarisaona, S. (1996) 'Can tourism make a major contribution to the conservation of protected areas in Madagascar', *Biodiversity and Conservation*, vol 5, pp345–353.
Fanamby (2002) *Proposition de Zonage pour les Forêts du Menabe Central*, Fanamby, Morondava.
Fanamby (2005) *Projet Fanamby Menabe Central, Plan de Travail Annuel 2005*. Fanamby, Morondava.
Favre, J.-C. (1996) 'Traditional utilization of the forest', in J. U. Ganzhorn and J. P. Sorg (eds) *Primate Report 46-1: Ecology and Economy of a Tropical Dry Forest in Madagascar*, Deutsches Primatenzentrum, Gottingen.
Ferraro, P. J. (2002) 'The local costs of establishing protected areas in low-income nations: Ranomafana National Park, Madagascar', *Ecological Economics*, vol 43, pp261–275.
Ganzhorn, J. U., Lowry II, P. P., Shatz, G. E. and Sommer, S. (2001) 'The biodiversity of Madagascar: one of the world's hottest hotspots on its way out', *Oryx*, vol 35, pp346–348.
Geist, H. J. and Lambin, E. F. (2001) *What Drives Tropical Deforestation? A Meta-analysis of Proximate and Underlying Causes of Deforestation Based on Subnational Case Study Evidence. LUCC Report Series No. 4*, Land-Use and Land-Cover Change Project, Louvain-la-Neuve.
Ghimire, K. B. (1994) 'Parks and people: livelihood issues in National Parks Management in Thailand and Madagascar', *Development and Change*, vol 25, pp195–229.
Harris, A. R. (2011) 'Out of sight but no longer out of mind: a climate of change for marine conservation in Madagascar', *Madagascar Conservation and Development*, vol 6, pp7–14.
Keller, E. (2008) 'The banana and the moon: conservation and the Malagasy ethos of life in Masoala, Madagascar', *American Ethnologist*, vol 35, pp650–664.
Kull, C. A. (1996) 'The evolution of conservation efforts in Madagascar', *International Environmental Affairs*, vol 8, pp50–86.
MAP (2007) *Madagascar Action Plan 2007–2012: A Bold and Exciting Plan for Rapid Development*, Government of Madagascar, Antananarivo.
Marie, C. M., Sibelet, N., Dulcire, M., Rafalimaro, M., Danthu, P. and Carrière, S. M. (2009) 'Taking into account local practices and indigenous knowledge in an

emergency conservation context in Madagascar', *Biodiversity and Conservation*, vol 18, pp2759–2777.

McGrath, M. (2005) 'Falling from the tree', *The Guardian*, London, 22 October.

MEP (1990) *Chartre de L'Environnement*, Ministère de l'Economie et du Plan, Antananarivo.

Mercer, E., Kramer, R. and Sharma, N. (1995) 'Rainforest tourism: estimating the benefits of tourism development in a new national park in Madagascar', *Journal of Forest Economics*, vol 1, pp239–269.

Ministère du Tourisme (2012) www.mtoura.gov.mg, accessed 30 December 2012.

Minten, B. and Méral, P. (2006) *International Trade and Environmental Degradation: A Case Study on the Loss of Spiny Forest in Madagascar*, World Wild Fund For Nature, Antananarivo.

Myers, N., Mittermeier, R. A., Mittermeier, C. G., da Fonseca, G. A. B. and Kent, J. (2000) 'Biodiversity hotspots for conservation priorities', *Nature*, vol 403, pp853–858.

Norris, S. (2006) 'Madagascar defiant', *BioScience*, vol 56, pp960–965.

Ormsby, A. and Mannie, K. (2006) 'Ecotourism benefits and the role of local guides at Masoala National Park, Madagascar', *Journal of Sustainable Tourism*, vol 14, pp271–287.

Peters, J. (1998) 'Transforming the integrated conservation and development project (ICDP) approach: observations from the Ranomafana National Park Project, Madagascar', *Journal of Agricultural and Environmental Ethics*, vol 11, pp17–47.

Peters, J. (1999) 'Understanding conflicts between people and parks at Ranomafana, Madagascar', *Agriculture and Human Values*, vol 16, pp65–74.

Pollini, J. (2011) 'The difficult reconciliation of conservation and development objectives: the case of the Malagasy Environmental Action Plan', *Human Organization*, vol 70, pp74–87.

Raharinjanahary, L. (2004) *Etude Socio-culturelle et Economique dans le Cadre du Processus de Mise en Place du Site de Conservation du Menabe Central*, Comité Régional du Développement Menabe, Morondava.

Reardon, T. and Vosti, S. A. (1995) 'Links between rural poverty and the environment in developing countries: asset categories and investment poverty', *World Development*, vol 31, pp1933–1946.

Robbins, P. (2004) *Political Ecology: A Critical Introduction*, Blackwell Publishing, Malden.

Scales, I. R. (2011) 'Farming at the forest frontier: land use and landscape change in western Madagascar, 1896 to 2005', *Environment and History*, vol 17, pp499–524.

Scales, I. R. (2012) 'Lost in translation: conflicting views of deforestation, land use and identity in western Madagascar', *The Geographical Journal*, vol 178, pp67–79.

Seddon, N., Tobias, J., Yount, J. W., Ramanampamonjy, J. R., Butchart, S. and Randrianizahana, H. (2000) 'Conservation issues and priorities in the Mikea Forest of south-west Madagascar', *Oryx*, vol 34, pp287–304.

Steneck, R. S. (2009) 'Marine conservation: moving beyond Malthus', *Current Biology*, vol 19, R117–R119.

Stephenson, P. J. (1993) 'The impacts of tourism on nature-reserves in Madagascar: Perinet, a case-study', *Environmental Conservation*, vol 20, pp262–265.

Stott, P. and Sullivan, S. (2000) *Political Ecology: Science, Myth and Power*, Arnold, London.

Tonganiriko, B. K., Rakotoarison, B. and Rivoarijaona, A. (2002) *Stratégie de Redressement du CFPF*, Centre de Formation Professionnelle Forestière, Morondava.

Twinning-Ward, L. (2009) *Sub-Saharan Africa Tourism Industry Research*, World Bank, Washington, DC.

Walsh, A. (2005) 'The obvious aspects of ecological underprivilege in Ankarana, Northern Madagascar', *American Anthropologist*, vol 107, pp654–665.

West, P., Igoe, J. and Brockington, D. (2006) 'Parks and peoples: the social impact of protected areas', *Annual Review of Anthropology*, vol 35, pp251–277.

12 Bioprospecting a biodiversity hotspot

The political economy of natural products drug discovery for conservation goals in Madagascar

Benjamin D. Neimark and Laura M. Tilghman

Introduction

> Yes, I feel that we [the International Cooperative Biodiversity Groups (ICBG)-Madagascar bioprospecting project] have done good work, even though at this point we do not have a drug to show for it. Probably the biggest benefit to date has been the establishment of the *Montagne des Français* as a protected area. There have also been numerous training activities and infrastructure improvements... in addition to the economic development projects funded by the upfront contribution.[1]
> Principle Investigator, ICBG-Madagascar project

> If they [ICBG-Madagascar] tell us that they get new drugs from the plants, and not hide it, maybe there will be a benefit for people in the village. Still, we didn't know why they had gone into the forest, since it was only after they arrived there that we knew. And we didn't know if they had permits or not.[2]
> Rural resident living adjacent to a plant collection site

In the early 1990s, there was a great deal of optimism in the conservation community that drug discovery from nature, known as 'bioprospecting,' could single-handedly accomplish the development goals set forth at the Earth Summit in Rio de Janeiro.[3] Principal among these goals included community participation in conservation and income generation in areas of high biodiversity in the tropics (Abelson, 1990). At the time, several large-scale bioprospecting ventures were launched. Most visible among these were the 1991 bilateral contract between Costa Rica's National Biodiversity Institute (INBio) and pharmaceutical giant Merck (Reid et al., 1993), and the US federally funded ICBG program (Rosenthal, 1999). Excitement garnered by these high-profile projects helped establish bioprospecting as the contemporary model for sustainable development.

Madagascar has long been an important destination for bioprospecting. During its early contact with Europeans, explorers and naturalists traveled the island's far-flung ecosystems collecting samples of its botanical riches and medicinal remedies. From the northern limestone outcrops (*tsingy*) to the extremely dry forests in the south-east and humid eastern forest corridor, Madagascar holds roughly 12,000–14,000 flowering plants species and some of the highest percentage of endemic mammals, earning the moniker – the world's '8th continent' (Tyson, 2000). It is this floral and faunal diversity that provides contemporary bioprospectors a wide breadth of unique samples to test for new pharmaceuticals.

Madagascar is also an ideal location for bioprospecting given the generally long-standing funding for conservation and development interventions. Its 15-year National Environmental Action Plan (NEAP), one of the first of its kind in Africa, was accompanied by immense funding from a host of multilateral and bilateral donors creating conditions for large environmental non-governmental organizations (NGOs) to flourish (see Chapter 7 by Kull). For example, Conservation International (CI)'s hotspot strategy identified Madagascar as one of its 'hottest hotspots' for biodiversity conservation – a place where conservation dollars may be best spent (Myers et al., 2000, p853).[4] This funding for environmental programs has resulted in some of the most intensive conservation and development interventions across Africa, including one of the largest drug discovery projects to date – the ICBG-Madagascar project.

However, is bioprospecting on the island too good to be true? This chapter's purpose is to investigate the winners and losers in Madagascar's rapidly expanding bioprospecting sector and lay out future implications for sustainability. Using a political–ecological lens, we survey the rhetoric surrounding bioprospecting, questioning if the practice is truly a way forward for sustainable development (see also Hayden, 2003; Neimark, 2012). First, we begin with a brief history of the practice of drug discovery from nature, laying out the legal and ethical dilemmas and debates that have continually plagued bioprospecting since its inception. In the second section, we give the programmatic framework for bioprospecting in Madagascar and discuss how access to biogenetic resources is mediated at various scales and by different stakeholders. In the third section, we provide a case study of the ICBG-Madagascar project, highlighting the benefits and burdens of bioprospecting and the spectrum of compensation mechanisms promised and delivered at different levels. We conclude by discussing sustainability within an environmental justice framework and address questions of bioprospecting as a tool for conservation.

History of natural products discovery

Even though bioprospecting is generally thought of as a modern practice, the reality is that humans have used natural resources for medicinal and functional purposes for thousands of years. Evidence of such use is remarkably well documented. For example, the Chinese *Materia Medica* (125 BC) is a landmark medicinal text that provides glimpses of many medicinal plant prescriptions and over 1,000 drugs utilized during ancient ruling dynasties (Cragg and Newman, 2005; Sneader, 2005). In India, *Atharvarvada*, a text thought to be the last of the Vedas or the Brahamanic constitutions of Hinduism and which dates back to 1000 BC, is filled with countless references to the use and preparation of medicinal plants for healing and spiritual purposes (Sneader, 2005). From the fifth to twelfth century, Arab civilizations became the center of medicinal plant use and knowledge. Physicians of this period, including the Abu Bakr al-Razi, the Abu Al-Qasim Al-Zahrawi, and the Persian philosopher, Avicenna, published some of the most influential medical practices using herbal remedies known at that time (Sneader, 2005).

European exploration in the New World proved very important for expanding the range and scope of medicinal plant use. Some sources claim that the discovery of the medicinal remedy for intermittent sickness originated with Jesuit priests who observed Indians in Quito, Peru, using cinchona bark (*Cinchona officinalis*) in a decoction to reduce shivering and cold spells. This medicinal remedy subsequently led scientists to isolate quinine to treat malaria. Another important discovery brought over from the New World to Europe was used for amoebic dysentery derived from ipecacuanha root (*Cephaelis ipecacuanha*). This compound is still used today as an emetic for respiratory infections (Sneader, 2005).

In the US, heightened interest in drug discovery from natural products came in the 1940s with the demand for antibiotics to treat wounded soldiers during WWII. Government contracts with Pfizer, Inc., for the mass production of penicillin, an antibiotic derived from *Penicillium* fungi, spurred rapid advancement in the science of drug discovery. This drive for drugs from nature was spearheaded by the Natural Products Branch of the National Cancer Institute (NCI), which began a massive program of collecting biological resources worldwide (Cragg and Newman, 2005). The first NCI plant screening program (1955–1982) included 14,000 crude natural products (plant, marine, microorganisms) sourced from 60 different countries (Aylward, 1995). In 1986, spurred by the discoveries of the anticancer drug paclitaxel (Taxol) from the bark of the Pacific yew tree (*Taxus brevifolia*), the NCI natural products program began a second phase of natural products research (Aylward, 1995).

In the 1990s, the science of natural products drug discovery began to change dramatically. The changes were mainly due to new advances in drug screening and information technology (Miller, 2007). While bioprospecting was still based on biogenetic resources collected from plant, marine, and microbial organisms, scientific innovations allowed easier isolation and more efficient analysis of chemical compounds and genetic material found in these organisms.

However, many changes in the process of bioprospecting that began in the 1990s can also be attributed to growing awareness of the possible commercial exploitation of local traditional knowledge and natural resources.

Many of these ethical issues that were addressed at the Earth Summit led to the subsequent signing of the Convention on Biological Diversity (CBD) in 1993. The concerns at the core of the CBD included not only the conservation of biodiversity but also the safeguarding of intellectual property rights for those engaged in research, the promotion of economic and social development, and, most noteworthy, the equitable distribution of benefits from the exchange or use of biodiversity and ethnobotanical knowledge (Schweitzer et al., 1991; see also Rosenthal, 1999; Brown, 2003).[5]

In response, bioprospecting schemes post-CBD, such as INBio and ICBG (described in more detail below), were founded on the logic that natural product discoveries should be monetarily rewarded in a pre-determined compensation deal known as an 'Access and Benefit-Sharing' (ABS) agreement (Eisner, 1992; Barrett and Lybertt, 2000). For many involved, this amounted to a 'win–win' providing a mechanism to finance conservation efforts in tropical ecosystems, where both biodiversity and traditional knowledge of medicinal usage were deemed to be the highest (Balick, 1990). To the US scientific community at the time, bioprospecting created an opportunity for global recognition, financing, and natural resource sourcing on a scale previously unforeseen.

One of the first bioprospecting projects designed to meet many of the CBD goals and provisions was based on an agreement signed in September 1991 which brought the US pharmaceutical giant, Merck and Co., and INBio together on a joint research-sharing platform. According to the terms of the agreement, Merck paid US$1.135 million for a two-year research and sampling project and royalties on products subsequently commercialized from plant, insect, and other biological samples (Reid et al., 1993). In return, INBio was to contribute 10 percent of the budget and 50 percent of any royalties to biodiversity conservation efforts in Costa Rica (Reid et al., 1993). This bioprospecting agreement, which ran from 1992 to 1997, was lauded by many, and heralded as delivering on many of the core tenets of the CBD (Reid et al., 1993). Moreover, it signified a major transformation in the way public–private collaborations between developed and developing countries operated, paving the way for subsequent similar bioprospecting initiatives (Aylward, 1995).

A more far-reaching in its geographic scale and long-term bioprospecting project post-CBD is the ICBG program. The ICBG was funded by the National Institutes of Health (NIH), the National Science Foundation (NSF), and the United States Agency for International Development (USAID).[6] Since its founding, the ICBG has expanded into one of the most ambitious bioprospecting projects ever attempted by the US government involving eight collaborative research groups which conduct research in 12 countries (Rosenthal and Katz, 2004). By 1999, the ICBG had reported the collection of 11,000 samples from approximately 5,800 species of plants, and over 500 insects and fungi, respectively. The ICBG had conducted up to 200,000 different types of therapeutic screens (Rosenthal and Katz, 2004), located 260 active compounds, in which at least 60 are reported as 'novel' (Rosenthal et al., 1999).

The main differences between the NCI collections of the mid-twentieth century and the INBio and ICBG of the 1990s and beyond are the projects' institutional structural complexity. The latter are collections of both private and public research laboratories, corporate partners, and a variety of civil society organizations and individuals, with each of these holding a special role in the bioprospecting process. Ironically, as we will see below, many of the roles of these groups evolved out of the ethical guidelines set under the CBD, but had the effect of building layers of regulatory barriers which decreased transparency. Due to these and other ethical questions surrounding the commercialization of traditional knowledge and fair compensation to participating local actors, critiques of the practice began to appear which eventually built to a wave of widespread re-assessment of bioprospecting.

'Biopiracy' and beyond: ethical and legal debates about bioprospecting

Drug discovery under bioprospecting is a difficult multi-step process that takes selected chemical compounds and structures from natural products and randomly screens them against a selected disease target (HIV/AIDS, cancer, etc.) (Aylward, 1995). Once bioactive substances are eventually found, the natural product must then be isolated through multiple fractionation and purification, requiring advanced knowledge and training in organic and inorganic chemistry, and access to libraries of known compounds (Weiss and Eisner, 1998). What was once a long and arduous process, advances in drug discovery science and genomics observed in the 1990s – including the advent of new screening technology – speeded up the process considerably, paving the way for researchers to run thousands of extracts of biological resources at rates commercially attractive to large-scale private laboratories and pharmaceutical firms (Miller, 2007). As private sector involvement increased, so did the concern for fair compensation for those who supplied the resources and intellectual property which led to the discovery. Concerns stemmed from the legacy of unjust colonial extraction of natural resources, such as rubber and timber extraction in the Congo; precious minerals, diamonds and gemstones in South and Southern Africa, respectively; and ongoing commercial exploitation in the developing world. These fears coincided with scientists' growing anxiety for the environment, including mass species extinction (Dorsey, 2003).

By the early 2000s, much of the initial euphoria of bioprospecting as a triple-edged panacea for illness, environmental destruction, and poverty was replaced with doubt and caution. While some of this was a natural process as high expectations met the sluggish reality of pharmaceutical development (Macilwain, 1998), it also was a result of a growing body of literature from both scholars and civil society groups that was deeply critical of bioprospecting. The Canadian activist group Rural Advancement Foundation International (RAFI) coined the term 'biopiracy' (Svarstad, 2005), which was then elaborated by the Indian scholar–activist Vandana Shiva in a book by the same name (Shiva, 1997). The main message of such criticisms was that bioprospecting was a reinvented and reinvigorated form of colonialism and theft of developing countries' natural resources and traditional knowledge by corporations and scientists of industrialized countries. The growing backlash against bioprospecting reached a new level when protests in the mid-1990s forced the early termination of an ICBG-funded project in southern Mexico before any real plant collecting had begun (Dalton, 2001; Rosenthal, 2006), demonstrating how salient critiques of the practice can be.

While some people involved in bioprospecting seemed blindsided by their vocal and media-savvy critics (e.g. Berlin and Berlin, 2004), a careful look at the political economic landscape of the late twentieth century reveals many clues to suggest that criticism of bioprospecting was if not unavoidable than at least unsurprising. The 'biopiracy' battles of the 1990s and 2000s involved many of same rhetoric of the 'seed wars' in the 1970s and 1980s, a period when activists and scholars debated the merits of collecting, privatizing, and genetically modifying agricultural germplasm. Both faced criticisms that had a common trope of theft by multi-national corporations and industrialized countries of the resources and traditional knowledge found in developing countries. The 'victims' changed from agricultural crop varieties developed by small-scale farmers, to genes of tropical plants guarded by indigenous healers, but the overall message was the same and indeed was conveyed by some of the same activist groups (Hamilton, 2006).

Furthermore, it should be noted that modern bioprospecting arose at a time of increasing activism in developing countries for indigenous rights and sovereignty (Hodgson, 2002), as well concern that traditional environmental knowledge (TEK) of these indigenous peoples was threatened (Bird and Sattaur, 1991; Plotkin, 1993). Thus it should be no surprise that bioprospecting is often the most controversial in countries with active indigenous rights organizations such as Mexico and Peru (Brown, 2003; Greene, 2004).

Underlying much of biopiracy literature and activism are deeper concerns about the commodification of nature (see Chapter 13 by Brimont and Bidaud) and in particular so-called 'patents on life,' or owning nature under market-led conservation projects as well as the messy international North–South relations concerning the trade of biodiversity and agricultural crops. Activism around biopiracy is thus part of a larger social movement against globalization and neoliberalism. Some of the biopiracy claims rely on over-simplification or even erroneous representation of the mundane process of bioprospecting research and product development, as we can see from the rosy periwinkle (Box 12.1; see also discussion in Moran et al., 2001; Svarstad, 2005). However, as Hamilton (2006) convincingly argues, we ignore biopiracy claims at our peril, since they raise important concerns about intellectual property and benefit sharing that must be addressed beyond simple modifications to the *status quo*.

Box 12.1 Madagascar's rosy periwinkle: fodder for the debate

Madagascar has played an interesting role in the debate around bioprospecting, being used as supporting evidence for both proponents and critics. However, it is not current bioprospecting activities that are discussed, but research that took place in the 1950s. Researchers looking at the rosy periwinkle (*Catharansus roseus*) (Figure 12.1) from Madagascar isolated alkaloids that led to the creation of blockbuster cancer drugs to treat childhood leukemia and Hodgkin's disease. Proponents of bioprospecting used Madagascar's rosy periwinkle as proof that tropical environments could indeed lead to cures for major illnesses (e.g. Swerdlow and Johnson, 2000), while critics noted that Eli Lilly, the pharmaceutical firm to develop these medicines, gained millions of dollars of profits while the country of Madagascar received nothing in return (Kadidal, 1993).

Figure 12.1 Rosy periwinkle: medicinal wonder plant or the global case of 'biopiracy'?

More careful scholarship shows that both proponents and critics alike have misrepresented the rosy periwinkle story (see discussions in Tilghman, 2004; Harper, 2005; Neimark, 2009). While the periwinkle does indeed originate from Madagascar, it has for centuries been propagated around the world. Nor is it a classic case of cures coming from endangered ecosystems, as the plant is a weedy shrub found in arid environments and secondary growth. While the periwinkle is indeed used in Malagasy traditional medicine, researchers at Eli Lilly were initially drawn to the plant from reports of its use as an herbal remedy by Caribbean and Filipino communities. Furthermore, while Madagascar received no royalties from the sale of Eli Lilly's cancer drugs, it did export large amounts of plant material and thus receive financial benefits for those involved in its collection and transport.

The difficulties in ensuring that all parties are informed and share in the benefits deriving from commercialization have caused confusion in the industry about the correct way to move forward with natural product research (ten Kate and Laird, 1999, 2000). These problems led many in the pharmaceutical industry to explore other options to discover drugs, such as computer-generated and synthetic-based compounds instead of natural sources. These new efforts are seen as 'rational' and 'scientific,' and have the advantage of leading to new drug discoveries without all the political entanglements that come with 'nature' (Parry, 2004).

Intellectual Property Rights

Intellectual Property Rights (IPRs) are goods that are intellectually derived and protected through various ways, such as copyrights, patents, trademarks, industrial design, trade secrets, and domain names. These protection mechanisms in effect provide the 'holder' the right to maintain exclusive control over the material (Walden, 1995). The economic rationale for IPRs asserts that allowing individuals or companies to have monopoly rights provides 'motivation' and 'remuneration' for the inventor having the desired effect of fending off competitors (Walden, 1995, p182). Bioprospectors use IPRs to make the most of their investments by restricting access to the compounds and processes they discover and thus being the sole provider of a good or service on the market.

Internationally, the World Trade Organization's (WTO) Agreement on Trade Related Aspects of Intellectual Property Rights (TRIPS) was the main force and fury behind the debate of IPRs. TRIPS emphasizes private property rights, especially concerning intellectual property, providing an easy way for companies and individuals to patent discoveries made from nature based on scientific or traditional knowledge.

One of the most contentious issues facing IPRs has to do with claims of 'novelty' involving biogenetic resources and traditional knowledge.[7] Critics claim that the patenting of biological life under biotechnology and bioprospecting breaches ethical boundaries, setting a damaging precedent for corporate control of life forms. For these critical scholars and activists, such as Vandana Shiva, 'traditional' knowledge was never formally accounted for under the TRIPS agreement, and this meant that there would be no protection and no compensation given for the hard work, creativity, and countless years of medicinal plant and agricultural crop selection through traditional 'trial and error' experimentation by diverse traditional communities (Shiva, 1997). Second, TRIPS made it easier to privatize knowledge under the framework of patent rights, potentially taking away the ability of local communities to now use the crop or plant medicinally and thus violating a company's IPRs. And last, many individuals and communities do not see themselves as owners of the biodiversity and associated knowledge per se, but rather 'keepers' of it and thus feel that is not correct to negotiate agreements to commercialize it (McAfee, 1999; GRAIN, 2004).

The TRIPS agreement was not the only international framework regarding intellectual property, and there are several agreements that have attempted to recognize traditional access rights and 'cultural' knowledge. Other examples include the 1988 International Conference of *Belém*, Brazil, and the second Code of Ethics of the International Society of Ethnobotany codified in 1991 in Kunming, China (Soejarto et al., 2005). These agreements were the first formal recognition that traditional knowledge was not only 'intellectual,' but also 'innovative,' and, thus, was protected under any formal patent rights (Soejarto et al., 2005, p16; see also Posey and Dutfield, 1996). Both of these agreements were codified under the CBD in 1992, which again established a number of regulatory hurdles that bioprospectors had to overcome in order to access and utilize traditional medicinal knowledge, including obtaining informed consent from all the parties involved in the exchange and sharing benefits with those who supplied the knowledge. The tension between TRIPS, which emphasizes privatizing knowledge and the needs of corporations, and the CBD, which recognizes the importance of traditional knowledge and the needs of nation states and communities, still has not been fully resolved.

IPRs have also become a touchstone of biopiracy activism, helped by several cases in which patents were granted to US-based individuals or institutions for plants with long histories of use by farmers and indigenous peoples, such as the neem tree in India (Burns, 1995) or the Amazonian *ayahuasca* (Dorsey, 2003). The main claim of biopiracy activists and scholars is that the international IPR regime favors industrialized countries, while disenfranchising local communities and indigenous peoples who do not have the resources to file or dispute patents. Furthermore, while centuries of experimentation and observation by people living in areas of biodiversity have allowed them to gain knowledge about the medicinal uses of plants, this knowledge cannot be protected under the international IPR regime because it is not novel. Critics say that bioprospectors who claim to 'discover' novel natural compounds are in fact benefitting from the local people who protected the resource at a cost to themselves, or worse using their knowledge to focus their plant collection and analysis without any acknowledgement.

Access and Benefit-Sharing (ABS)

The second aspect of bioprospecting that has received a great deal of attention from critics is Access and Benefit-Sharing (ABS) arrangements, the goal of which are to ensure that access to natural resources is given with full informed consent, and that the benefits of commercializing those natural resources is shared with all necessary stakeholders.

Of the different types of benefits that may arise from a bioprospecting project, *royalties* and *milestone payments* are the monetary benefits that have been most analyzed, but least realized (ten Kate and Laird, 1999). In a bioprospecting project, 'milestone' payments are usually generated when significant discoveries are made at successive stages of the research process, whereas 'royalties' only come following the full commercialization of a natural product (ten Kate and Laird, 1999; see also Miller, 2007). There have been only a few reported cases where cash payments in the form of royalties were shared by rural actors incorporated into a bioprospecting project (Barrett and Lybbert, 2000; Laird, 2002; Lybbert et al., 2002). Beyond these milestone payments and royalties, monetary benefits can also include financing conservation or development projects in the country generally or targeted to the bioprospecting collection sites. Non-monetary benefits that may arise from a bioprospecting project could include technology transfer and capacity building to reinforce the capacity of developing nations to conduct their own scientific research, or agreements for bioprospectors to focus on diseases of high priority for the country where biogenetic resources are collected.

ABS is explicitly required by the CBD in Article 8j, but critics of bioprospecting express concern that international power dynamics put both developing countries and local communities within their borders at a disadvantage in negotiating agreements with bioprospectors from industrialized countries. As a prominent biopiracy activist stated: 'The current socio-economic environment makes mutually-beneficial contracts unlikely or impossible' (Mooney, 2000, p37). In other words, the power wielded by wealthy industrialized countries and multinational corporations and their greater influence in developing international treaties and regulations puts them at a distinct advantage at the negotiating table. Second, it costs thousands of dollars in legal fees and significant time and knowledge of the legal process to draft a mutually beneficial contract. Access to such resources is just not possible for small communities and, thus, it is not really an a option to engage in such procedural agreements without considerable support from outside groups.[8]

Mediating access to Madagascar's biogenetic resources

Bioprospectors seek access to biogenetic resources so that they can be analyzed for potential useful compounds that may have commercial applications. Some bioprospectors are also interested in TEK, the know-how of healers, farmers, and rural residents of how to use biogenetic resources. Beyond the international agreements that have set the general rules of the game, principally the CBD and TRIPS, bioprospectors must also negotiate various rules and interest groups at different scales that mediate access to biological and cultural resources. In this section, we will discuss the principal mediating factors impacting bioprospecting in Madagascar, as well as some ways in which bioprospectors have sought to sidestep these controls.

Accessing land and plants in Madagascar

Land tenure in Madagascar, particularly in rural areas of interest to bioprospectors, mixes Western property rights introduced by the French with traditional Malagasy customary laws usually emphasizing inheritance of individual plots of land through clans or lineages and larger spaces of communal use governed by local authorities (Healy and Ratsimbarison, 1998). Bioprospectors working in Madagascar seem thus far to have sidestepped the complexities of negotiating access in this patchwork of different land tenure regimes by negotiating directly with the national government and collecting in areas that are at least nominally state-managed.

The protected areas code[9] does not place legal limits on bioprospecting within state-managed protected areas, nor does it specifically address spatial limits to where bioprospecting may take place. According to a leading administrator of the system of protected areas, national parks and other reserves are meant to be off-limits to bioprospectors (Quansah, 2003). However, associated scientific research activities for explicit commercial outputs – considered by some to be bioprospecting – have been conducted in protected areas including national parks in the past (Quansah, 2003). Furthermore, the expansion of protected areas under new protected area schemes in Madagascar has created conservation sites that may now be managed by communities (see Chapter 8 by Pollini et al.), but are also open to bioprospecting by companies or private individuals. In effect, some sites, such as *Montagne des Français* in northern Madagascar, are in danger of essentially becoming protected extractive reserves, an outcome that has raised concerns all the way up to the highest levels of national office (Neimark and Schroeder, 2009).

Accessing medicinal plant knowledge in Madagascar

Malagasy TEK,[10] particularly of medicinal plants, is prominent throughout the island as is its commercialization, as evidenced by a booming national trade in raw plants (Dauphiné, 2002) and many Malagasy companies producing processed medicinal plant remedies for the national market (Rasoanaivo, 1990). Medicinal plant knowledge can be generally divided into two categories (Tilghman, 2004):

1   Specialist/spiritual plant knowledge is the domain of traditional healers and diviners (called *ombiasy* generally). Traditional healers treat illnesses of a spiritual nature that are the result of human agency, intentional or accidental, in the form of spirit possession or broken taboos (*fady*). The traditional healer invokes and consults ancestral spirits for diagnosis; thus, knowledge of how to use plants is actually the domain of the ancestors and not the traditional healer per se.

2 Non-specialist/secular medicinal plant knowledge is the domain of the general populace and acquired via family and social networks throughout one's life. It is used to treat non-spiritual diseases that are caused by imbalances in hygiene, physical activity, and diet. This type of knowledge is part of the common domain, and its use is mediated by differences in personal preference as well as depth and range of social networks (see Chapter 14 by Kaufmann for more on indigenous environmental knowledge).

Bioprospectors seeking to mine the wealth of Malagasy TEK face several hurdles. First, conferring with traditional healers who have more specialized knowledge of plants is complicated by the fact that they see themselves as vessels for transferring (or protecting) ancestral knowledge rather than as knowledgeable individuals in their own right. Additionally, specialist plant knowledge is inextricably linked to other aspects of the Malagasy healing system such as divination, astrology, and sprit possession (Sharp, 1993; Harper, 2002), which are largely not of interest to bioprospectors. While conferring with non-specialists would seem to avoid some of these issues, bioprospectors may be dismayed to learn that many Malagasy folk remedies rely on plants from disturbed areas close to habitations rather than the more unique (and patentable) primary forest plants (Lyon and Hardesty, 2005).

Avoiding or attenuating regulation

Madagascar has yet to develop a comprehensive system that regulates bioprospecting directly and at the scale of the nation-state. However, given the history of bioprospecting initiatives by both national research laboratories and Malagasy independent organizations, there is a framework by which researchers must adhere to to extract resources for drug discovery. Most notably, it allows regional and local authorities the power to regulate access to genetic resources policy and, as noted by Quansah (2003), allows large-scale NGOs to actively promote community-level, genetic, resource-based activities.

Bioprospectors, once they have gained access to land and/or knowledge, are basically left to create their own protocols regarding ABS, IPR, and other ethical dimensions of the bioprospecting process. Yet government institutions in Madagascar have been seen to have inadequate capacity to understand these frameworks, to disseminate information to national, regional, and local stakeholders, and to effectively enforce individual agreements as well as any pertinent national laws or regulations.

Case study: the ICBG-Madagascar project

The ICBG-Madagascar project has been conducting research on Madagascar's unique flora and fauna since 1998. Phase I of the project

(1998–2003) collected plants in the area of the Zahamena National Park, in the eastern forests of the Analanjirofo Region to the northwest of the city of Toamasina. Phase II of the project (2003–2008, extended to 2013) collected plants in the forests of the Diana Region to the south of the city of Antsiranana, as well as marine specimens in the northern coastal areas of the country near Nosy Be (Figure 12.2).

As ICBG-Madagascar developed and expanded its operations from Phase I to Phase II, the project had to make many adjustments. Phase I collected plant material in a relatively small area of eastern montane rainforest, while Phase II collected plant material over a much larger area with great biogeographic scope ranging from dry *xenotrophic* to moist, high-altitude and montane forests.[11] Phase I set up operations in an area where one of the partner organizations (CI) had already been working for over a decade, and thus the project was largely fused with the preexisting social relationships that the organization had developed with local authorities and communities. Phase II moved to an area where the core ICBG organizations had not previously worked, requiring them to develop new partnerships with other organizations and build community relations from the ground up, if at all. With this expanded geographic focus and new working relationships, the project's dynamics changed. First, project staff in Phase II needed to hire many disparate groups that did not necessarily live in the villages adjacent to the forest collection sites, and thus the people the project came into contact with did not always know the project's goals or mission. Local development projects as part of the project's ABS mechanisms in Phase II were scattered over a much greater area rather than only adjacent to plant collection sites, in essence decoupling biogenetic resource extraction and benefit sharing through community development. Second, the large range of collection in Phase II presented many logistical challenges, including the distance that needed to be covered to get to plants and the time it took to get the plant material back to the laboratory before it began to decompose. Ultimately, this meant working with a number of smaller groups over a much larger area and, in return, the project sacrificed long-term relationships as seen in Phase I.

Financial structure of the ICBG-Madagascar project[12]

According to the ICBG website,[13] since 2000, total funding derived from US inter-agency sources topped approximately a little over US$9.3 million. According to David Kingston, the Principle Investigator of ICBG-Madagascar, the total annual costs for running the project are roughly US$700,000. Yet it is not a total dollar amount that can be used to evaluate the equitability of a bioprospecting project, but rather who controls how that money is spent and what mechanisms are in place for transparency and collaboration (Table 12.1).

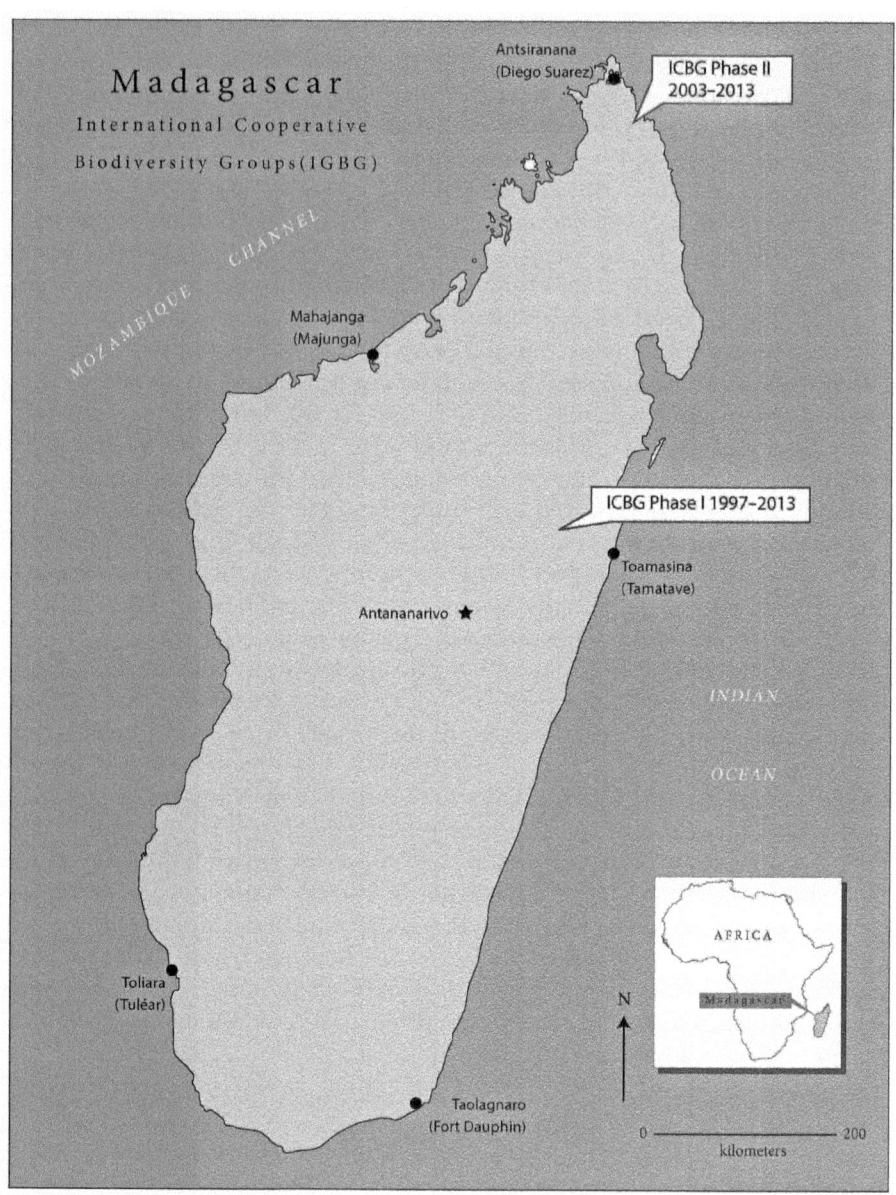

Figure 12.2 ICBG-Madagascar project plant collection locations

Table 12.1 Role and function of US- and Madagascar-based funding institutions and research organizations in the ICBG-Madagascar

US-based funding institutions and research organizations in the ICBG

Institution/organization name	Acronym	Role and function for ICBG
Fogarty Institute and other National Institutes of Health	NIH	Lead administrative body and major funder for the ICBG programs worldwide
National Science Foundation	NSF	Major funding body
US Department of Agriculture	USDA	Major funding body
Virginia Polytechnic Institute and State University	VPISU	US lead for research on drug discovery (laboratory analysis and testing of samples)
Dow AgroSciences	DAS	Funding for upfront compensation/private research for agrochemicals
Eisai Pharmaceuticals	EISAI	Funding for upfront compensation/research for pharmaceuticals
Missouri Botanical Garden	MBG	Lead botanical agency (plant collection)
Conservation International	CI	Administers conservation projects

Madagascar-based funding institutions and research organizations in the ICBG

Institution/organization name	Acronym	Role and function for ICBG
Centre National d'Applications des Recherches Pharmaceutiques	CNARP	Lead research institution conducting drug discovery in Madagascar (plant processing, analysis, and laboratory testing on all samples)
Centre National de Recherches Sur l'Environnement	CNRE	Leads research on marine species
Centre National de Recherches Océanographiques	CNRO	Conducts research and collection on marine species
Office National pour l'Environnement	ONE	Head Malagasy administration for conservation trust fund
Malagasy Teknisiana Mivondronaho Aro sy TEzan'i Zahamena ary ny Ala atsinanana	MATEZA	Local conservation and development projects (Phase I)
Service d'Appui à la Gestion de l'Environnement	SAGE	Local conservation and development (Phase II)

While the ICBG-Madagascar project is multilateral, funds are not distributed to or controlled equally by all partners. The operating budget for the project is centralized at Virginia Polytechnic Institute and State University (VPISU), so that the other organizations act as subcontractors. For the two Malagasy partners who participate in biological collection and research, *Centre National de Recherches Sur l'Environnement* (CNRE) and *Centre National d'Applications des Recherches Pharmaceutiques* (CNARP), the money for their operating costs rests in a 'parked' account and is distributed to each partner depending on their activities and involvement. However, CNARP's funds are managed by the US-based Missouri Botanical Gardens (MBG); whereas CNRE's funds are dispersed directly to them. Although the initial decision to have MBG manage their money was made in concert with the previous director of CNARP, it now causes considerable anxiety among those now leading the project. This loss of autonomy is coupled with criticisms from some that the ICBG is selling Malagasy resources 'on the cheap.' Many of the US-based private partners, such as MBG and CI, travel to ICBG meetings in SUVs and are able to continually tap into pools of money for their research, while CNARP scrapes by to cover its administration costs. CNRE, on the other hand, has the ability to source a large amount of government and private funding due to its wider environmental research focus, placing the institution in a much better position than CNARP in terms of equipment and materials, despite also being a Madagascar-based institution.

The ICBG program at the NIH funds the operating costs of the project, while funding for ABS mechanisms is provided directly by the private commercial partners (Eisai Pharmaceuticals and Dow Agrosciences), at roughly US$50,000 annually each. This money goes directly into a fund that is managed by the MBG and sent once a year to the National Environmental Office (*Office National pour Environnement*, or ONE). Two Malagasy partner institutions are co-signatories, the Director of CNARP and the Secretary General of ONE. Half of this money goes to support small-scale rural development projects (called 'Upfront Compensation') and the other half is placed into a Biodiversity 'Trust-Fund' which will gain interest to support ongoing conservation activities. It is unclear how or when this money, if ever, will be filtered back to Malagasy research laboratories so that they can conduct their work, nor to those target communities adjacent to collection areas to foster a biodiversity conservation ethic amongst residents – two of the main goals of the project. Either way, the state institutions of Madagascar specifically charged with organizing environmental programs are thus provided a subsidy for development and conservation activities that they can maneuver as they see fit, again placing considerable power in the hands of particular actors to manage and control action in the rural landscape.

Local benefits and perceptions of ICBG-Madagascar

ICBG-Madagascar, through its private commercial partners Eisai Pharmaceuticals and Dow Agrosciences, financed a number of development projects to share benefits locally as part of its ABS mechanisms. This tactic of providing upfront compensation to people living near plant collection sites is now a widely used practice in the ICBG program and elsewhere, as it ensures that benefits to developing countries and local communities do not rest entirely on the risky and lengthy process of pharmaceutical development.

In Phase I, upfront compensation took the form of construction projects: two large grain storehouses, one community center, one bridge, and one primary school. Visits to the region two years after the completion and inauguration of the projects revealed that they had met with varied success. Both grain storehouses were barely in use and the bridge had been destroyed in a cyclone. Some of the upfront compensation programs included in Phase II of the ICBG included two wells, a watering trough, a small dam, up to 20 horticultural gardens, and two new hen houses. Time will tell how these projects fare as sustainable ABS mechanisms compared to Phase I efforts.

Not only was success of ICBG-Madagascar's benefit-sharing mechanisms mixed at best, access as defined by informed consent of all participating stakeholders is uncertain. This is illustrated by the large variety of Malagasy words that were used to describe the Upfront Compensation development projects of Phase I in interviews in 2003 (Table 12.2).

Table 12.2 Malagasy terms used to describe upfront compensation development activities (Phase I)

Malagasy word	English translation	Value (positive/negative/neutral)
Tombon-tsoa	Benefit; something gained or received	Positive
Valisoa	Reward for something done well	Positive
Tamby	Wages; recompense	Neutral
Takalony	Exchange for something of equal value	Neutral
Fanasoavana	Gift; an act of generosity	Positive
Asasoa	Literally 'good work,' usually refers to charity or development activities	Positive
Tamberim-bidy	Money given to compensate for the value of something that has been taken	Neutral

Source: interviews conducted in 2005 in villages along the periphery of the Zahamena protected area by Tilghman.

These words reveal that local people perceived the projects in very different ways, from a special reward, to an equal exchange, to a payment. Moreover, interview responses revealed that very few individuals understood that these projects were meant by the ICBG to compensate for plants collected in nearby forests for commercial research. Some people recognized the link between development projects and plant collection, but not in the intended way. For example, grain storage buildings were sometimes interpreted as a place for researchers to store plants or hold meetings rather than as a development project to ease local food insecurity and reduce slash-and-burn within the Zahamena protected area. This variation reflects the failure of ICBG-Madagascar to fully educate local citizens about the project, and therefore casts doubt on whether local communities could give fully informed consent. Unfortunately, the lack of dissemination of vital project information is symptomatic of some larger problems of NGOs non-engagement with local communities commonly observed across environmental sectors in Madagascar (see also Chapter 8 by Pollini et al. and Chapter 9 by Corson).

Discussion: justice and equality in the ICBG-Madagascar project

Policy surrounding bioprospecting up to this point has mainly developed around the idea that environmental inequality should be measured mainly in distributive forms (i.e. compensation provided to individuals of groups in return for their participation). Recently, however, scholars have begun to view questions of environmental equality under a more procedural or democratic decision-making framework. According to Lake (1996, p165), procedural equity entails 'full democratic participation not only in decisions affecting distributive outcomes but also, and more importantly, in the gamut of prior decisions affecting the production of costs and benefits to be distributed'. Following Lake, we hold that balancing the environmental inequalities found in bioprospecting requires incorporating a more complete definition of justice that includes both distributive and procedural mechanisms. To illustrate this concept, we describe two main bioprospecting activities in which ethical questions are raised and possible outcomes to ameliorate inequality are addressed: first, resource collection and community involvement and, second, analysis and product development.

During the resource collection phase of bioprospecting, which takes place in rural areas, Malagasy residents of nearby communities have very little knowledge about the projects and what any type of benefits they may receive from the discovery of a drug for both phases of the project. Considering the financial windfall that potentially may come from the discovery of a new pharmaceutical product, this ignorance may be seen as a purposeful attempt by bioprospectors to hold back information (i.e. their goals of drug discovery) so that rural actors will not restrict access to forested sites if they feel they are not being fairly compensated and continue to participate as manual laborers.¹⁴ This lack of knowledge about bioprospecting also questions just how Malagasy are participating in the decision-making process of drug discovery and related conservation activities. Phase II of ICBG-Madagascar led to the creation of the new conservation site of *Montagne des Francais*, located to the east of the city of Antsiranana. Although there have been some reserved optimism from rural Malagasy living nearby for the protection of local resources, there still seems to be quite a bit of confusion on their part as to just what 'protection' means and that the creation of a new conservation site will restrict some of their livelihood activities. And in some cases, residents question their ability to restrict access to foreigners coming in to collect biogenetic resources.

Many of these access dynamics that are taking shape within areas of bioprospecting collection need to be addressed head-on if significant areas of forest are to be placed under some form of protection by bioprospecting projects. To do this, bioprospectors must find creative ways to inform rural inhabitants about the project's goals and possible benefits of their activities, and devise ways that rural inhabitants can participate in the decision-making process of bioprospecting and associated conservation activities. Furthermore, conservation projects must occur in the context of a more democratic process, with input from inhabitants who are potentially most affected by the projects themselves. This process must include participation by the Malagasy state, the legal owner of the forested sites of collection. State agencies and institutions that provide bioprospectors access to these sites must be willing to hold back collection permits unless a democratic process by which rural Malagasy are informed and participate in the process of decision making.

In the analysis and product development phase of bioprospecting, uneven partnerships between collaborating laboratories have shifted power in favor of foreign scientists and laboratories. These uneven power relations have skilled Malagasy scientists conducting menial tasks such as exporting of ready-made extracts to high-tech labs in the US where American scientists conduct drug discovery. In return for the source material, host-country laboratories should be provided with current drug discovery equipment and materials so that they can conduct parallel drug discovery research using their own scientific knowledge and skills. Second, compensation for participating Malagasy scientists needs to be levied on the ability to discover new molecules and not new drugs, which can take up to 10–15 years. Finally, actors along the natural products pharmaceutical chain need to be paid a fair price, not only for their labor, but also for their skills and ability to conduct comparable research within their host-country. Furthermore, compensation for bioprospecting can also be delivered in the form of health care, either technical capacity (training of doctors, nurses, or medical technicians) or the return of pharmaceutical products to Madagascar. This idea has been discussed by some bioprospectors; however, up to this point it has continued to be dismissed by many of the commercial and research partners as not economically or politically feasible.¹⁵

Conclusion: bioprospecting and conservation of natural resources

Much of the criticism of bioprospecting focuses on its social impacts, particularly the redistributive justice of IPRs and ABS. It is surprising that much less attention has been devoted to assessing whether bioprospecting is an effective conservation mechanism, given that one of its three major goals is to help stem environmental destruction. It is to this question that we will now turn. How in theory should bioprospecting work as a market-based conservation mechanism? Bioprospecting proponents have argued that the practice would have a positive impact on the environment in two different ways.

First, it was argued that bioprospecting would be a viable alternative to more environmentally destructive activities such as logging and mining. The logic behind this argument was that environmental destruction in tropical regions was driven by economic need. As a newspaper article in the British newspaper *The Independent* put it, 'Madagascar is the unique home of perhaps 5 per cent of the world's species. It is the biological equivalent of an Arab oil sheikdom. Yet, without an income from its huge biological wealth, it has chopped down most of its forests to feed its people' (quoted in Boyle, 1996, p128). Bioprospecting would be a sustainable alternative that would provide at least as much income as environmentally destructive activities. It was argued that bioprospecting would have minimal environmental impact, since research required the collection of comparatively little biological material, and modern technology such as bioengineering and chemical synthesis would allow drugs or other products to be made with little additional impact on forests. Furthermore, equitable bioprospecting contracts that followed the guidelines of the CBD would provide immediate financial resources for additional secondary activities to protect the environment beyond the actual research and product development. These activities could range from establishment of protected areas to sustainable income-generation activities for rural individuals as an alternative to slash-and-burn agriculture.

Second, it was argued that bioprospecting would change the hearts and minds of people living in tropical countries, from government officials to peasant farmers. Once bioprospecting had been demonstrated as a sustainable but economically profitable alternative to more destructive activities, they would come to see intact forests as valuable investments that would pay far into the future in the form of additional income earned through sustainable activities (bioprospecting, ecotourism, etc.), cures for diseases afflicting developing countries (malaria, HIV/AIDS, etc.), and innumerable as-yet-undiscovered products and services. Thus, bioprospecting's impact on the conservation of natural resources was hoped to have more diffuse but long-lasting impacts, by helping to change how people value and perceive of the environment itself.

But how has bioprospecting actually fared as a conservation mechanism? In reality, it is difficult to say. Research published by individuals or groups involved in bioprospecting ventures tend to portray the practice in glowing terms all around, including its role in biodiversity conservation (e.g. Kingston et al., 1999; Rosenthal et al., 1999; Soejarto et al., 2006; Kursar, 2007). Analytical scholarship has overwhelmingly focused on the social dimensions of the practice, analyzing issues around IPR and ABS (Moran et al., 2001). The few studies that have actually tried to empirically assess the environmental impact of bioprospecting give mixed or negative reviews.

To begin with, there is the environmental impact of the research and product development itself. Assumed to be benign, especially when compared to logging or mining, it turns out that bioprospecting is not inherently sustainable. The collection of plants for bioprospecting research actually requires large amounts of biomass which can have negative environmental impacts. Scholars reviewing bioprospecting activities all over the world found that few had implemented plant collection policies to take into account conservation principles, such as changing collection procedures for rare plants (Dhillion et al., 2002). Regarding the INBio project, often used as a model of best-case bioprospecting methods, the researchers found that the project nevertheless could not be considered ecologically sustainable per se, since its collection was not sensitive to endemic or rare species, specimen selection processes were not always followed, cultivation to reduce impact on wild populations was only observed for one lead, and little was done to study or monitor impact of research activities overall (Dhillion et al., 2002).

Furthermore, bioprospecting can continue to have a negative environmental impact once a commercial product has been developed, particularly if cultivation or chemical synthesis is not possible and wild resources form the base for the natural product. If market demand is high for the product, unsustainable collection of wild resources may occur, leading to species decline and secondary ecological impacts resulting from the reduced presence of the target species and harvesting activities. This has been demonstrated clearly in southern Africa where the Hoodia cactus (*Hoodia gordonii*) has been collected at unsustainable levels for herbal remedies to treat obesity by reducing appetite (Wynberg and Laird, 2007). While regulation may in some cases help curb the negative impacts of wild harvesting of plants, policy implementation and enforcement is often weak in tropical developing countries that are of interest to bioprospectors. In Madagascar, high demand for bark of the African cherry tree (*Prunus africana*), used in medications to treat urinary ailments and benign enlargement of the prostrate (Benign prostatic hyperplasia – BPH), has led to unsustainable collection of wild stocks, despite the plant being listed on the Convention of International Trade in Endangered Species of Wild Fauna and Flora (CITES) (Neimark, 2010). Even when cultivation of the species that forms the basis of a natural product is possible, if not properly managed this may still have a negative environmental impact. Dorsey (2003, p148) describes the replacement of Amazonian rainforest with large 'agroforestry plantations' of the Sangre de Drago plant (*Croton lechleri*) in Ecuador as a result of a bioprospecting firm's demand for the plant.

While the actual environmental impact of bioprospecting research and production is not without its faults, it could be argued that it pales in comparison to the wholesale destruction of landscapes from logging and mining. Additionally, bioprospecting proponents would argue that it provides funding and incentives for secondary conservation activities, such as environmental education, protected area creation, and alternate income-generating activities for poor rural residents. Yet little outside analysis of the empirical environmental impacts of such secondary activities has been done to date, making it hard to judge whether or not they are effective at reducing pressure on natural resources. One must also note that that funding for conservation that is tied to bioprospecting is usually only found in multi-lateral bioprospecting projects such as the ICBG program, while they are absent or superficial in bioprospecting that is conducted by businesses with a purely market-driven approach. As Wynberg and Laird conclude,

> Despite the early rationale that bioprospecting would enable biodiversity conservation to 'pay its way,' the reality is that the *high technology industries that engage in this field are not interested in supporting biodiversity conservation as a way of protecting their research interests*. For many, natural products and genetic resources are only one part of a complex research strategy, that must compete with approaches that require fewer resources and are less legally ambiguous; for others, numerous ex situ sources of material exist, for example in private collections and seed banks, and, increasingly, in a company's backyard. Some of the more carefully crafted bioprospecting partnerships have included payments to conservation funds or parks and support research on biodiversity, but *incentives have never existed for these industries to invest in conservation as part of their business model.*
>
> (2007, pp29–30, emphasis added)

What of bioprospecting's immaterial impact on conservation: changing the hearts and minds of the stakeholders whose activities most influence tropical resources? Again, analysis shows that its impacts are at best mixed and often difficult to measure. Suriname established a four-million acre protected area after its involvement in the ICBG program, which proponents deemed a result in how bioprospecting had changed the perspectives of local scientists and government officials (Kingston et al., 1999; Rosenthal et al., 1999). Yet grand examples such as these are few and far between. Research conducted by the authors in Madagascar found that while convincing Malagasy people of the value of natural resources was often a stated goal of those involved in the ICBG program, translating this into results on the ground proved difficult. We found that in both Phase I (Tilghman) and Phase II (Neimark), residents of local communities that were beneficiaries of development projects funded through ICBG had little knowledge of ICBG-Madagascar, and thus were unlikely to associate any personal benefits from the development projects with plant collections and natural resources more generally. To further illustrate how difficult it was for ICBG-Madagascar to change environmental values, our interviews with rural residents near plant collection sites sometimes revealed that local people suspected that plant collectors were in fact there to secretly dig for gold or sapphires. While ICBG-Madagascar had hoped to convince people that trees are more valuable standing and worth more than precious metals and stones, for now that lesson has yet to sink in.

We should note in conclusion that much of the promotion of bioprospecting as a conservation tool relies on simplified explanations of the causes of environmental change that have been discussed elsewhere in this book (see, in particular, Chapter 5 by Scales). Poverty and irrationality are often seen as the reasons why tropical countries have thus far ignored the potential value of forests for the development of new medicines, cosmetics, and industrial products in favor of immediate profits from logging and mining concessions, or immediate sustenance from slash-and-burn farming. As we unpack the set of assumptions behind this common narrative, we must in turn question the logic behind the justifications for bioprospecting as a sustainable alternative.

Acknowledgments

The authors would like to thank James Bano, Micky Martial Robson, Yvette Harisoa Randrianatody, and Ania Rakotondrazaka for translation and research assistance, and Etienne Rakotobe and Rabodo Andriantsiferana of CNARP and Mamitiana Rakotozafy of MATEZA for institutional support. Research in Madagascar was facilitated by the wonderful staff of Parc National Zahamena, MATEZA, CNARP, and the Missouri Botanical Gardens. Laura Tilghman was supported by the University of Vermont Environmental Program Merck Senior Thesis fund, a Fulbright/IIE research award, and the NSF Research Experience for Graduate Students program. Benjamin Neimark was supported by a Fulbright/IIE research award, and Rutgers University and Bevier Graduate Fellowship award.

Notes

1 Anonymous interview (February 11, 2005).
2 Anonymous interview (March 5, 2006).
3 Contemporary bioprospecting is generally defined as the search for biogenetic resources and their development into commercial products. While our focus is on the investigation of plants and their constituent compounds for new medicines, as this makes up the great majority of the bioprospecting taking place in

Madagascar as well as more globally, it is important to note that many programs work on other life forms such as fungi, microbes, or marine organisms, and commercializing natural products for the nutritional, agricultural, and cosmetic industries.

4 This strategy 'prioritizes endemic-rich areas' where global conservation dollars would be best spent.
5 Provisions for the fair and equitable sharing of benefits from genetic resources, whereas language of access to genetic resources is provided in Article 15. Other articles that also address bioprospecting include traditional knowledge (Article 8(j)); technology transfer (Article 16); exchange of information (Article 17); and scientific cooperation (Article 18). Access and Benefit-Sharing (ABS) provisions were adopted at the CBD's 10th Conference of the Parties (Nagoya Protocol) (CBD, 2012).
6 USAID has been replaced by the Foreign Agriculture Service of the United States Department of Agriculture.
7 This precedent began with the 1980 Supreme Court ruling that upheld a lower Federal US court's decision that life forms were patentable after the plaintiff, Amanda Charkrabarty, attempted to patent a bacterium he developed which was capable of breaking down the structural components of petroleum while working for General Electric. The case *Diamond vs. Charkrabarty* was a landmark decision that opened the way for the patenting of life forms (Shiva, 1997).
8 For more on this and practical ideas for reaching equitable agreements, see Laird (2002).
9 As shown under the *Code des Aires Protégées* or COAP – Loi No. 2001/05. The protected area system in Madagascar is administered by Madagascar National Parks (MNP), and was formerly called *Association National de la Gestion des Aires Protégées* (ANGAP).
10 We discuss TEK because some bioprospecting projects use medicinal plant knowledge to help guide research efforts by collecting organisms that have traditional uses in local communities. However, we should note that the ICBG-Madagascar project does not base its research operations on TEK and uses a 'random' plant collection model.
11 In addition to its terrestrial activities, Phase II of ICBG-Madagascar also included marine collections.
12 The majority of this section is based on information provided by individuals involved directly in the ICBG-Madagascar project, including principal investigators as well as others who chose to remain anonymous. These interviews and questionnaires were conducted in 2005 and 2007 by Neimark.
13 Total annual ICBG financial data since 2000 was accessed on September 2012 at: http://projectreporter.nih.gov/project_info_description.cfm?aid=7538540&icde=6660918 (accessed August 15, 2012).
14 This development that has been observed in contemporary bioprospecting sites in Mexico and Peru where bioprospectors were denied access to collecting sites by locally organized resistance groups (see Hayden, 2003; Berlin and Berlin, 2004; Greene, 2004).
15 Gordon Cragg, personal communication, 2005.

References

Abelson, P. H. (1990) 'Medicine from plants', *Science*, vol 247(4942), p513.
Aylward, B. (1995) 'The role of plant screening and plant supply in plant conservation, drug development and health care', in T. Swanson (ed.) *Intellectual*

Property Rights and Biodiversity Conservation: A Multidisciplinary Analysis of the Values of Medicinal Plants, Cambridge University Press, Cambridge.

Balick, M. J. (1990) *Ethnobotany and the Identification of Therapeutic Agents from the Rainforest*, John Wiley & Sons, New York.

Barrett, C. B. and Lybbert, T. J. (2000) 'Is bioprospecting a viable strategy for conserving tropical ecosystems?', *Ecological Economics*, vol 34, no 3, pp293–300.

Berlin, B. and Berlin, E. A. (2004) 'Community autonomy and the Maya ICBG project in Chiapas, Mexico: how a bioprospecting project that should have succeeded failed', *Human Organization*, vol 63, no 4, pp472–486.

Bird, C. and Sattaur, O. (1991) 'Medicines from the rainforest', *New Scientist*, vol 131 (August 17), p34.

Boyle, J. (1996) *Shamans, Software, and Spleens: Law and the Construction of the Information Society*, Harvard University Press, Cambridge, MA.

Brown, M. F. (2003) *Who Owns Native Culture*, Harvard University Press, Cambridge, MA.

Burns, J. F. (1995) 'Tradition in India vs. a patent in the U.S.', *The New York Times*, September 15, 1995.

CBD (2012) 'ABS provisions in the convention: the adoption of the Nagoya Protocol', www.cbd.int/abs/background, accessed January 30, 2012.

Cragg, G. M. and Newman, D. J. (2005) 'International collaboration in drug discovery and development from natural sources', *Pure and Applied Chemistry*, vol 77, no 11, p1923.

Dalton, R. (2001) 'The curtain falls', *Nature*, vol 414, p685.

Dauphiné, N. (2002) *Medicinal Plant Trade, Use, and Habitat in the Highlands of Madagascar*, Master's Thesis, Ithaca, Cornell University.

Dhillion, S. S., Svarstad, H., Amundsen, C. and Bugge, H. (2002) 'Bioprospecting: effects on environment and development', *AMBIO*, vol 31, no 6, pp491–493.

Dorsey, M. K. (2003) 'The political ecology of bioprospecting in Amazonian Ecuador: history, political economy, and knowledge', in S. R. Brechin, P. R. Wilshusen, C. L. Fortwangler and P. C. West (eds) *Contested Nature: Promoting International Biodiversity Conservation with Social Justice in the Twenty-first Century*, State University of New York Press, Albany, NY.

Eisner, T. (1992) 'Chemical prospecting: a proposal for action', in F. H. Bormann and S. R. Kellert (eds) *Ecology, Economics, and Ethics: The Broken Circle*, Yale University Press, New Haven, CT.

GRAIN (2004) 'Community or commodity: what future for traditional knowledge', *Seedling*, July 29, 2004, www.grain.org/article/archive/categories/37-seedling-july-2004, accessed April 1, 2008.

Greene, S. (2004) 'Indigenous people incorporated?', *Current Anthropology*, vol 45, no 2, pp211–237.

Hamilton, C. (2006) 'Biodiversity, biopiracy and benefits: what allegations of biopiracy tell us about intellectual property', *Developing World Bioethics*, vol 6, no 3, pp158–173.

Harper, J. (2002) *Endangered Species: Health, Illness and Death among Madagascar's People of the Forest*, Carolina Academic Press, Durham, NC.

Harper, J. (2005) 'The not-so rosy Periwinkle: political dimensions of medicinal plant research', *Ethnobotany Research & Applications*, vol 3, pp295–308.

Hayden, C. (2003) *When Nature Goes Public: The Making and Unmaking of Bioprospecting in Mexico*, Princeton University Press, Princeton, NJ.

Healy, T. and Ratsimbarison, R. (1998) *Historical Influences and the Role of Traditional Land Rights in Madagascar: Legality versus Legitimity*, Proceedings of the International Conference on Land Tenure in the Developing World, Cape Town, South Africa.

Hodgson, D. L. (2002) 'Introduction: comparative perspectives on the indigenous rights movement in Africa and the Americas', *American Anthropologist*, vol 104, no 4, pp1037–1049.

Kadidal, S. (1993) 'Plants, poverty, and pharmaceutical patents', *Yale Law Journal*, vol 103, no 1, pp223–258.

Kingston, D. G. I., Abdel-Kader, M. and Zhou, B.-N. (1999) 'The Suriname International Cooperative Biodiversity Group program: lessons from the first five years', *Pharmaceutical Biology*, vol 37, pp22–34.

Kursar, T. A. (2007) 'Linking bioprospecting with sustainable development and conservation: the Panama case', *Biodiversity and Conservation*, vol 16 no 10: 2789–2800.

Laird, S. A. (2002) *Biodiversity and Traditional Knowledge*, Earthscan, London.

Lake, R. W. (1996) 'Volunteers, NIMBYs, and environmental justice: dilemmas of democratic practice', *Antipode*, vol 28, no 2, pp160–174.

Lybbert, T. J., Barrett, C. B. and Narjisse, H. (2002) 'Market-based conservation and local benefits: the case of argan oil in Morocco', *Ecological Economics*, vol 41, no 1, pp125–144.

Lyon, L. M., and Hardesty, L. H. (2005) 'Traditional healing in the contemporary life of the Antanosy people of Madagascar', *Ethnobotany Research & Applications*, vol 3, pp287–294.

McAfee, K. (1999) 'Selling nature to save it? Biodiversity and green gevelopmentalism', *Environment and Planning D: Society and Space*, vol 17, no 2, pp133–154.

Macilwain, C. (1998) 'When rhetoric hits reality in debate on bioprospecting', *Nature*, vol 392, no 6676, pp535–540.

Miller, J. S. (2007) 'Impact of the convention on biological diversity: the lessons of ten years of experience with models for equitable sharing of benefits', in C. McManis (ed.) *Biodiversity and the Law: Intellectual Property, Biotechnology and Traditional Knowledge*, Earthscan, London.

Mooney, P. R. (2000) 'Why we call it biopiracy', in H. Svarstad and S. S. Dhillion (eds) *Responding to Bioprospecting: From Biodiversity in the South to Medicines in the North*, Spartacus Forlag AS, Oslo.

Moran, K., King, S. R. and Carlson, T. J. (2001) 'Biodiversity prospecting: lessons and prospects', *Annual Review of Anthropology*, vol 30, pp505–526.

Myers, N., Mittermeier, R. A., Mittermeier, C. G., da Fonseca, G. A. B. and Kent, J. (2000) 'Biodiversity hotspots for conservation priorities', *Nature*, vol 403, pp853–858.

Neimark, B. (2009) *Industrial Heartlands of Nature: the political economy of biological prospecting in Madagascar*, PhD thesis, Rutgers University, New Brunswick, NJ.

Neimark, B. (2010) 'Subverting regulatory protection of "natural commodities": the Prunus Africana in Madagascar', *Development and Change*, vol 41, no 5, pp929–954.

Neimark, B. (2012) 'Industrializing nature, knowledge, and labour: the political economy of bioprospecting in Madagascar', *Geoforum*, vol 43, no 5, pp580–590.

Neimark, B. and Schroeder, R. (2009) 'Hotspot discourse in Africa: making space for bioprospecting in Madagascar', *African Geographical Review*, vol 28, pp43–70.

Parry, B. (2004) *Trading the Genome*, Columbia University Press, New York.
Plotkin, M. J. (1993) *Tales of a Shaman's Apprentice: An Ethnobotanist Searches For New Medicines in the Amazon Rain Forest*, University of California Press, Berkeley, CA.
Posey, D. A. and Dutfield, G. (1996) *Beyond Intellectual Property: Toward Traditional Resource Rights for Indigenous Peoples and Local Communities*, International Development Research Center, Ottawa.
Quansah, N. (2003) 'Access to genetic resources in Madagascar', in K. Nnadozie, R. Lettington, C. Bruch, S. Bass and S. King (eds) *African Perspectives on Genetic Resources: A Handbook on Laws, Policies, and Institutions Governing Access and Benefit Sharing*, Environmental Law Institute, Washington, DC.
Rasoanaivo, P. (1990) 'Rain forests of Madagascar: sources of industrial and medicinal plants', *Ambio*, vol 19, no 8, pp421–424.
Reid, W., Laird, S. A., Mayer, C. A., Gamez, R., Sittenfeld, A., Janzen, D., Gollin, M. and Juma, C. (1993) *Biodiversity Prospecting: Using Genetic Resources for Sustainable Development*, World Resources Institute, Washington, DC.
Rosenthal, J. P. (1999) 'Combining high risk science with ambitious social and economic goals', *Pharmaceutical Biology*, vol 37, pp6–21.
Rosenthal, J. P. (2006) 'Politics, culture, and governance in the development of prior informed consent in indigenous communities', *Current Anthropology*, vol 47, no 1, pp119–142.
Rosenthal, J. P. and Katz, F. N. (2004) 'Natural products research partnerships with multiple objectives in global biodiversity hot spots: nine years of the international cooperative biodiversity groups program', in A. T. Bull (ed.) *Microbial Diversity and Bioprospecting*, ASM Press, Bull Washington, DC.
Schweitzer, J. H. F., Edwards, J., Harris, W. F., Grever, M. R., Schepartz, S. A., Cragg, G., Snader, K. and Bhat, A. (1991) 'Summary of the workshop on drug development, biological diversity and economic growth', *Journal of the National Cancer Institute*, vol 83, no 18, pp1294–1298.
Sharp, L. A. (1993) *The Possessed and the Dispossessed: Spirits, Identity, and Power in a Madagascar Migrant Town*, University of California Press, Berkeley, CA.
Shiva, V. (1997) *Biopiracy: The Plunder of Nature and Knowledge*, South End Press, Boston, MA.
Sneader, W. (2005) *Drug Discovery: A History*, Wiley, New York.
Soejarto, D. D., Fong, H. H., Tan, G. T., Zhang, H. J., Ma, C. Y., Franzblau, S. G., Gyllenhaal, C., Riley, M. C., Kadushin, M. R. and Pezzuto, J. M. (2005) 'Ethnobotany/ethnopharmacology and mass bioprospecting: issues on intellectual property and benefit-sharing,' *Journal of Ethnopharmacology*, vol 100, nos 1–2, pp15–22.
Soejarto, D. D., Gyllenhaal, C., Regalado, J. C., Pezzuto, J. M., Fong, H. H. S., Tan, G. T., Hiep, N. T., Xuan, L. T., Binh, D. Q., Hung, N. V., Bich, T. Q., Thin, N. N., Loc, P. K., Vu, B. M., Southavong, B. H., Sydara, K., Bouamanivong, S., O'Neill, M. J., Lewis, J., Xie, X. M. and Dietzman, G. (2006) 'Studies on biodiversity of Vietnam and Laos, 1998–2005: examining the impact,' *Journal of Natural Products*, vol 69, no 3, pp473–481.
Svarstad, H. (2005) 'A global political ecology of bioprospecting', in S. Paulson and L. L. Gezon (eds) *Political Ecology Across Spaces, Scales, and Social Groups*, Rutgers University Press, New Brunswick, NJ.

Swerdlow, J. L. and Johnson, L. (2000) 'Nature's Rx: growing importance of plant-based pharmaceuticals,' *National Geographic*, vol 197, no 4, p98.
ten Kate, K. and Laird, S. A. (1999) *The Commercial Use of Biodiversity: Access to Genetic Resources and Benefit-sharing*, Earthscan, London.
ten Kate, K. and Laird, S. A. (2000) 'Biodiversity and business: coming to terms with the grand bargain', *International Affairs*, vol 76, no 2, pp241–264.
Tilghman, L. M. (2004) 'Bioprospecting: perspectives from Madagascar,' B.A. thesis, University of Vermont at Burlington, VT.
Tyson, P. (2000) *The Eighth Continent: Life, Death, and Discovery in the Lost World of Madagascar*, Perennial, New York.
Walden, I. (1995) 'Preserving biodiversity: the role of property rights', in T. Swanson (ed.) *Intellectual Property Rights and Biodiversity Conservation: An Interdisciplinary Analysis of the Values of Medicinal Plants*, Cambridge University Press, Cambridge.
Weiss, C. and Eisner, T. (1998) 'Partnerships for value-added through bioprospecting', *Technology in Society*, vol 20, pp481–498.
Wynberg, R. and Laird, S. A. (2007) 'Bioprospecting: tracking the policy debate', *Environment*, vol 49, no 10, pp20–32.

13 Incentivising forest conservation

Payments for environmental services and reducing carbon emissions from deforestation

Laura Brimont and Cécile Bidaud

Introduction

It is widely acknowledged that conservation strategies cannot succeed without taking into account the support of local populations and their livelihood concerns (Adams *et al.*, 2004; Sunderland *et al.*, 2008). One of the biggest challenges for conservation in Madagascar has been finding ways to create alternative sources of income for forest-dependent households. Promising new tools have recently emerged that attempt to increase the financial value of ecosystems. These 'economic' or 'incentive-based' instruments[1] are based on putting a price on the functions performed by ecosystems, and getting those who benefit from such functions to pay for them. The idea is that this creates a financial incentive to maintain the ecosystems that provide the key services.

Contrary to state-enforced coercive instruments, incentive-based instruments rest upon individual economic choices. Such instruments have been praised as 'win–win', providing alternative livelihoods for those living in and around key ecosystems and thereby reducing poverty (Landell-Mills and Porras, 2002; Pagiola *et al.*, 2002; Grieg-Gran *et al.*, 2005). Furthermore, incentive-based instruments provide new funding sources for conservation that are potentially larger than traditional funding from governments and international donors (Ferraro, 2011).

While incentive-based instruments look promising, there is a lack of research on how such instruments might work in different places and with different ecosystems, as well as a lack of knowledge of their possible environmental and social effects (Ferraro and Pattanayak, 2006; Pattanayak *et al.*, 2010). This chapter looks at how incentive-based approaches generally, and Payments for Ecosystem Services (PES) more specifically, have developed in Madagascar over the past ten years. The first part of this chapter explores the basic principles of incentive-based approaches and reviews emerging critiques. We then look at key Malagasy case studies that highlight some of the major challenges of incentive-based approaches. We argue that incentive-based instruments have shown limited success in integrating conservation and poverty reduction and have in many cases increased economic inequalities within local communities.

Economics and the environment

Global environmental politics have evolved significantly over the past 30 years, seeing a shift from 'fortress conservation' (based on national parks) to the greater involvement of communities living around protected areas (see Chapter 7 by Kull and Chapter 8 by Pollini et al.). The challenge of community-based approaches to conservation in low-income nations has been to create viable livelihood alternatives for rural households whose activities often impact on ecosystems such as tropical forests.

Unfortunately, the first generation of tools designed to achieve synergy between conservation and local livelihood concerns, which were based primarily on various forms of tourism and attempts to valorise non-timber forest products, often failed to meet expectations (see Chapter 11 by Scales and Chapter 12 by Neimark and Tilghman). Many authors have described the difficulties of integrating conservation and development goals (Wells and McShane, 2004; Hockley and Andriamarovololona, 2007; Blom et al., 2010). These include the inadequate scope and timeframe of conservation projects; the complex political and economic realities of such projects; a lack of local participation in the planning process; the unrepresentative nature of the local community associations created; and over-ambitious goals and unkept promises from conservation organisations.

In response to such failures, economists have proposed a new kind of conservation tool, commonly referred to as incentive-based instruments. Incentive-based instruments are a product of developments in economic theory during the twentieth century. Economists began taking an interest in environmental matters in the 1960s and 1970s, when industrial pollution arose as a major issue (Meadows et al., 1972; Daly, 1977). Up to that point, standard economic theory considered the economy and the environment as two separate entities, ignoring the economic value of nature.

In the language of economics, when nature does not have a price the environment is treated as an 'externality', i.e. something whose cost or benefit is not transmitted to markets through prices. As the cost of pollution is not borne by the polluter, not only is there little incentive to reduce pollution but the cost of pollution is not reflected in the cost of goods and services. Thus both the production and consumption of goods tend to be environmentally damaging. However, as environmental economists pointed out, pollution imposes a range of costs. For example, an industry polluting a river without any restrictions imposes a cost for other river users. Industry thus 'externalises' the cost of pollution and makes others suffer its consequences.

Alfred Pigou popularized the idea that pollution could be considered as a 'negative externality', caused by the incapacity of the market to take into account the social cost of pollution – that is the cost imposed on society as a whole (Pigou, 1932). Pigou's solution was to impose a governmental tax on polluting activities to 'internalize' the social cost of pollution. By making pollution a costly activity and making the polluter pay, producers would be incentivised to minimise it.

By contrast, Ronald Coase (1960) did not believe state enforced regulation was necessary. Instead, negative externalities could be internalized through negotiations between individual actors. In this model those carrying out activities with negative impacts negotiate a level of compensation with the individuals whose well-being is being impacted on. The amount of compensation should be equal or superior to the opportunity costs of the abandoned activity and equal or inferior to the costs previously borne by the second individual.

A new branch of economics, environmental economics, extended the scope of cost–benefit analysis to natural resource use, with the aim of internalising environmental impacts in economic decisions. A new concept of 'ecosystem services' emerged, based on a utilitarian framing of nature and highlighting societal dependence on ecosystems. Ecosystem services are defined as benefits obtained from nature that satisfy human needs (MEA, 2005). The term was initially used as a metaphor to stress human reliance on the functions performed by ecosystems and increase public interest in biodiversity conservation (Ehrlich and Ehrlich, 1981), and then became mainstream in the sustainability literature (Daily, 1997). The Millennium Ecosystem Assessment (MEA, 2005) defined four categories of ecosystem services: i) provisioning services (such as food, water and timber); ii) regulating services (that influence climate, floods, disease, waste and water quality); iii) cultural services (recreational, aesthetic and spiritual benefits); and iv) supporting services (e.g. soil formation, photosynthesis and nutrient cycling).

A further step towards the incorporation of nature into economic systems was the promotion of incentive-based instruments. The early 2000s saw the emergence, both in academic and policy circles, of direct payments to incentivise conservation practices (Gomez-Baggethun *et al.*, 2010). Such market-based instruments (MBIs) differ from the first generation of tools designed to link development and conservation (for example ecotourism) on two main points:

1 Directness: Integrated Conservation and Development Projects (ICDPs) promoted activities to divert local communities from extracting activities, what Ferraro and Simpson (2002) named 'conservation by distraction'. With MBIs, the payment is directly linked to a behavioural change resulting in an environmental outcome.
2 Performance based: while ICDP benefits were not linked to the realisation of environmental outcomes, MBI payments are conditional to the realisation of such outcomes.

Among MBIs, PES have received the most interest in both academic and political circles. Wunder (2005) has defined PES as: i) voluntary transactions; ii) involving a well-defined ecosystem service (or a land use likely to secure that service); iii) 'bought' by a (minimum of one) buyer; iv) from a (minimum of one) provider; and v) if and *only* if the provider secures the provision of the ecosystem service (i.e. there is conditionality).

Since the first Costa Rican national programme experiments in 1996 aimed at preserving forest cover, PES have spread worldwide. Some countries – such as Costa Rica, Mexico and Peru – have implemented national PES programmes, but the majority of PES projects have been on a case-by-case basis. These projects have involved a variety of ecosystem services, from watershed protection, to carbon storage and landscape preservation.

In low- and middle-income countries, the popularity of PES is due to four main factors (Pattanayak *et al.*, 2010): i) weak state institutions make the enforcement of regulations difficult and therefore state-led coercive methods ineffective; ii) PES fit well with the conditionalities imposed by donors and their desire for measureable performance indicators to make aid more efficient; iii) PES link poverty alleviation with ecosystem protection; and iv) PES create funds to pay for conservation activities.

With regards to forest conservation and management, the most significant incentive-based mechanism to have emerged is the United Nations 'Reducing Emissions from Deforestation and Forest Degradation' (REDD+) mechanism. Tropical forests have a major role in climate regulation through the absorption and sequestration of carbon dioxide and deforestation is a significant source of carbon dioxide (Pan *et al.*, 2011). REDD+ aims to use financial incentives to reduce emissions of greenhouse gases from deforestation and forest degradation. The basic principle of REDD+ is that industrialised high-income nations that produce large quantities of greenhouse emissions pay tropical low-income nations to maintain forest cover and therefore reduce emissions from forest loss and degradation.

REDD+ has stimulated keen interest from policymakers and researchers, as shown by the huge financial amounts invested (about US$5 billion between 2010 and 2012)[2] and the large literature emerging on the subject. REDD+ is generally presented as a 'win–win' mechanism: 'REDD+ has the potential to deliver large cuts in emissions at a low cost within a short time frame and, at the same time, contribute to reducing poverty and sustainable development' (Angelsen, 2008, pviii). It is presented as a new opportunity for financing forest conservation and 'green' economic development in low-income nations and emerging economies in the tropics. Moreover, REDD+ is expected to be an efficient mechanism for mitigating against global warming and climate change, as reducing deforestation is seen as a cheaper way of decreasing CO_2 emissions compared to other options such as reducing fossil fuel consumption (Eliasch, 2008; Stern, 2008; McKinsey and Company, 2009).

Incentive-based instruments such as REDD+ fit well with the holy grail of the 'win–win' principle of sustainable development, which is based on the assumption that economic growth is possible at the same time as conserving biodiversity and maintaining ecosystem function. Unfortunately, economic theory supporting incentive-based instruments is based on a certain set of hypotheses and assumptions: i) that there is pure and perfect information available to enable both service providers and service users to make informed decisions; ii) that there are clearly defined property rights; and iii) that different forms of capital are substitutable. This is often very far from reality, especially in low-income nations. The difference between the theory and practical implementation of incentive-based instruments is at the root of a series of critiques (Muradian et al., 2010).

The first problem is that our understanding of ecosystems, both in terms of the links between their different components and the measurement of the services they provide, is still marred by uncertainty. It is difficult to define a causal relationship between a change in a specific human activity and the delivery of a complex ecosystem service such as watershed protection. In Madagascar, for example, research has shown that a PES scheme for watershed protection has no measurable scientific basis to prove that the conservation project has effectively delivered the expected service (Bidaud et al., 2011). There is also the potential for competition between different ecosystem services. For example, while afforestation or reforestation with fast-growing tree species delivers carbon storage services, it has less benefit from a biodiversity perspective (Kosoy and Corbera, 2010; Vatn, 2010). The wide range of different functions performed by ecosystems also means that there is scope for different stakeholders valuing and prioritising different services (Vatn, 2010). People living in or around forests value provisioning services (honey, fishing, hunting) whereas Western tourists might prioritise aesthetic services and those involved in REDD+ would prioritise carbon sequestration services.

As well as these conceptual challenges, there are a number of technical issues that undermine the overall efficiency of such instruments. The performance of PES schemes such as REDD+ is rooted in one concept specific to carbon projects – that of additionality. A carbon project is 'additional' when a reduction in emissions recorded after project implementation is directly the result of the project, and would not have been possible without it. For example, if a carbon PES scheme is implemented in a forested zone that experiences massive rural emigration and where, as a result, deforestation is likely to decrease in the future, the project is not additional. The reduction in forest loss and emissions of carbon dioxide would have occurred regardless.

Measuring additionality therefore requires foreseeing how carbon emissions would have developed without the project. It is thus necessary to establish a baseline, i.e. the emissions that would have occurred under 'business as usual'. Several solutions exist to build such models. The project planner can take the historical trend of deforestation in the country or region where the REDD+ project is to be implemented and assume that the deforestation rate would stay the same without intervention. In Madagascar, the Makira REDD+ project's scenario is built on historical trends of deforestation in the Makira forest zone (WCS, 2011). The problem is that deforestation rates are not constant through time (see for example Scales, 2011) and depends on a complex set of factors (e.g. demography, the political context, economic growth, support to the agricultural sector) specific to each region and country (see Chapter 5 by Scales for more on the drivers of deforestation).

To cope with the limitations of using historical scenarios for estimating future emissions, the different drivers of deforestation can be modelled. In Madagascar, the *Projet Holistique de Conservation de la Forêt* (PHCF) used a predictive model based on population density and spatial factors such as the altitude of forests, distance from the forest edge and fragmentation (Grinand *et al.*, 2013) to model deforestation. Although predictive models have stronger scientific basis than historical models, they are unable to address all the uncertainties surrounding deforestation. For example, while political crisis in Madagascar generally leads to an increase in deforestation (Allnutt *et al.*, 2013), it is extremely difficult or even impossible to predict the occurrence of a *coup d'état*. Furthermore, modelling deforestation in developing countries is constrained by the lack of socio-economic data.

Another technical issue concerning the functioning and efficiency of forest carbon projects is the risk of 'leakage', i.e. emissions moving to another location or sector where they remain uncontrolled or uncounted (Murray, 2008). For example, reduction in forest loss in one area might simply lead to more deforestation elsewhere as rural migrants or commercial plantations shift. Such displacements might occur at both national and international scales. To reduce the potential for leakage, REDD+ was initially designed as a national-level mechanism, with each country implementing a national REDD+ programme (Murray, 2008). With measurement at the national level, any displacement within a country means that emissions would not be counted as avoided. However, as REDD+ has moved towards a project-level approach (Dahan *et al.*, 2011), the risk of national leakage has increased. Furthermore, international leakages are likely to occur if some forested countries do not participate in the global effort to reduce deforestation and activities that lead to deforestation (for example the clearance of rainforest for cultivation) move to those areas. The issues of additionality and leakage have led to strong criticism of the likely efficiency of forest carbon PES instruments.

Other critiques of PES have focused on the unequal distribution of benefits. Following Coase (1960), any individual who has clear defined use rights over a natural resource can provide an environmental service by giving up a damaging activity, in exchange of a compensation at least equal to his opportunity cost. In practice, a lack of information and insecure tenure over natural resources often raises difficult questions about the distribution of payments. First, there is the issue of the adverse selection effect – the more an individual is dependent on consumptive uses of natural resources, the higher the opportunity cost. By contrast, an individual whose revenues are not based on environmental damaging activities would accept a lower payment to protect natural resources. In the first case, the payment has to be high to convince the individual to supply an environmental service, and lead to high additionality. In the second case, the payment is lower, and so is the additionality.

An adverse selection effect is especially likely to occur with fixed payments programmes (Alix-Garcia et al., 2008 ; Wünscher et al., 2008). For example, economists have shown that the Costa Rica national programme has not been fully additional because it has mainly attracted large landowners who can easily set aside unused land for conservation, rather than small landowners who are directly dependent on land (Pagiola, 2008). To avoid an adverse selection effect, payments have to be flexible and adapt to individual opportunity costs, although this increases data purchase transaction costs.

The second issue with regard to payment distribution relates to tenure rights. In low-income nations, rights over natural resources are often complex, mixing legal and customary laws, with a superposition of different rights held by different people and groups on the same territory (Roy et al., 1996). Land is often viewed as a common resource, where different people share different usufruct rights. Land users therefore do not necessarily have exclusive rights. This clearly creates problems for PES schemes, which need to be based on clear ownership rights in order to function. The choice of who will benefit from a PES is likely to be a highly political issue (Corbera et al., 2007). The poorest people tend to be the most directly dependent on forests for their livelihoods; have the most insecure tenure rights; and have little political power to claim their rights over natural resources during the implementation of PES (Pagiola et al., 2005; Zbinden and Lee, 2005).

Last but not least, there is a possibility that incentive-based instruments fail to address the drivers of degradation. In its most limited form, a PES can be viewed as a use-restricting instrument – a buyer pays an environmental service supplier to give up an activity. This version of PES has been criticized as being a form of annuity for the poorest (i.e. a simple fixed payment for a specific period of time), without providing real livelihood alternatives to get themselves out of poverty (Karsenty, 2007). In Madagascar, one of the major threats to forests is swidden cultivation (see Chapter 5 by Scales). By simply stopping forest clearance, a PES does not offer livelihood alternatives to farmers who do not have the financial capital or technical skills to invest in other agricultural practices (Ducourtieux, 2009; Karsenty et al., 2010). Some PES projects try to provide alternatives, by funding investments in intensive agricultural techniques, for example, but such efforts cannot succeed if they are not undertaken in line with national policies aimed at creating favourable economic conditions to realize agricultural transition.

To summarise, incentive-based instruments have arisen from sustainable development ideology, whose basic principle is to promote economic growth without degrading ecosystems. This is based on the assumption that changing the relative prices of degrading versus preserving natural resources is the best way to manage ecosystems. Many criticisms have been raised, mainly with regards to the technical issues determining the efficacy of incentive instruments and the political issues surrounding the distribution of benefits. However, much of the work to date has been theoretical and conceptual and there is a lack of empirical evidence. The second part of this chapter addresses this gap by analysing several case studies of incentive-based conservation in Madagascar.

Incentive-based mechanisms in practice: Madagascar case studies

Since the 1980s, Madagascar has been the focus of considerable attention from international conservation organisations (see Chapter 7 by Kull). It is therefore not surprising that incentive-based conservation has played an increasingly important role in Madagascar's environmental discourse over the past decade. The trend began with the Durban World Parks Congress in 2003, which announced a tripling of the island's protected area network (see Chapter 9 by Corson). In an attempt to justify the extension of protected area networks and to find other funding mechanisms than simply donors, economists from the World Bank carried out a cost–benefit analysis of the existing protected area network (Carret and Loyer, 2003). This analysis was based on the monetary evaluation of three main environmental services (biodiversity conservation, ecotourism and watershed protection), compared to the costs of conservation (made up of the costs of managing the network as well as the opportunity costs of those dependent on forests). The authors concluded that based on the overall benefits, the protected area network was economically beneficial for Madagascar.

The World Bank report perfectly illustrates the growing trend promoting conservation as a means of economic development through assigning economic value to natural resources and ecosystem functions. However, PES was not put forward as a tool for financing conservation. Rather than considering ecosystem services more broadly, the authors only considered watershed protection and concluded that 'downstream' urban water users were too poor to pay rural farmers to stop forest clearance and maintain watershed services.

Since then, the potential for a wider range of PES mechanisms has increased, most notably with the emergence of REDD+ in international forest conservation discourse. Carbon credits are now considered as a major source of funding for forested protected areas (World Bank, 2011). The Malagasy government has been engaged in the REDD+ mechanism since 2008, when a technical committee was established to put together a REDD+

national strategy. This committee was made up of representatives of various ministries, research institutes and universities, together with international conservation non-governmental organisations (NGOs) involved in shaping conservation policy in Madagascar. This committee was established in order to obtain a grant from the World Bank Forest Carbon Partnership Facility (FCPF), aimed at helping nations develop the necessary policies and systems to make REDD+ possible. However, because of the 2009 political crisis, this funding (US$3.6 million) has not been released, even though the draft document produced by the committee has been unofficially accepted by the FCPF commission (Bidaud, 2012).

In parallel to the unsuccessful governmental efforts to engage in UN-REDD+, four REDD+ projects led by international NGOs have emerged over the past few years. These REDD+ projects fit with the Durban Vision (see Chapter 9 by Corson) while integrating new elements linked to the sale of carbon credits, such as reduction emission scenarios and all the technical processes directed at getting the certification necessary to sell carbon on voluntary markets.

Incentive-based conservation can also be seen in other PES-like schemes that have emerged under different names, for example the 'Participatory Ecological Monitoring and Community Based Competitions' (Sommerville et al., 2010) implemented by the Durrell Wildlife Conservation Trust (DWCT; a British NGO); the 'Conservation Agreements' supported by Conservation International (CI; a major American NGO); and the experiments in 'Alternatives to Slash-and-Burn' of the French *Centre de Coopération Internationale en Recherche Agronomique pour le Développement* (CIRAD). However, such PES-like schemes are a long way away from Wunder's (2005) definition of PES.

Finally, another kind of incentive-based instrument has appeared in the Malagasy conservation landscape, namely the 'eco-certification' scheme. The purpose of eco-certification is to provide consumers with information on the social and ecological context of the goods they purchase. It is part of broader trend in 'ethical' consumption. Underlying eco-certification is the belief that the decisions of individual consumers might help to correct some of the negative impacts of consumption (Brockington and Duffy, 2010). Several labelling schemes have been developed in Madagascar for products such as vanilla and silk (Pierre, 2011).

The Makira REDD+ project

The Makira REDD+ project, supported by Wildlife Conservation Society (WCS), is the most advanced REDD+ project in Madagascar. It has also been the first new protected area created in the wake of the Durban conference. The Makira forest is located in north-eastern Madagascar, the most forested part of the island. With a total area of 707,643 hectares, Makira is the largest protected area in Madagascar. It is divided into a core protected area of 372,470 hectares and a protection zone of 335,173 hectares that forms a belt around the core protected area and encompasses the adjacent community areas (WCS, 2011).

The first carbon evaluation studies ever conducted in Madagascar were conducted in the Makira region in 2001 (Meyers and Berner, 2001). These studies, financed by the United States Agency for International Development (USAID), aimed to establish the feasibility of funding Malagasy rainforest conservation through the sale of carbon credits (Bidaud, 2012). In 2003, the first boundaries of the protected area were designed and local communities were informed of the creation of the park. In 2005, the Malagasy government decreed the temporary classification of a National Park (IUCN Category II), and in 2011 WCS was designated as the manager of the new protected area. In 2012, a definitive decree was granted. So although the project is presented as a PES scheme by project planners, the Makira REDD+ project is based on a traditional conservation approach consisting of a protected area surrounded by community management units (Holmes et al., 2008).

In 2008, WCS created the Makira Carbon Company to market the emission reductions generated by the Makira conservation project.[3] The sale of carbon credits has raised US$700,000,[4] which has been used to finance project implementation (Christopher Holmes, personal communication). At the time of writing, the process of carbon funding has not made significant headway. While the Makira REDD+ project received certification in 2012 allowing it to sell 830,000 verified carbon units on voluntary markets for the first five-year monitoring period, it has not yet concluded a sale agreement. The difficulties in selling carbon credits seem to lie in the drying out of voluntary carbon markets due to the stagnation of international negotiations on mitigating against climate change (Christian Burren, personal communication). Hence, the funding capacity of incentive-based instruments such as REDD+ relies heavily on political decisions that shape the demand for carbon credits. If industrialised countries do not show a clear commitment to reducing emissions, a market-financed REDD+ mechanism, either voluntary or regulatory, is no longer conceivable.

Another issue at stake in the Makira REDD+ project is the distribution of benefits. When the WCS made an agreement with the Malagasy government to create the Makira Carbon Company, they agreed on a distribution of revenues (Figure 13.1).

The distribution of carbon benefits, particularly the share allocated to local communities, is similar to the share of the revenues received from national parks' entry fees. However, the precise mechanism for deciding how carbon revenues will be distributed to local communities is still uncertain.[5] One option being considered is that the share allocated to local communities could be placed in environmental funds managed by a Malagasy environmental foundation, which would use returns on the fund to finance local development projects proposed by community

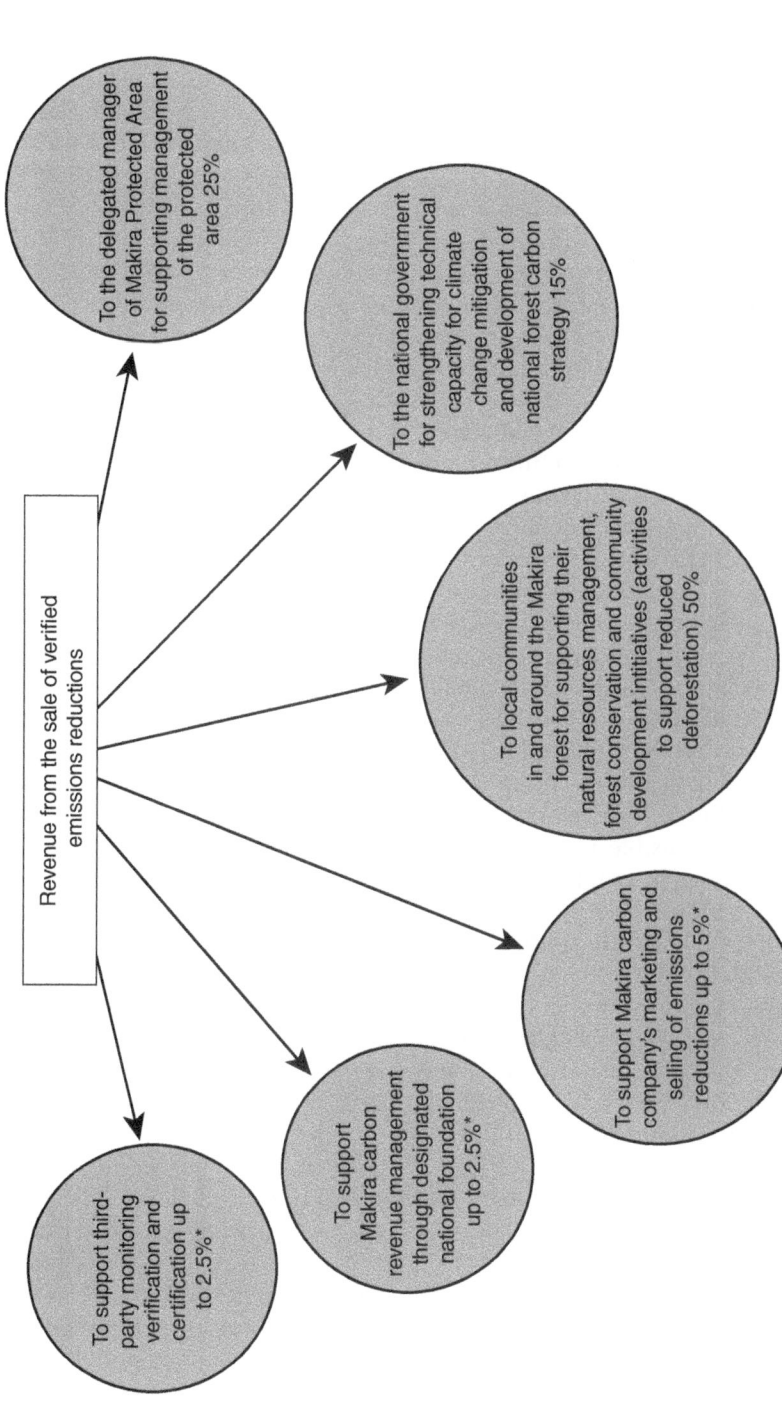

Figure 13.1 The distribution of carbon revenues from the Makira REDD+ project

Source: WCS, Makira Forest Protected Area Project Design Document 2011, p73.

management organisations (for more on community resource management institutions, see Chapter 8 by Pollini *et al.*). Applications would have to propose a community-based project intended to promote local development while preserving forest ecosystems (Monique Andriamananoro, personal communication). Such a financing scheme raises several issues:

- Representativeness – many scholars have described how newly established community organisations have often failed to provide an adequate representation of the rural communities (see Chapter 8 by Pollini *et al.* and Chapter 14 by Kaufmann). As in many African countries, community-based resource management in Madagascar has been undermined by corruption as well as elite capture of benefits. The danger is that development projects funded by carbon revenues will only benefit a marginal part of the local community.
- Linking benefits to livelihood activities – forest clearance is usually a household decision rather than a group one. For projects to lead to land use change, they must filter down to households. Benefits that accrue at the community level are less likely to lead to behavioural change at the individual or household level (see also Chapter 11 by Scales). A household survey carried out by one of the authors in different parts of the Makira protected area showed that few households have concrete ideas of how much compensation they could ask for to stop swidden agriculture, and what levels of investment they require to improve their agricultural activities and living conditions.[6]
- The level of funding – an a priori distribution of benefits based on fixed sharing between conservation projects actors (e.g. the government, NGOs responsible for management of the project, third party monitoring) does not guarantee that REDD+ provides sufficient money to create genuine livelihood alternatives and therefore ensure a decrease in forest loss in the middle and long term. The risk of a shortfall is particularly high due to the fact that problems with voluntary carbon markets have resulted in the price of carbon credits falling to near zero. In the case of the Makira REDD+ project, the price of carbon dioxide in the 2011 negotiations between the Malagasy government and a potential buyer was around US$3 per tonne (Christian Burren, personal communication).

The issue of adequacy between carbon revenues and local needs to adapt to land use restrictions leads us back to the question of the capacity of incentive-based instruments to address the drivers of deforestation. Environmental economics and the 'greening' of capitalism is based on the assumption that a change in the relative prices of resource degradation and resource preservation will induce behavioural changes and lead to ecosystem protection. However, the efficacy of incentive-based instruments depends on institutional and economic factors such as access to markets, opportunities for off-farm employment or agricultural policies. In a country such as Madagascar, where poverty rates are extremely high and where government intervention is limited by general political instability, bad governance and a lack of financial means, incentive-based instruments are likely to be seriously hampered.

Conservation agreements in the Ankeniheny-Zahamena corridor

Another interesting case study of experimentation with incentive-based conservation can be found in CI's attempt to establish conservation agreements in two municipalities in the Ankeniheny-Zahamena (CAZ) corridor in the eastern rainforests of Madagascar. The general objective of conservation agreements is to 'engage communities in activities to support management of the protected areas around which they lived ... The social objectives [are] to provide economic opportunities to benefit communities as incentives for conservation' (CI, 2011, p2). In this case, the conservation agreements consisted of a grant of US$5,000 paid directly to the community resource management organisation and which covers three types of expenses:

1   Remunerating local patrollers in charge of monitoring the territory managed by the community. Patrols are paid 5,000 Ariary per day (approximately US$2 at the time of writing), which is around twice the local daily agricultural salary. The grant also provides some equipment for patrolling (tents, mattress, binoculars, etc).
2   Funding for the functioning costs of the community organisation, including per diems for attending meetings or writing reports, transport costs and equipment.
3   Funding for development activities, which are decided during the community organisation's annual general assembly (for example, improving rice cultivation, rice storage, beekeeping, livestock).

Figure 13.2 summarises the share of the 2010/2011 grant between the different budget items for one specific community organisation in Didy municipality (Ambatondrazaka district, Alaotra-Mangoro region).

After some years of experimentation, these conservation agreements have struggled to fulfil their objectives. The biggest challenge is local monitoring. PES schemes require monitoring in order to assess whether a service is being delivered. Local patrolling has tended to be supported in the REDD+ and PES literature because it is seen as a low-cost way of monitoring (Böttcher *et al.*, 2009; Danielsen *et al.*, 2011). However, in Madagascar the number of forest officers available for such monitoring is too low: in 2012, there were 294 forest officers and six million hectares of protected area, roughly one agent for 20,000 hectares.[7] The delegation of monitoring to local communities seems to be unavoidable. Unfortunately, community patrols often fail to provide effective control. In the case of

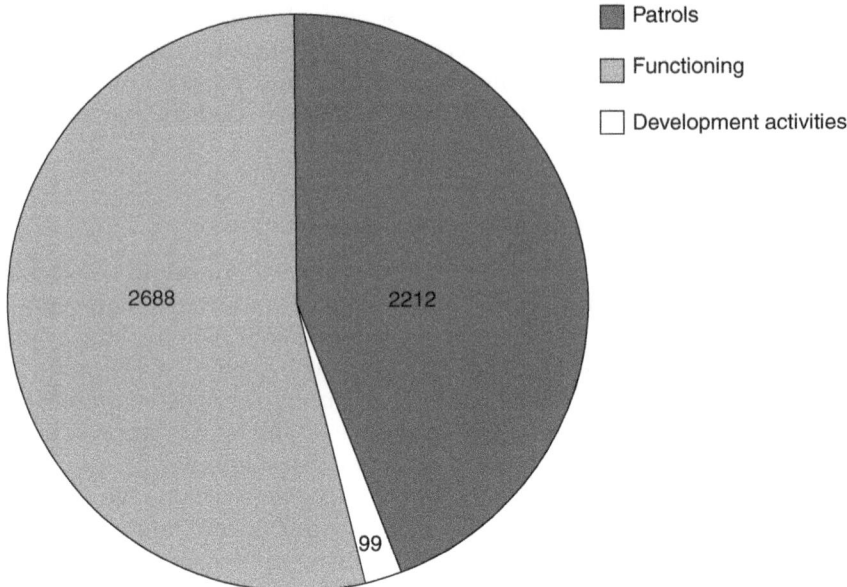

Figure 13.2 The share of conservation agreement's budget in US$ (VOI Taratra, Didy municipality, CAZ, 2011)

Source: CI (2011).

the CAZ conservation agreements, members of the community management organisation are supposed to make patrols every week in groups of four or five. However, patrols are tiring (involving between two and eight hours of walking), and members do not consider patrol salaries sufficient to cover their opportunity costs. As a result, they often ignore patrol duties and fake patrol reports (CI representative in Didy, personal communication). Furthermore, there is the issue of social pressure. During interviews led by one of the authors, many people involved admitted their reluctance to denounce members of their own communities for fear of reprisals (for example through black magic). Chapter 14 (Kaufmann) highlights the moral and ethical problems created by conservation programmes that expect members of communities to report each other for infringements. Ultimately, such schemes must also face the reality that community groups have no legal power to arrest or penalise those who break rules.

As well as issues of monitoring and enforcement, conservation agreements also face a problem common to any incentive-based mechanism – that of benefit distribution. In Didy municipality, customary land tenure over forest is held by pastureland (*kijana*) owners (Charbonnier, 1998). These pastureland owners live outside the forest and use it to graze zebu from December to March. Others live in the forest and use it all year round for their subsistence activities. Forest dwellers generally belong to the same families as pastureland owners but have set up livelihoods in forests because they do not have the necessary capital (zebus or land) to work irrigated rice fields. There are also migrants attracted by land availability. This traditional land tenure scheme has been respected in the CAZ conservation project, as community management territories are based on the geographical boundaries of pasturelands in forests, and pastureland owners are represented in the management groups. As a consequence, the benefits from conservation agreements are largely accrued by pastureland owners who do not bear the costs of forest conservation because zebu penning is still allowed in the forest and their livelihoods do not depend on the extraction of forest resources. The issue of distribution has also been observed in the Maroseranana municipality (another part of the CAZ), where the people who are most dependent on forest resources do not benefit from conservation agreements as their migrant status does not allow them to claim rights over forest land (Karsenty et al., 2009).

In addition to these distributional issues, conservation agreements have struggled to provide community management organisations with financial autonomy. This has been highlighted as a major constraint for local community management in Madagascar more generally (Hockley and Andriamarovololona, 2007). A key aim of conservation agreements was to fill the funding gap by providing start-up funds for future financing activities. In Didy village, after three funding campaigns, only one of eight community organisations has succeeded in generating collective benefits. For the others, benefits have been entirely monopolised by a small group of members, and virtually nothing has been reinvested in group activities.

Alternatives to Slash-and-Burn (ASB)

The final experiment in incentive-based conservation to be discussed in this chapter involves CIRAD's 'Alternatives to Slash-and-Burn' (ASB). The basic idea involves providing forest farmers with alternative agricultural practices, based on sowing under forest cover, to encourage them to stay on permanent agricultural plots and reduce forest clearance. The hope is that farmers will switch from a more extensive swidden cultivation system requiring 5–6 hectares of land per household (and forest clearance) to a more intensive form of agriculture using only 1–2 hectares per household, and at the same time increasing production from 0.9–1.3 to 2.0–2.5 tonnes per hectare (Raharison, 2012). This would be financed by the selling of un-emitted carbon from avoided deforestation.

ASB has been tried with a dozen or so farmers in Didy forest with limited success.[8] As with the two previous case studies discussed in this chapter, ASB highlights some major constraints for incentive-based instruments in a country such as Madagascar. First and foremost, such a scheme is incapable of dealing with the structural drivers of deforestation, such as demographic growth or agricultural policy. ASB cannot be expected to deal with the huge demographic pressure in the CAZ area, which is experiencing a 4.2 per cent annual growth in population. The integration of new agricultural techniques is long and complex, and migrants place large and urgent demands on forests. Agricultural intensification is hindered by the risk management strategies of forest households. In the context of annual cyclones and droughts, farmers cannot afford to pay for agricultural inputs to intensify their agriculture without an insurance system to protect them from the risk of losing their harvest (see Chapter 5 by Scales for a discussion of the risk averse nature of swidden cultivation). As private insurance is absent from the agricultural sector in Madagascar, state intervention is needed to provide public insurance to support agricultural intensification.

Once again, the ASB case study shows the challenge of creating genuine livelihood alternatives for rural households. The cost of the ASB project amounts to US$1,573 per household. Considering all the impacted households living in the CAZ forest (a total of 2,101 households), the costs of providing agricultural support for the CAZ would be US$3,305,520 for a period of five years (Desbureaux, 2012). It is predicted that deforestation in the CAZ area will result in the emission of 21,279,632 tonnes of CO_2 between 2010 and 2030 (Ramaroson, 2012). Assuming that the conservation project halves deforestation and that deforestation is linear in time, avoided emissions for a five-year period would be 2,659,954 tonnes. In order to fund the support to agricultural intensification through carbon credits, the price of carbon would have to be US$1.24 per tonne. However, this price does not take into account the other costs involved in a REDD+ project, such as monitoring and operating costs. Such costs are expected to be very high, as CAZ forests are also impacted by illegal mining, which is complex and costly to monitor and control. Ultimately, the cost of saving Madagascar's forests while reducing poverty is higher than the money that can be currently generated through carbon markets.

Conclusion

Incentive-based instruments are the most recent tool to emerge in efforts to conserve tropical forests. They stem from environmental economic theory, which is based on the idea that the pricing of ecosystem services creates incentives for individuals to exchange environmental goods or services and is the best way to protect natural resources. Many criticisms have emerged concerning the capacity of incentive-based instruments to address the drivers of deforestation. In this chapter, we have illustrated some of the major challenges of using these instruments in Madagascar. Three principle lessons emerge from our analysis:

1   Incentive-based instruments implemented at the community level face serious problems with the distribution of costs and benefits and are likely to increase inequality *within* local communities. The community resource management institutions that are created generally fail to represent the common interests of the whole community and are hampered by the monopolisation of power, as well as corruption and bad governance. Land tenure complexity also contributes to create winners and losers, with some individuals taking advantage to enrich themselves, to the detriment of others who are generally the most dependent on natural resources.
2   Incentive-based instruments by themselves cannot solve the problem of deforestation. Such instruments are presented as more efficient than state-based coercive methods. However, without adequate political, institutional and economic support, they are currently incapable of generating sufficient income and alternative livelihoods for rural households. Reducing swidden cultivation will require households and individuals to invest in more intensified agriculture or move to off-farm economic sectors. Incentive-based instruments are not a miracle solution to cope with the problem of deforestation. They must be considered as just one option, taking into account their limitations.
3   Incentive-based instruments cannot be seen as sufficient sources of funding for protecting Madagascar's forests. The expectation is that incentive-based instruments will not only create alternative livelihoods but create funding for conservation. However, the financing potential of incentive-based instruments is shaped by political commitments concerning demand for ecosystem services and factors such as the price of carbon, which is itself subject to complex political and economic dynamics. In Madagascar, the demands of reducing both deforestation and rural poverty go far beyond the finances currently provided by carbon markets.

Acknowledgements

The authors' research referred to in this chapter was financed by two research programmes funded by the French agency ANR Systerra: Serena, (ANR-08-STRA-13) and Pesmix (ANR-10-STRA-008.01).

Notes

1 We prefer to use the term 'incentive-based instruments', which refers to instruments based on a change of relative prices, rather than the term 'market-based instruments', which refers more specifically to a transfer of property rights, which doesn't necessarily happen in incentive-based mechanisms. For more discussion of this debate, see Karsenty, A. and Ezzine-de-Blas, D. (2013) 'Are PES "market-based instruments" for commodifying nature?', *Ecological Economics* (forthcoming).

2 For more on this see Butler, R. A. 'What is the current status of REDD+?', mongabay.com, 23 March 2011.
3 The Makira Carbon Company has been allowed to sell forest carbon credits by the Malagasy government, which is the legal owner of forests in Madagascar and therefore the legal owner of carbon credits generated from conservation projects.
4 These cannot be referred to as proper carbon credits since they do not have the necessary certification to be sold on voluntary markets. These carbon transactions have involved private companies such as Dell and Mitsubishi, and the rock group Pearl Jam.
5 This revenue distribution scheme has not been applied for first carbon credit sales in 2005 and 2008.
6 Brimont, survey conducted in the East and North East management transfer of the Makira protected area in June and September 2012.
7 General secretary of environment in Madagascar, communication in PHCF Day, 18 September 2012.
8 See Raharison (2012) for more details.

References

Adams, W. M., Aveling, R., Brockington, D., Dickson, B., Elliott, J., Hutton, J., Roe, D., Vira, B. and Wolmer, W. (2004) 'Biodiversity conservation and the eradication of poverty', *Science*, vol 306, no 5699, pp1146–1149.

Alix-Garcia, J., De Janvry, A. and Sadoulet, E. (2008) 'The role of deforestation risk and calibrated compensation in designing payments for environmental services', *Environment and Development Economics*, vol 13, pp375–394.

Allnutt, T. F., Asner, G. P., Golden, C. D. and Powell, G. V. N. (2013) 'Mapping recent deforestation and forest disturbance in northeastern Madagascar', *Tropical Conservation Science*, vol 6, no 1, pp1–15.

Angelsen, A. (2008) *Moving Ahead with REDD: Issues, Options and Implications*, Center for International Forestry Research, Bogor Barat.

Bidaud, C. (2012) *Le Carbone qui Cache la Forêt. La Construction Scientifique et la Mise en Politique du Service de Stockage du Carbone des Forêts Malgaches. Etudes de Développement*, PhD thesis, Institut de Hautes Etudes Internationales et du Développement (IHEID), Geneva.

Bidaud, C., Serpantié, G. and Méral, P. (2011) *Knowledge Mobilization in Water and Carbon PES Projects Implementation in Madagascar*, BIOECON, Geneva.

Blom, B., Sunderland, T. and Murdiyarso, D. (2010) 'Getting REDD to work locally: lessons learned from integrated conservation and development projects', *Environmental Science & Policy*, vol 13, no 2, pp164–172.

Böttcher, H., Eisbrenner, K., Fritz, S., Kindermann, G., Kraxner, F., McCallum, I. and Obersteiner, M. (2009) 'An assessment of monitoring requirements and costs of reduced emissions from deforestation and degradation', *Carbon Balance and Management*, vol 4, no 7.

Brockington, D. and Duffy, R. (2010) 'Capitalism and conservation: the production and reproduction of biodiversity conservation', *Antipode*, vol 42, no 3, pp469–484.

Carret, J.-C. and Loyer, D. (2003) *Comment Financer Durablement le Réseau d'Aires Protégées Terrestres à Madagascar? Apport de l'Analyse Economique*, World Parks Congress, Durban.

Charbonnier, B. (1998) *Limites et Dynamique Coutumières dans la Forêt Classée d'Ambohilero, à l'Intérieur de la Cuvette de Didy, S.E. d'Ambatondrazaka*, ENGREF, Montpellier.
CI (2011) *Conservation Agreements in Madagascar: An Update from the Conservation Stewards Program*, Conservation International, Washington, DC.
Coase, R. H. (1960) 'The problem of social cost', *Journal of Law and Economics*, vol 3, pp1–44.
Corbera, E., Brown, K. and Adger, N. (2007) 'The equity and legitimacy of markets for ecosystem services', *Development and Change*, vol 38, no 4, pp587–613.
Dahan, A., Buffet, C. and Viard-Crétat, A. (2011) *Le Compromis de Cancun: Vertu du Pragmatisme ou Masque de l'Immobilisme?*, Koyré Climate Series no 3, Centre Alexandre, Koyré.
Daily, G. C. (1997) *Nature's Services: Societal Dependence on Natural Ecosystems*, Island Press, Washington, DC.
Daly, H. E. (1977) *Steady State Economics*, W.H. Freeman, San Francisco.
Danielsen, F., Skutsch, M., Burgess, N. D., Jensen, P. M., Andrianandrasana, H., Karky, B., Lewis, R., Lovett, J. C., Massao, J., Ngaga, Y., Phartiyal, P., Poulsen, M. K., Singh, S. P., Solis, S., Sørensen, M., Tewari, A., Young, R. and Zahabu, E. (2011) 'At the heart of REDD+: a role for local people in monitoring forests?', *Conservation Letters*, vol 4, pp158–167.
Desbureaux, S. (2012) *L'insertion des Instruments Incitatifs dans les Politiques de Préservation des Ressources Naturelles. Etude de Cas: Enjeux de la Mobilisation d'Instruments PSE pour la Gestion de la Nouvelle Aire Protégée du Corridor Ankeniheny-Sahamena à Madagascar*, Masters thesis, Université Paris X Nanterre La Défense, Ecole des Mines ParisTech, ESCP EUROPE, Master 2: 102.
Ducourtieux, O. (2009) *Du Riz et des Arbres: L'interdiction de l'Agriculture d'Abattis-brûlis, une Constante Politique au Laos*, IRD, Karthala, Paris.
Ehrlich, P. R. and Ehrlich, A. H. (1981) *Extinction: The Causes and Consequences of the Disappearance of Species*, Random House, New York.
Eliasch, J. (2008) *Climate Change: Financing Global Forests: The Eliasch Review*, Routledge, London.
Ferraro, P. J. (2011) 'The future of payments for environmental services', *Conservation Biology*, vol 25, pp1134–1138.
Ferraro, P. J. and Pattanayak, S. K. (2006) 'Money for nothing? A call for empirical evaluation of biodiversity conservation investments', *PLoS Biology*, vol 4, e105.
Ferraro, P. J. and Simpson, R. D. (2002) 'The cost-effectiveness of conservation payments', *Land Economics*, vol 78, pp339–353.
Gomez-Baggethun, E., De Groot, R., Lomas, P. L. and Montes, C. (2010) 'The history of ecosystem services in economic theory and practice: from early notions to markets and payments schemes', *Ecological Economics*, vol 69, pp1209–1218.
Grieg-Gran, M., Porras, I. and Wunder, S. (2005) 'How can market mechanisms for forest environmental services help the poor? Preliminary lessons from Latin America', *World Development*, vol 33, no 9, pp1511–1527.
Grinand, C., Vieilledent, G., Rakotomalala, F. and Vaudry, R. (2013) 'Estimating past deforestation from 2000 to 2010 in Madagascar using multi-date Landsat satellite images and the Random Forests classifier', *Remote Sensing for Environment*, vol 139, pp68–80.

Hockley, N. J. and Andriamarovololona, M. M. (2007) *The Economics of Community Forest Management in Madagascar: Is There a Free Lunch? An Analysis of Transfert de Gestion*, United States Agency for International Development, Washington, DC.

Holmes, C., Carter Ingram, J., Meyers, D., Crowley, H. and Ray, V. (2008) *Case Study: Forest Carbon Financing for Biodiversity Conservation, Climate Change, Mitigation and Improved Livelihoods: The Makira Forest Protected Area, Madagascar*, Wildlife Conservation Society, New York.

Karsenty, A. (2007) 'Questioning rent for development swaps: new market-based instruments for biodiversity acquisition and the land-use issue in tropical countries', *International Forestry Review*, vol 9, pp503–513.

Karsenty, A., Randrianarison, M., Andrianjohaninarivo, T., Ranoarisoa, P. and Randriamavo, L. (2009) *Les Contrats de Conservation à Madagascar: Enquête Socio-économique dans 3 Villages de la Commune de Maroseranana*, CIRAD, Montpellier.

Karsenty, A., Sembres, T. and Randrianarison, M. (2010) 'Paiements pour services environnementaux et biodiversité dans les pays du sud: le salut par la "déforestation évitée"?', *Revue Tiers Monde*, vol 202, pp57–74.

Kosoy, N. and Corbera, E. (2010) 'Payments for ecosystem services as commodity fetishism', *Ecological Economics*, vol 69, pp1228–1236.

Landell-Mills, N. and Porras, I. T. (2002) *Silver Bullet of Fools' Gold? A Global Review of Markets for Forest Environmental Services and Their Impact on the Poor*, International Institute for Environment and Development, London.

McKinsey & Company (2009) *Pathways to a Low-Carbon Economy. Version 2 of the Global Greenhouse Gas Abatement Cost Curve*, McKinsey & Company, London.

MEA (2005) *Ecosystems and Human Well-Being*, Millenium Ecosystem Assessment, United Nations Environment Programme, Nairobi.

Meadows, D. H., Randers, J. and Meadows, D. L. (1972) *The Limits to Growth*, Universe Books, New York.

Meyers, D. and Berner, P. O. (2001) *Carbon Sequestration: Maroantsetra Carbon Project Progress Report*, United States Agency for International Development, Washington, DC.

Muradian, R., Corbera, E., Unai, P., Kosoy, N. and May, P. H. (2010) 'Reconciling theory and practice: an alternative conceptual framework for understanding payments for environmental services', *Ecological Economics*, vol 69, pp1202–1208.

Murray, B. C. (2008) *Leakage from an Avoided Deforestation Compensation Policy: Concepts, Empirical Evidence, and Corrective Policy Options*, Nicholas Institute for Environmental Policy Solutions, Duke University, Durham.

Pagiola, S. (2008) 'Payments for environmental services in Costa Rica', *Ecological Economics*, vol 65, pp512–524.

Pagiola, S., Arcenas, A. and Platais, G. (2005) 'Can payments for environmental services help reduce poverty? An exploration of the issues and the evidence to date from Latin America', *World Development*, vol 33, no 2, pp237–253.

Pagiola, S., Landell-Mills, N. and Bishop, J. (2002). 'Making market-based mechanisms work for forests and people', in S. Pagiola, J. Bishop, and N Landell-Mills (eds) *Selling Forest Environmental Services: Market-based Mechanisms for Conservation and Development*, Earthscan, London.

Pan, Y., Birdsey, R. A., Fang, J., Houghton, R., Kauppi, P. E., Kurz, W. A., Phillips, O. L., Shvidenko, A., Lewis, S. L., Canadell, J. G., Ciais, P., Jackson, R. B., Pacala, S. W., McGuire, A. D., Piao, S., Rautiainen, A., Sitch, S. and Hayes, D. (2011) 'A large and persistent carbon sink in the world's forests', *Science*, vol 333, no 6045, pp988–993.

Pattanayak, S., Wunder, S. and Ferraro, P. J. (2010) 'Show me the money: do payments supply environmental services in developing countries?', *Review of Environmental Economics and Policy*, vol 4, no 2, pp254–274.

Pierre, R. (2011) 'La prise en compte de la notion de Service Environnemental dans les Labels: L'exemple de Madagascar', *UFR Sciences des Territoires et de la Communication*, 35. Bordeaux: Université Michel de Montaigne Bordeaux 3.

Pigou, A. C. (1932) *The Economics of Welfare*, Macmillan, London.

Raharison, T. (2012) *Rapport de Mission d'Evaluation des Itinéraires Techniques Alternatifs au Tavy*, Volet Paiement pour Services Environnementaux, PSE Tavy, COGESFOR.

Ramaroson, N. (2012) *Analyse Historique de la Déforestation par Télédétection et Modélisation de la Déforestation à Madagascar: Cas du Corridor Ankeniheny-Zahamena*, Télédétection & Risques Naturels, Antananarivo, Master 2: 38.

Roy, E. L., Karsenty, A. and Bertrand, A. (1996) *La Sécurisation Foncière en Afrique: Pour une Gestion Viable des Ressources Renouvelables*, Karthala, Paris.

Scales, I. R. (2011) 'Farming at the forest frontier: land use and landscape change in western Madagascar, 1896 to 2005', *Environment and History*, vol 17, pp499–524.

Sommerville, M., Milner-Gulland, E. J., Rahajaharison, M. and Jones, J. P. G. (2010) 'Impact of a community-based payment for environmental services intervention on forest use in Menabe, Madagascar', *Conservation Biology*, vol 24, pp1488–1498.

Stern, N. (2008) *Key Elements of a Global Deal on Climate Change*, London School of Economics and Political Science, London.

Sunderland, T. C. H., Ehringhaus, C. and Campbell, P. M. (2008) 'Conservation and development in tropical forest landscapes: a time to face the trade-offs?' *Environmental Conservation*, vol 34, no 4, pp276–279.

Vatn, A. (2010) 'An institutional analysis of payments for environmental services', *Ecological Economics*, vol 69, pp1245–1252.

WCS (2011) *Makira Forest Protected Area Project Design Document*, Wildlife Conservation Society, New York.

Wells, M. P. and McShane, T. O. (2004) 'Integrating protected area management with local needs and aspirations', *Ambio*, vol 33, no 8, pp513–519

World Bank (2011) *Project Paper on a Proposed Additional IDA Credit in the Amount of SDR26 Million and a Proposed Additional Grant from the Global Environment Facility Trust Fund in the Amount of US$10.0 Million to the Republic of Madagascar for the Third Environmental Program Support Project (EP3)*, World Bank, Washington, DC.

Wünscher, T., Engel, S. and Wunder, S. (2008) 'Spatial targeting of payments for environmental services: a tool for boosting conservation benefits', *Ecological Economics*, vol 65, pp822–833.

Wunder, S. (2005) 'Payments for environmental services: some nuts and bolts', CIFOR Occasional Paper 42, Center for International Forestry Research, Bogor Barat.

Zbinden, S. and Lee, D. R. (2005) 'Paying for environmental services: an analysis of participation in Costa Rica's PSA program', *World Development*, vol 33, no 2, pp255–272.

14 Contrasting visions of nature and landscapes

Jeffrey C. Kaufmann

Conservation and culture

'Culture' has begun to matter to conservationists who realize that among the beliefs, meanings, and practices shared by a local population there are some that can aid in the conservation of nature. In New Zealand, the Maori concept of *kaitiakitanga*, which conservationists considered an indigenous resource management term, formed the basis of the Resource Management Act of 1991 (Kawharu, 2000; cf. Roberts et al., 1995). In Malaysia, conservationists supported the National Geographic explorer and ethnobotanist Wade Davis' urgings to publicize the Penan people's resistance to logging within their territories by setting up blockades against powerful timber corporations as 'icons of resistance for environmentalists worldwide' (Davis and Henley, 1990).[1] In Madagascar, conservationists gauged the Malagasy notions of *fady* and *dina* (taboos and community laws) as instruments of an indigenous conservation ethic drawn to serve a global agenda (Andriamalala and Gardner, 2010; Jones et al., 2008). In each of these cases, conservationists have begun to plumb the mission of 'saving nature' out of local cultural ideas and institutions, moving away from a victim–victimizer model that considers nature being torn asunder by culture.

While finding ways to include human inhabitants and their cultural ideas in possible solutions to local conservation problems is an improvement over excluding people from the very contexts of which they are part (Kaufmann, 2006), conservationists are not anthropologists. As it turns out, *kaitiakitanga* means much more than 'resource management.' It has numerous meanings for Maori people, not just one generic notion such as 'guardianship,' meanings that run in both philosophical and pragmatic directions, depending on the kin group (Kawharu, 2000). Its main structural idea is social, serving to manage relationships across a world layered in time and space. The term aims to balance the 'genealogical layering' of descent-time (linear) and kinship-space (lateral) in an ethic of reciprocity between humans, ancestors, the spirit world, and the natural environment (Kawharu, 2000; Roberts et al., 1995). *Kaitiakitanga* trains under *whakapapa*, 'to place in layers' (Roberts et al., 2004, p1). This latter term refers to the

Maori's classification scheme, their folk taxonomy, an ordering of many layers of related beings, both spiritual and material (Roberts et al., 2004). 'Resource management' and 'guardianship' do not get to the multidimensional nature of maintaining 'kinship links between themselves and their environment' (Kawharu, 2000, p366). Insisting that they do flattens the layers of meanings into a category with which conservationists might be comfortable but which does not represent Maori cultural thinking.

If this brief introduction to the way Maori navigate their system of layered connections among all sorts of related beings bewilders the reader, it should. Such a complex theory of knowledge cannot be understood with ready-made substitutions that take out the unfamiliar ideas and replace them with concepts familiar to a Western audience. There is more at stake than paring down ideas to familiar territory, to categories of thought that help readers half a world away feel at ease. Power plays a role. The original ideas can be transformed to do work for which they were not intended, to carry meanings with a different power behind them. Such is the case with the translations of Penan ideas of resisting the destruction of their rainforest by timber companies.

According to one of the Penan's ethnographers, there are two senses of knowledge that apply in the Penan case. The first is the 'objectivist' sense, which has to do with Penan knowledge of their world, the forest, its waterways, and natural resources; this is the sense that cultural anthropologists work in. The other is the 'environmentalist' sense that does not come from knowing the language, which borrows from ethnographies to 'make an argument and mobilize support,' and which links cultural knowledge to keywords such as 'sacred' or 'spiritual insights' that carry emotional weight among Western audiences, thus transforming Penan knowledge about their environment into 'wisdom'—the cornerstone of Davis' and Henley's environmental campaign to save cultural diversity (Brosius, 1997, pp53–55). Conserving cultural diversity is not the problem. The problem is with Davis' and Henley's embellishments of Penan ideas that transformed them into serviceable categories for an environmentalist agenda but removed them from a Penan cultural context and sense of reality. For instance, Brosius wrote about the *molong* concept of 'to preserve' in order to demonstrate that the nomadic Penan were not wayward wanderers on the landscape but had a system of land tenure that gave people different rights to the land (Brosius, 1997, p56). But Wade and Henley transformed Brosius' descriptions of *molong* into 'sacred,' giving souls to features in the landscape that Penan do not reckon, romanticizing their notion of preserving their place in the landscape into inaccuracies that Brosius labels 'ecological etherealism,' not recognizable as Penan thought but Western fantasies about tribal wisdom (Brosius 1997, pp58–59).

The process of making cultural differences legible to, and thus available for appropriation by, conservation organizations is called 'cultural generification' (Errington and Gewertz, 2001). The process, which fits under the umbrella term 'acculturation,' involves a translation 'such that the cultural particular either has become translated into the cultural general or into a general example of the cultural particular' (Errington and Gewertz, 2001, p510). Interestingly, conservationists are not the only people who reduce esoteric cultural information to something close and familiar. Missionaries and colonialists paved the way in this regard for conservationists. Moreover, members of indigenous populations are nowadays participating in the process of translating their esoteric knowledge into a generic form that fits into the conservation discourse (Brockington, 2001, 2005; Brosius, 1999; Errington and Gewertz, 2001). They even put anthropologists in the uncomfortable position of helping them translate their culture in ways that 'soften the edges of differences,' that 'erase the vernacular,' and that 'conform to models and ideologies of the powerful' (West, 2005, p633). Anthropologists tend to resist the invitation because helping indigenous cultures acculturate to the dominant culture by making esoteric ideas understandable to powerful conservationists would result in local populations losing power via translation (Errington and Gewertz, 2001).

In this chapter I bring out some of the contrasts between conservationists and Malagasy people by formulating, in their own terms, Malagasy approaches to nature and landscape. The reader should recognize that there are different 'ways of being in the environment' than those suggested in conservation discourse. In the next two sections I discuss nature and landscape with an eye on remaining faithful to the vernacular categories of thought that shape and give these concepts heft. I consider how the practice of making history with the help of trees gets at Malagasy conceptions of nature. Next, I discuss how Malagasy inform their landscapes with taboos (*fady*) and community conventions (*dina*), concepts that conservationists in Madagascar consider as keys to an 'indigenous conservation ethic,' as ways to make Malagasy cultural ideas instrumental in their efforts to save nature. My discussion shows that taboos and community conventions are cultural terms directing social behavior by instructing people on how to conduct themselves in a social environment with other people, and not elements of a conservation ethic. Conservationists are embellishing Malagasy ideas for their own designs when they refashion social principles into environmental rules. I finish by considering this embellishment through the lens of the fifteenth-century French political philosopher, La Boétie. I discuss briefly his concept of 'voluntary servitude' to compare past and present forms of dominance. La Boétie's lens clarifies the link between conservation power and colonial power. It helps one see the persistence of colonial institutions in Madagascar. I conclude by offering some anthropological recommendations for conservationists to think about.

Nature in things past

The Malagasy practice of using trees to make history informs their attitudes toward nature and plays an important role in how rural households relate to and use forests. I am not referring to the process of turning trees into paper and writing down their stories of things past in books. While Madagascar has produced some excellent academic historians who have written remarkable histories of the island (see, for example, Esoavelomandroso, 1979; Rakoto, 1997; Rakotoarisoa, 1998; Ramarolahy, 1972), most Malagasy rely on oral history to transmit their knowledge of the past. What interests me about their brand of historicism—their particular way of knowing themselves by knowing their past—is their use of trees to locate significant human ancestors.

Malagasy consider their ancestors a very important part of the past. For most Malagasy, 'the idea of the ancestors ... encompasses and expresses all that is considered morally desirable or appropriate in social relations' (Mack, 1986, p64). *Ny razana*, the ancestors and the spirits that, according to Malagasy belief, stay in contact with their descendants after their bodies have perished, occupy various spaces in *ny tanindrazana*, the ancestral lands, within reach of their living kin. For example, Merina and Betsileo *razana* stay in underground tombs; Bara in aboveground caves; Sakalava *razan'olo* (ancestor people) have their trees, as do Mahafale ancestors and many other Malagasy ethnic groups whose ancestral homelands are 'coastal' and turned away from the interior highlands. Most Merina of the central highlands perform the ritual of *famadihana*, or 'turning the bones,' wherein every few years the shrouded bones of ancestors are brought out of the tomb and into the light of day, reshrouded, fêted and turned a little more into dust and a little further into the past (Bloch, 1971; Graeber, 2007b).

Malagasy are adept at applying rituals, like the *famadihana*, to remember people past by the work of people present. They find trees helpful in making history by placing ancestors in a relationship with their descendants via trees. Trees are effective agents for several reasons (Feeley-Harnik, 1991): they are close at hand, the hardwoods last many generations, and the trees are upstanding—an ambiguous, hence powerful, quality found in some people too. After the French colonial era ended in 1960, Sakalava, who have their ancestral homeland in the west and northwest, used trees to tell the story of the history of royal dynasties ruling the territories before the French colonized them (Feeley-Harnik, 1991; see also Lambek, 2002). Putting together the royal funeral (*menaty*) and in particular carving down certain trees to get at their heart or core (*teza*), then erecting the core poles around the tomb amounted to erecting 'tree-people'—a category of thought, somewhat like the classification 'Christmas tree' as representative of Christ—whereby Sakalava see and treat certain trees as royal ancestors. As Feeley-Harnik explains:

> In short, workers in the menaty service handled the trees like royal bodies, and they handled the teza like royal corpses. The transformation of trees that had died on the mainland into posts reburied in the

ground around the royal tomb thus paralleled the transformation of the fleshy corpse into a skeleton, enclosed in a tree trunk and buried in the tomb itself.

(1991, p445)

Antankaraña, who live in the far north of the island, employ a mast-raising ceremony, the *Tsangantsainy*, as a way to make history (Walsh, 2001). As workers placed the tree mast in the ground, religious leaders invoked the ancestors who had ruled in the past. By associating various personalities and characteristics with the tree, the mast became, like the second person plural 'You,' polysemic in that it meant different things to different people involved in the ceremony. It marked authority and legitimacy to the Antankaraña political institution; it symbolized the allegiance of the workers in service to the institution; it created a memorable event to the audience observing the performance; and it created an ethnic event perpetuating Antankaraña custom to the invited dignitaries. Madagascar might be rich in species that live in trees, but Walsh (2001) demonstrates that trees host a wealth of Antankaraña meanings.

Burial of the deceased and erecting 'tree-people' poles around tombs or 'tree-people' masts in the center of local government are elaborate displays of making history in trees. A more common or widespread way to pay respect to one's ancestors is to use trees in cattle sacrifices to ancestors, in which the living ask blessings from the dead. Betsimisaraka, who live in the wet eastern part of the island and do not raise many cattle themselves, spend large sums of money to purchase their sacrificial animals (Cole, 2001). At a circumcision ceremony that made several youth into men, a bull was sacrificed next to a prayer post (*jiro*), 'a tree cut in the shape of zebu horns' (Cole, 2001, p178). As the ritual leader invoked the names of ancestors, he held a wand of a plant the Betsimisaraka associate with growth and efficacy they call *hasina*, meaning 'ancestral power' (Cole, 2001, p116). Boys soon to be of age climbed the 'haughty tree,' a tree in particular favor with the ancestors, who received coins as 'symbols of ancestral blessing' after ascending the tree (Cole, 2001, p181). Finally, the sacrificial meal was served on the large fronds of the traveler's palm (*Ravenala madagascariensis*; for a social history of this tree, see Feeley-Harnik (2001)). By comparison, people in the south and west use the deciduous *kily* (*Tamarindus indica*) fruit tree in ancestor invocations.

Trees also encapsulate sacred grounds, usually cemeteries, and individual trees considered the habitat of active ancestral spirits. Sacred groves are 'clean' and 'good' places, which their spiritual occupants are said to covet, and which they protect by draping them in taboos (more on taboos in the next section). Outside of the village of Analakely, near the Bezaha Mahafaly Reserve in Toliara province, I accompanied a herder through a sacred forest as he took his cattle to a watering hole on the other side. Before entering, he told me all of the things I could not do, including even picking up a single dead branch. My guide explained that the ancestors guard jealously their habitats, some of which were *doany* or the tree abodes of powerful ancestors who entered people, usually women, through *tromba* or spirit possession. Villagers knew the ancestor spirits as if they were living people, learning about them from spirit mediums during séances in which the medium became apparently possessed by the spirit and exhibited a different personality, tastes, and wants, told stories about himself, and so on.

Trees help to connect many Malagasy with what is most important to them in the past, their ancestors. This does not mean that Malagasy are obsessed with the past, caring more about the past than about the present or the future.[2] Malagasy are instead a people of many historians in pursuit of knowing the past and knowing their history. Trees help them gather this knowledge. They practice historicism by recognizing that knowledge of history is a sure way of knowing themselves. Perhaps George Orwell, in his classic book *1984*, had them or someone like them in mind when he wrote his dictum, 'Who controls the past controls the future: who controls the present controls the past' (Orwell, 1961). Ancestors are revised and renewed in every invocation of their blessings and in every situation that calls for their participation.

Nature, then, can be thought of in a Malagasy way as an historical method, as a way to understand the past. It attests to the powers of human imagination and expression. This way of considering nature should not be thought of as 'strange' or 'superstitious' or 'exotic' because it is different from our own. Anthropology, as a method of knowing ourselves through knowing others, demands its practitioners to suspend their ethnocentric attitudes of superiority over others in order to apprehend, from the inside out, the reasons for the differences (and eventually a recognition of the similarities) of others.

In order to bring out Malagasy interactions with, and constructions of landscapes out of, nature, it will help to look at two kinds of rules that position their behaviors toward nature. These rules change in perspectives depending on village and lineage, thus they get to landscape theory.

Landscapes as a mosaic of perspectives

Madagascar consists of a mosaic of landscapes and cultural practices that have arisen there over time (Esoavelomandroso, 1988). The landscape concept gives details and nuance to blocky labels such as 'tropical island' for Madagascar. Landscapes are different perspectives upon the land or nature or built spaces, each having its own stories. Landscapes reveal what one wants to emphasize about a place, about certain spaces that have significance and meaning to human narrators, often closely connected to identity and how one sees oneself and others like them as belonging to a location, to a place, and how that place gives meaning to life and sometimes to living on Earth. A place, then, gets to feelings, to emotions. Getting people to talk about place, to describe a location on a landscape of meaning, is one way that anthropologists can get at another person's feelings.

Southern Madagascar has vast forests of mainly xerophytic trees, short in height and well equipped to protect their water holdings in a part of Madagascar that tends to receive around only 30 cm of rain a year. As one might expect, Malagasy farmers, pastoralists, and fishers who live around trees have their own knowledge base about the various trees. I do not intend to discuss a list of categories and cultural practices involving trees. I want instead to get to a broader point, one that helps the reader see that competing landscapes, different ethnoecologies or ways of knowing and perceiving nature, can lead to the breakdowns in communication that Nora Haenn (1999) identified between government-sponsored conservation practices and the wishes of Campeche populations in Mexico's Calomel Biosphere Reserve.

Mahafale herders see trees functioning as health providers by providing shade for their livestock. They see land as 'open or spacious' (*lalake*) and suitable to farming after cutting trees and preparing the soil for crops. They also see trees in economic terms, picking out some species for house building, some for chairs, some for spears, and still other varieties for thorny enclosures around economic spaces, and so on. We have already seen how they use trees as anchors to the past, as ways of making history by interacting with nature in ways that make sense to them. I turn next to the ways different rules, namely taboos and community laws, inform their landscapes and their perspectives toward nature.

Fady: *forbidden behavior*

If 'landscape' gets at the idea of different points of view toward nature and the land, then 'taboos' help people organize their behaviours in a moral landscape. Among the Maori, *tapu* are items 'set apart from normal use' (Kawharu, 2000, p357). The most widespread taboo among humans is the incest taboo, which dictates to not have sexual relations with close relatives. This taboo also provides a map of potential marriage partners out of those to whom the taboo does not apply (anyone not related too closely). Taboos tell subscribers what not to do by proscribing unfit behavior, and they also prescribe what to do in terms of proper rule-following behavior. Taboos help to carve out, Michael Lambek reminds us, a space for the construction of self-identity (Lambek, 1992). They provide people with an assurance of who they are in contrast to other people with a different set of taboos and a different identity. Taboos also convey status to their subscribers. A descendant of royalty in Madagascar, for example, will have discerning taboos that a descendant of slaves will not.

Mahafale people know taboos. Their name means, after all, 'having the power to taboo.' Yet the people do not use the name to refer to themselves as having the power to taboo, even if they do make the rules. Like most things Malagasy, their language takes the spotlight off of the speaker and moves it onto a more ambiguous, and therefore powerful, canvas: in this case, landscape. Notice I did not say 'vague,' which clouds meaning with uncertainty. Ambiguity comes from the Latin meaning 'wanderer,' which, in the Malagasy case, draws not on the idea of an aimless roamer but of someone who gets to certainty, to the meaning, in an indirect way. For speakers of other languages that pride themselves on direct speech, on providing clarity with as little effort for the listener or reader as possible, the Malagasy way of demanding an effort on the other's part can be disconcerting.

When early French colonialists asked residents what name should go in the record book for the local population, the people responded that they were the people of the tabooing land (*tane mahafale*), which the French interpreted as having the same meaning as in Merina (the predominant dialect on the island) where *faly* means 'happy' (Eggert, 1981). Foreigners called them, erroneously, 'the happy people.' Even today, guidebooks make the same mistake, failing to recognize that the Mahafale dialect shifts the consonant <d> to <l>, and the near silent ending vowel <y> to <e> (among many other differences) rendering a word with a different lexicon.

Being a people of a tabooing land should not be thought of imparting a special non-human agency to land. It is my understanding that Mahafale use the idiom of taboo to name the land of which they are part for its central characteristic: a highly unpredictable environment subject to changes in rainfall from year to year, a region prone to drought and the occasional famine, where pastoralists delight in raising zebu cattle in the southwest's slight but famously healthy grass (*ahitse*). To my knowledge, Mahafale do not interpret rainfall variability as being caused by transgressions of taboos.

Both French colonial administrators and conservation non-governmental organizations (NGOs) make the same mistake with regards to ethnicity, identity, and culture—they treat culture (whether it is taboos, *dina*, or ethnic identity) as something concrete (see also Box 6.2 on page 138).[3] They both try to categorize and reify culture. They want to label different kinds of people and the marks of their distinct identity. However, the eighteen ethnic groups officially recognized by the colonial government ethnicity is largely a colonial administrative construct. Many of these people see themselves belonging clearly to lineages and clans but not so clearly to an ethnic group called, for example, 'Mahafale.' A performative sense of identity—you are what you do—informs a lineage structure that helps people decide, for example, who potential marriage mates might be (Astuti, 1995; Eggert, 1981; Larson, 1996; Poyer and Kelly, 2000; Rakotondrabe, 1993). Although there are boundaries among the various populations—for example, their homelands, dialects, and taboos—trying to describe these in ethnic terms is not always as productive as missionaries, colonialists, and conservationists have assumed. So the view that these 'Sakalava' or those 'Tandroy' people are 'indigenous' and have such and such 'traditional' practices is both wrong and potentially highly damaging. Wrong because in reality all these cultural ideas are fluid, messy, and context specific. Dangerous because it can play into power struggles, into the power dynamics behind taboos consisting of the religious specialists (*olobe* and *ombiasy*) whose charge it is to uphold the taboos set down by an influential ancestor.

Moreover, taboos can be very localized and specific—they can apply to all people or just some people, some areas but not all—as I illustrate with the following scene from my fieldwork around Androka Vaovao in the far southwest. One day, I was asked by my hosts to help one of my local consultants gather kindling for the local gendarme office. We hitched up the steers to the ox-cart and headed north, away from the sea. After about an hour of hot rodding, of racing the ox-cart by whipping the steers into a gallop, then steadying them at a trot, then resting them at a walk, only to repeat the pattern three or four times, we arrived at our destination. A thick grove of euphorbia trees banked the base of a long hill (a long-dead sand dune), with large brown and round shrubs swathing the sandy soil. We gathered these shrubs, which made the dense and rapid fires that pastoralists used to burn off the thorns of the prickly pear cactus before feeding them to their cattle, and loaded them into the ox-cart. After a while I ventured into the euphorbia forest, to enjoy a bit of shade. I walked into a killing field where hundreds of Malagasy land tortoise shells lay. Some had been freshly cooked in their shells, others had been there for years.

I was stunned. For I had thought that the one animal, besides ring-tailed lemurs (*maki*, *Lemur catta*), that Mahafale held in such high esteem that their ancestors proclaimed them taboo was the radiated tortoise (*sokatse*, *Geochelone radiata*). They linked their long history of survival in the region to their being aware and respectful of the land's special demands and the animals that, according to oral traditions, at one time or another led their ancestors to water and out of danger. The trope of 'being shown the way' (*toro lala*), of learning how to survive in the mahafale region (the land that had the power to taboo), merged in innumerable creation stories of various taboos involving animals (Kaufmann, 2003, p113, n4). People interpreted taboos not as barriers to individual freedom but as guides to family survival in a land of difficulty where conditions fluctuated from year to year (Kaufmann, 2004, 2008).

My hosts explained that while *sokatse* was taboo to Mahafale people, others had no such prohibition. The rural policemen (gendarmes) came from all over the island, with their own sets of taboos. When they conducted their monthly walking tours of villages in their jurisdiction, the non-Mahafale among them would pick up tortoises they came upon and bring them to the campsite I had stumbled upon. My contacts saw no offense to the practice, since it did not involve anybody bound by a rule that forbade harming the animals. One informant, for whom *sokatse* was taboo, even said that she would have no problem eating a tortoise if she had no other food. She would ask forgiveness for the transgression by offering a blood sacrifice to the ancestors.

My experience in the killing field changed my understanding of how taboos worked in the tabooing land. I realized that taboos, in a sense, are meant to be broken. They gain articulacy and power in the minds of their followers, not by being unmindfully followed, but by a conscious effort, perhaps like a silent prayer, on the part of the person to honor the wishes of the ancestors as well as the object being separated from the person via the taboo. In Andrew Walsh's (2002, p466) words, taboos always suggest 'the freedom to do otherwise.' Transgressing taboos creates a dialogue with the ancestors, at once challenging their authority and then righting it again (Walsh, 2002, p455). They map out a moral landscape of right and wrong.

Mahafale see nothing sacred about resource conservation. Tabooing radiated tortoises and ring-tailed lemurs does, indirectly, conserve those animals (in an ideal world, in which the only people in the tabooing land are the Mahafale). Malagasy taboos are directed at something very different from conservation; namely, at pursuing a structured relationship with their ancestors (Lambek, 1992). David Graeber reminds us that tabooing is 'one of the most basic ways of demonstrating authority' (Graeber, 1995, p265). Malagasy ancestors keep them busy enough attending to the sacred realm while trying to make ends meet in the mundane realm of their ordinary lives. Conservation projects are things that foreigners do (*zavatse vazaha*), which are outside the social and cultural realm of Mahafale.

This has not hindered conservation organizations, especially those that aim to pursue a collaborative conservation methodology, from exploiting the informal institution (non-state level) of tabooing as a way to regulate resource extraction. The logic seems to be that conservationists can acculturate local populations by incorporating a sacred element into conservation. Acculturation means exploiting the similarities between cultures in an effort to close the distance between them (Ratsimbazafy and Kaufmann, 2008). Conservationists assume that by going through the local taboos, through the sacred groves, through a sacredness concept attributed to the landscape, they will arrive at collaborative conservation. Yet, as I have shown, the context is much more intricate than they think (for similar results, see Cinner, 2007; Jones et al., 2008).

If acculturation does not lend itself to the sacred, how does it fare with a non-sacred informal institution, such as the *dina*, which, among other things, regulates resource extraction? Might conservation organizations have more success appropriating secular local laws to serve resource conservation?

Dina: *community regulations*

Dina are community conventions that help to regulate the social environment at the village level. Such rules are located among the power base in Malagasy rural communities: the elders, sorcerers, and ceremonial leaders in charge of family, lineage, and clan alliances. Community laws have legitimacy among community members because the power base that the community has legitimized vets those rules and sees that they are enforced.

There are as many kinds of *dina* as there are Malagasy communities who use the tool to regulate their living together. For example, pastoralists have *dina* to help them sort out grazing commons, common-pool watering rights, and livestock exchanges. Fishers regulate, for instance, octopus no-take zones (Langley, 2006, p28). Horticulturalists have codes to help them identify use rights in the forest (Henkels, 1999, p42). Only Mikea foragers seem not to codify publicly their dealings with one another (Tucker et al., 2011),[4] though to my mind their use of reciprocity is itself a *dina*—a community convention transmitted by oral tradition (though some Malagasy communities write them down and register them with the village president or the town mayor). Rakotoson and Tanner (2006, p862) have identified three types of *dina*:

> The first type of *Dina* does not require a control of legality and needs no legal entity for approval. This is the case with most oral traditions that remain mostly unwritten, especially in the remote rural communities. This type of *Dina* is applicable immediately without legal procedure, and is the most efficient for local traditional communities. The second type of *Dina* requires a judicial review. These *Dinas* have to be approved by the Ministry of Justice and tested first to ensure that they are not in contradiction to official laws and the Constitution. For instance, if a *Dina* sanctions the death penalty, it will not be approved legally and is essentially forbidden since the Malagasy Constitution and laws prohibit the death penalty. The third category of *Dina* are those which are legal by nature and created by legal and official institution.

The varieties of customary law presents a problem for conservationists, who tend toward formalizing them into the third type, into legally binding social codes (Bérard, 2009), while most Malagasy villages in the countryside use the first type (see Chapter 8 by Pollini et al. for the problems that this division creates for attempts at community-based resource management). In addition, rules and institutions created through conservation institutions such as *Gestion Locale Sécurisée* (GELOSE) and community-based natural resource management (CBNRM) face the problem of legitimacy in Malagasy rural communities (see also Chapter 8 by Pollini et al.). *Dina* are legitimized, followed, and enforced because they emerge through local institutions identified as the seat of community power.

The same is true for tenure and customary land rights. Malagasy conceive of ownership in ways different than most conservationists. Some land can be deeded but most is held as communal use rights. Cropped land, where investments of time and labors are evident to the community, carry a stronger sense of ownership. Individual fruit trees and livestock are private property. Different tenure systems matter since conservation projects increasingly push privatization, enclosure, and profit as a way of incentivizing conservation (see Chapter 11 by Scales, Chapter 12 by Neimark and Tilghman, and Chapter 13 by Brimont and Bidaud for discussion of such schemes). This does not fit easily with Malagasy views of ownership.

In 1996, the Malagasy government enacted Law 96025, which aimed to integrate development and conservation by transferring to local communities the authority to manage natural resources (Rakotoson and Tanner, 2006, p860). Most villagers were not prepared to work with the new bureaucratic jargon. They required help and that fell to various conservation NGOs. Out went oral tradition as the bearer of the laws and in came literacy and bureaucratese. Hence, Law 96025 linked conservation NGOs to the communities in order to help them to write new *dina*, which would conserve various natural resources and species, in such a way that the local *de facto* rules—rules that exist in fact, though not by legal establishment or by official recognition—could be recognized as Malagasy *de jure* law—rules recognized by legal establishment (see, for example, Andriamalala and Gardner, 2010, p450; Rarivoson, 2007, p312).

The Madagascar government had experimented with the idea of enacting a conservation agenda in local *dina* before Law 96025 came about. In 1991, the Ministry of Waters and Forests (*des Eaux et Forêts*) joined forces with Bara herders in south central Madagascar and wrote the *Dinan'ny Mpanao Hatsaka* (Horning, 2004, p180). This law was targeted at farmers who were encroaching on pastoralist land by deforesting and planting crops. The *dina* read as follows: 'It is strictly forbidden to clear new parcels [of forest]. Or else, rule-breakers will have to abandon their land to the state and their crops to the community. In addition, they will be fined' (Horning, 2004, p180). The convention went through a number of modifications trying to better the monitoring capacity, the penalties, and the specific natural resources. The World Wide Fund for Nature (WWF) eventually took over the *dina* and put resource management in the hands of the eight main ethnic leaders, essentially transforming the power dynamic on forest protection from community convention to elite control (Horning, 2004, p185), which exacerbated divisions among Bara villagers and eroded village authority, leading to an erosion of legitimacy and trustworthiness. In the end, Bara considered the *Dinan'ny Mpanao Hatsaka* as doing the bidding of outsiders and not looking out for their own interests (Horning, 2004, p186). In other words, the rules had 'become instruments of power acquisition and maintenance,' of appropriation, domination, and exclusivity over the matter of natural resources (Horning, 2004, p241).

Malagasy communities have limited answers to the problem of power and domination, which we might, in this case, sum up as the problem of acculturation—the problem of narrowing the distance between cultures. Conservationists tend to see acculturation as a good thing, as a way to lessen cultural distance between the cultures in contact, but anthropologists know that the exchange is never equitable: the weaker side is asked to give in to the demands of the more powerful side, not the other way around (Ratsimbazafy and Kaufmann, 2008, pp41–42).

One common Malagasy response to *zavatse vazaha*, foreigner's things such as their acculturation pushiness, is to back away from compliance with the *dina* that they see as not theirs. Horning (2004, p150) noted that Bara choose to not comply with *dina* that did not have the legitimacy of the ancestors, in other words that were not local de facto conventions. They did not fear the consequences since foreigners rather than their ancestors had articulated the *dina* as a state-level rule.

Another reply to acculturation is to submit to the new power dynamic. This has deeper consequences than the Bara's avoidance technique that I referred to above. Marcus (2008, pp94–95) argued that a systemic erosion of social cohesion has occurred as rural Malagasy *dina* are usurped by foreign-made *dina*. Giving up, for instance, the social contract of *titike*, 'a moral commitment that individuals make to each other to respect traditions and the common good' (Horning, 2004, p186 citing Randriatavy, 1994) in order to strengthen the monitoring and enforcement of a foreign conservation *dina*, leads Malagasy villagers to lose trust and commitment to one another. Conservation, however, does not have to come, necessarily, at the expense of social cohesion and good will among the local population (see, for example, Ratsimbazafy et al., 2008).

Conservationists are suggesting that conservation success requires the elimination of social contracts such as *titike* that reinforce social cohesion at the expense of monitoring violators of the *dina*. They recommend, in other words, that forests trump rural folk and that dehumanizing Malagasy forest dwellers will stop deforestation. A conservation study among Vezo fishers who belonged to the Velondriake conservation association in the Morombe district in west Madagascar pointed out that social cohesion (*fihavanana* in the dialect of the central highlands or *filongoa* in the west regional dialect) works against the enforcement of the conservation *dina*, since 'being family' (*filongoa*) among related clans transcended 'betraying' somebody close to them who had broken the *dina* and not followed the rules (Andriamalala and Gardner, 2010, p247). The proposed solution to the 'problem' of social cohesion was to offer monetary rewards ('appealing to their financial interests') to those who turned in relatives that had broken the conservation rules (Andriamalala and Gardner, 2010, p247). How did we get to the point of sanctioning bounty hunting on one's relatives as a way of reaching conservation goals?

Beside the above moral argument against conservationists undermining social cohesion in an effort to encourage the regulation and policing of resource use, there is also a pragmatic argument worth noting.[5] There is a real danger that, in an effort to promote conservation, this sort of policy might actually do the very opposite, by undermining communal thought and action and encouraging self interest. CBRNM is based on the fundamental premise that stronger communities are better for conservation and yet such policies do the very opposite by turning households against each other. La Boétie, to whom I turn next, warned that corrupting our personal integrity and honesty with one another leads to a hideous result—to giving up one's liberty to serve a tyrant.

Voluntary servitude

Étienne de la Boétie wrote *Le Discours sur la Servitude Volontaire* to counter moral resignation—to submitting, voluntarily, to servitude to a sovereign, forfeiting one's freedom, and creating a tyrant if enough people submit. La Boétie sets out his thesis as follows:

> I should like merely to understand how it happens that so many men, so many villages, so many cities, so many nations, sometimes suffer under a single tyrant who has no other power than the power they give him; who is able to harm them only to the extent to which they have the willingness to bear with him; who could do them absolutely no injury unless they preferred to put up with him rather than contradict him. Surely a striking situation! Yet it is so common that one must grieve the more and wonder the less at the spectacle of a million men serving in wretchedness, their necks under the yoke, not constrained by a greater multitude than they ...
>
> (La Boétie and Bonnefon, 2007, p112)

La Boétie did not declare that people have a natural inclination toward being dominated and enslaved. What he said was that humans have the capacity to give to an external power the greatest gift that nature has bestowed upon us—our freedom (Abensour, 2011). Humans are not pre-determined to slavery, but must work hard for their liberty and not resign our integrity for something of far less value (cottoning the ruler's favor). The concept of 'gifting,' which anthropologists have studied long and hard since La Boétie's time, is key: 'The common thread of La Boétie's analysis is generated by the verb *to give*. It is the gift of these multiple freedoms that establishes the power of the tyrant' (Abensour, 2011, p333, quoting J.-M. Rey, 1968, pp199, 202n3).

Voluntary servitude dissolves what La Boétie considered to be the glue of social cohesion, friendship—the greatest gift of nature. La Boétie knew what he was talking about, as he had formed a great friendship with the younger Michel de Montaigne, the French thinker who would go on to write about the subject, using La Boétie as his example, in his famous *Essais* (Montaigne, 1946),[6] which were precursors to anthropology texts, enlightening readers since their first printing in 1580 on how humans live (Bakewell, 2010). Friendship, La Boétie stressed, opposes voluntary servitude and helps, as oppositions often do, to understand the enigma of tyranny. Moreover, La Boétie identified friendship with nature (Conley,

1998, p67). Giving up the gift of friendship, which voluntary servitude does, depraves humans of a key part of their nature.

La Boétie's redress against tyranny was not violence or revolution, but passive resistance—by using the moral weapon of refusing obedience to the tyrant. For La Boétie, people make tyrants by giving them the power to abuse them. The collective action that made the tyrant can, therefore, just as easily unmake such a perversion of nature by taking away the power of servitude (La Boétie and Bonnefon, 2007, p116).

For La Boétie, the key to domination was profiting at the expense of another, which corrupts one's integrity and collapses people into enslavement like a domino effect. La Boétie identified the 'mainspring and the secret of domination' to arise with an inner circle of five or six who sell their freedom for profit, who then hire six hundred, who hire six thousand, until a collective movement supports the tyrant (La Boétie and Bonnefon, 2007, p140). People who make a tyrant, he points out, are worse than slaves—are worse than people who have unwillingly had their freedom taken from them—because they have become co-dependent with the tyrant and have given away their liberty.

But profiting from conservation is exactly the answer that conservationists have proposed to solve the 'problem' of social cohesion. They seek to appeal to Malagasy's financial interests in an attempt to break down their loyalty toward one another. If this idea catches on, the conservation *dinas* would have more success. But consider the cost. Conservation organizations' attempts to exploit customary devices, such as *dina*, taboos, and religious beliefs, to get local people to internally give up their power and freedom, attempts to turn them into willing servants of a master.

If one knows only one thing about Malagasy people, it is that they cherish freedom and stigmatize servitude and the slave position (Graeber, 2007a, pp272–274; Rakoto, 1997; Randrianja and Ellis, 2009, pp226–228). The Malagasy ethos emphasizes, on the one hand, their similarity with one another in having a natural right to be free, and, on the other hand, their differences brought out by ranking themselves in a hierarchy of unequals marked not so much by the haves and have-nots but by ascendency and descendancy. The tension between the two, free and unequal, keeps the two categories alive. No wonder, then, that rural Malagasy do not excel at participating, voluntarily, in the 'new conservationism' that attempts to involve local communities in biodiversity conservation by exploiting as tools of servitude some local customary devices that help Malagasy live in their environments.

The irony here would be funny if it weren't so serious. Malagasy now face the possibility of having no way out but to submit to a dominating power, forfeiting their freedom to gain more control of natural resources. Law 96025 aimed to give Malagasy more freedom in the control of managing natural resources. But they receive just the opposite. Rather than giving them more rights, it takes rights away—if they let it. They face a losing situation: if they volunteer to erode their social institutions, then they will have more control in managing their natural resources. They might have more control but with fewer choices. CBNRM supposedly gives people more control over resources but at the same time narrows what they can actually do with it, for example, limiting forest use to non-consumptive uses. If we continue to go down this road, conservation will have become voluntary servitude. So far, not many Malagasy have submitted to this. Their resistance to partnering with conservationists is tied very much to their idea of freedom and to their respect for their ancestors and the social institutions they represent.

The worst of this appropriation of local beliefs is not conservationists' misappropriation of anthropology; of trying to use anthropology, not as a methodological way to understand people in their own terms and for the general scientific good of exploring human diversity, but as a tool for discovering central features of local cultures and shaping them for the service of conservation. At least they see some value of anthropology as a way to explore cultural relativity, of trying to understand other ways of life on their own terms rather than in relation to our way of life. Yet their clumsiness with the methodology results from not managing their ethnocentrism—the view that their way of life is superior to other cultures—when they begin pottering with the method of cultural relativism. Social and cultural anthropologists are trained to suspend their ethnocentrism as we try to get a handle on the culture afoot, which is why anthropologists need long periods of time doing ethnographic fieldwork. Conservationists demonstrate that they are not very good at anthropological methodology.

Nor is the worst of this usurping of Malagasy ideas for environmental conservation coming from a new 'environmental elite' of Malagasy (Moreau, 2008, p54), who work for various conservation NGOs. One of the earliest publications that offered the Malagasy institution of *dina* to the conservation organizations working in Madagascar was a United States Agency for International Development (USAID) report (Razafindrabe and Thompson, 1994). This set the stage for numerous publications, many with Malagasy authors, that have considered turning local governance in the form of *dina* into a broad mandate of conservation policies (see, for example, Andriamalala and Gardner, 2010; Henkels, 1999; Langley, 2006; Pronk and Evers, 2007; Rakotondrasoa and Evers, 2010; Rakotoson and Tanner, 2006; Ralalarimanga, 2010; Randrianarison and Karpe, 2010; Rarivoson, 2007; Razanaka, 2000).

The real tragedy of this appropriation is that by latching onto local rules toward nature and grafting them into explicit wide-ranging conservation rules, the conservation organizations are asking local people to submit to voluntary servitude, the basis of tyranny according to La Boétie. By applying La Boétie's insight to conservationists who hope to conjure a far-reaching conservation ethic from cultural ideas and practices that have quite different intents and social referents, we see that this powerful group is trying to lure Malagasy to join them in making nature into a tyrant. Such appropriation of culture is consistent with conservation being a self-identified 'mission driven discipline' (Soulé, 1985). So it is little surprise that it often behaves like missionaries and colonialists, who in many ways took a similar view of indigenous culture and institutions—something to be co-opted where possible or preferably replaced with something 'better' (more 'rational' in the case of conservation). The persistence of colonial thinking, of controlling and dominating local populations by appropriating their local institutions and fitting them into colonial institutions, is in evidence among conservation organizations working in Madagascar.

Conclusions

Across Madagascar, case studies demonstrate the lack of success in integrating local socio-cultural beliefs and practices into a Western conservation ethos (Scales, 2012). On its surface, the idea of drawing local stakeholders into a conservation narrative, which aims, in the words of the esteemed primate conservationist Alison Jolly (2008, pxii), 'to benefit the rest of the world,' seems prima fascia a good idea. But on closer analysis, especially from an anthropological perspective, we learn that the challenges of involving communities in foreign conservation are deep seated. When conservationists play with anthropology—come to it with little academic training to get beyond their engrained ethnocentrism, lacking a sensitive understanding of different worldviews for what they are—as if toying with the idea of there being different value systems about the environment that would bring the results they hope for, they make clumsy attempts at wedding local socio-cultural values to an alien worldview that prefers biodiversity to Malagasy people.

Colonial empires collapsed from the weight of their oppressive mission; might conservationism be headed for the same fate? Perhaps it is. Conservation is increasingly aligning itself with market environmentalism (discussed in Chapter 13 by Brimont and Bidaud). If forests are privatized and enclosed in the name of Payments for Ecosystem Services (PES), rural Malagasy will most likely feel robbed. They may retaliate against the strangeness and the power inequities inherent in market environmentalism by harming the forests. Their culture will not be the only sufferer, so will conservation.

But conservation need not go the way of colonialism. There is a way forward. Brosius (1999) noted that the present day's conservation is quite different from the conservation of the previous decade. Conservation can root out its colonial legacies, starting with its mission. Holding instrumentalist interests in cultural practices will not do, especially when cultures are made to do conservationist work that has no legitimacy in the eyes of the people. The end need not justify the means; conservation ethics need not trump cultural vitality. Conservation can change to include local people and to do them no harm. Perhaps as conservation matures and changes itself from its many mistakes and failures, it can drop the 'mission' stance and develop a more collaborative approach to conserving nature with the help of culture. I am not so naïve to think that conservationists will become better anthropologists. But they can become better conservationists by being less colonialist.

Notes

1 See also the Ted Talks featuring Wade Davis, which perhaps have an impact on a wider audience than his books: 'Saving a Pristine Backyard Wilderness: Wade Davis at TED2012' [blog.ted.com/2012/02/29/wade-davis-at-ted2012/]; 'Wade Davis: Dreams from Endangered Cultures' [www.ted.com/talks/wade_davis_on_endangered_cultures.html]; 'Wade Davis: The Worldwide Web of Belief and Ritual' [www.ted.com/talks/wade_davis_on_the_worldwide_web_of_belief_and_ritual.html].
2 Nor does this mean that they have more history than others, or a deeper history, or a better relationship with the past. It means only that things past, especially things that gave life to them and life to good stories that remember the dead, are relevant to them in their lives.
3 I thank the editor for suggesting I stress many of the points in this paragraph.
4 The authors go on to note, 'in southwestern Madagascar, *dina* refers exclusively to payment of bloodwealth, cattle paid to families to compensate for raiding-related injuries' (Tucker et al., 2011, p301). I did not find such exclusive restrictions around Androka Vaovao.
5 I thank the editor for asking me to elaborate this argument here.
6 Montaigne expanded on friendship as depending on integrity and sincerity in his essay 'On Friendship,' in which he recounts—first as a letter to his father and published subsequently in his essay 'On Friendship'—staying at La Boétie's deathbed until his friend's death, at the age of thirty-two (Montaigne, 1946, vol 1).

References

Abensour, M. (2011) 'Is there a proper way to use the voluntary servitude hypothesis?', *Journal of Political Ideologies*, vol 16, pp329–348.
Andriamalala, G. and Gardner, C. J. (2010) 'L'utilisation du *dina* comme outil de gouvernance des ressources naturelles: leçons tirés de Velondriake, sud-ouest de Madagascar', *Tropical Conservation Science*, vol 3, pp447–472.
Astuti, R. (1995) '"The Vezo are not a kind of people": identity, difference and "ethnicity" among a fishing people of western Madagascar', *American Ethnologist*, vol 22, pp464–482.
Bakewell, S. (2010) *How to Live, or, a Life of Montaigne in One Question and Twenty Attempts at an Answer*, Other Press, New York.
Bérard, M.-H. (2009) 'Légitimité des normes environnementales et complexité du droit: l'exemple de l'utilisation des *dina* dans la gestion locale de la forêt à Madagascar', Docteure en Droit (LL.D.), Faculté de droit, Québec, de l'Université Laval.
Bloch, M. (1971) *Placing the Dead: Tombs, Ancestral Villages, and Kinship Organization in Madagascar*, Seminar Press, London.
Brockington, D. (2001) 'Communal property and degradation narratives: debating the Sukuma immigration into Rukwa Region, Tanzania', *Les Cahiers d'Afrique de l'Est*, vol 20, pp1–22.

Brockington, D. (2005) 'The politics and ethnography of environmentalisms in Tanzania', *African Affairs*, vol 105, pp97–116.

Brosius, J. P. (1997) 'Endangered forest, endangered people: environmentalist representations of indigenous knowledge', *Human Ecology*, vol 25, pp47–69.

Brosius, J. P. (1999) 'Analysis and Intervention: anthropological engagements with environmentalism', *Current Anthropology*, vol 40, pp277–309.

Cinner, J. E. (2007) 'The role of taboos in conserving coastal resources in Madagascar', *SPC Traditional Marine Resource Management and Knowledge Information Bulletin*, vol 22, pp15–23.

Cole, J. (2001) *Forget Colonialism? Sacrifice and the Art of Memory in Madagascar*, University of California, Berkeley.

Conley, T. (1998) 'Friendship in a local vein: Montaigne's servitude to La Boétie', *South Atlantic Quarterly*, vol 97, pp65–90.

Davis, W. and Henley, T. (1990) *Penan: Voice for the Borneo Rainforest*, Western Canada Wilderness Committee, Vancouver.

Eggert, K. (1981) 'Who are the Mahafaly? Cultural and social misidentifications in southwestern Madagascar', *Omaly Sy Anio*, vols 13–14, pp149–176.

Errington, F. and Gewertz, D. (2001) 'On the generification of culture: from Blow Fish to Melanesian', *Journal of the Royal Anthropological Institute*, vol 7, pp509–525.

Esoavelomandroso, M. (1979) *La Province Maritime Orientale du Royaume de Madagascar à la Fin du XIXe Siècle*, FTM, Antananarivo.

Esoavelomandroso, M. (1988) 'La destruction de la forêt par l'homme Malgache: un problème mal posé', *Recherches pour le Développement*, vol 2, pp183–186.

Feeley-Harnik, G. (1991) *A Green Estate: Restoring Independence in Madagascar*, Smithsonian Institution, Washington, DC.

Feeley-Harnik, G. (2001) '*Ravenala madagascariensis* Sonnerat: the historical ecology of a "Flagship Species" in Madagascar', *Ethnohistory*, vol 48, pp31–86.

Graeber, D. (1995) 'Dancing with corpses reconsidered: an interpretation of *famadiaha* (in Arivonimamo, Madagascar)', *American Ethnologist*, vol 22, pp258–278.

Graeber, D. (2007a) *Lost People: Magic and the Legacy of Slavery in Madagascar*, Indiana University Press, Bloomington.

Graeber, D. (2007b) *Possibilities: Essays on Hierarchy, Rebellion, and Desire*, AK Press, Oakland.

Haenn, N. (1999) 'The power of environmental knowledge: ethnoecology and environmental conflicts in Mexican conservation', *Human Ecology*, vol 27, pp477–491.

Henkels, D. M. (1999) 'Une vue de pres du droit de l'environnement Malgache', *African Studies Quarterly*, vol 3, pp37–57.

Horning, N. R. (2004) *The Limits of Rules: When Rules Promote Forest Conservation and When They Do Not—Insights from Bara Country, Madagascar*, PhD thesis, Political Science, Cornell University, Ithaca.

Jolly, A. (2008) 'Foreword', in J. C. Kaufmann (ed.) *Greening the Great Red Island: Madagascar in Nature and Culture*, Africa Institute of South Africa, Pretoria.

Jones, J. P. G., Andriamarovololona, M. M. and Hockley, N. (2008) 'The importance of taboos and social norms to conservation in Madagascar', *Conservation Biology*, vol 22, pp976–986.

Kaufmann, J. C. (2003) 'Cactus pastoralism: on its origin and growth in Madagascar', *Michigan Discussions in Anthropology*, vol 14, pp104–126.
Kaufmann, J. C. (2004) 'Prickly pear cactus and pastoralism in Southwest Madagascar', *Ethnology*, vol 43, pp 345–361.
Kaufmann, J. C. (2006) 'The sad opaqueness of the environmental crisis in Madagascar', *Conservation and Society*, vol 4, pp179–193.
Kaufmann, J. C. (2008) 'The non-modern constitution of famines in Madagascar's spiny forests: "water-food" plants, cattle, and Mahafale landscape praxis', *Environmental Sciences*, vol 5, pp73–89.
Kawharu, M. (2000) 'Kaitiakitanga: a Maori anthropological perspective of the Maori socio-environmental ethic of resource management', *The Journal of the Polynesia Society*, vol 109, pp349–370.
La Boétie, É. de and Bonnefon, P. (2007) *The Politics of Obedience and Étienne de La Boétie*, Black Rose, Montréal.
Lambek, M. (1992) 'Taboo as cultural practice among Malagasy speakers', *Man*, vol 27, pp245–266.
Lambek, M. (2002) *The Weight of the Past: Living with History in Mahajanga, Madagascar*, Palgrave Macmillan, New York.
Langley, J. M. (2006) *Vezo Knowledge: Traditional Ecological Knowledge in Andavadoaka, Southwest Madagascar*, Blue Ventures Conservation Report, London.
Larson, P. M. (1996) 'Desperately seeking "the Merina" (Central Madagascar): reading ethnonyms and their semantic fields in African identity histories', *Journal of Southern African Studies*, vol 22, pp541–560.
Mack, J. (1986) *Madagascar: Island of the Ancestors*, British Museum Publications, London.
Marcus, R. R. (2008) 'Tòkana: the collapse of the rural Malagasy community', *African Studies Review*, vol 51, pp85–104.
Montaigne, M. de (1946) *The Essays of Montaigne in Three Volumes*, The Heritage Press, New York.
Moreau, S. (2008) 'Environmental misunderstandings', in J. C. Kaufmann (ed.) *Greening the Great Red Island: Madagascar in Nature and Culture*, Africa Institute of South Africa, Pretoria.
Orwell, G. (1961) *1984: A Novel*, Signet Classics, New York.
Poyer, L. and Kelly, R. L. (2000) 'Mystification of the Mikea: constructions of foraging identity in Southwest Madagascar', *Journal of Anthropological Research*, vol 56, pp163–185.
Pronk, C. and Evers, S. J. T. M. (2007) 'Complexité de l'accès à la terre dans le sud-est de Madagascar', also published as 'The complexities of land access in southeast Madagascar', *Taloha*, vol 18, www.taloha.info/document.php?id=568.
Rakoto, I. (1997) *L'esclavage à Madagascar: Aspects Historiques et Résurgences Contemporaines*, Institute de Civilisations, Musée d'Art et d'Archéologie, Antananarivo.
Rakotoarisoa, J. A. (1998) *Mille Ans d'Occupation Humaine dans le Sud-Est de Madagascar: Anosy, une Ile au Milieu des Terre*, L'Harmattan, Paris.
Rakotondrabe, T. D. (1993) 'Beyond the ethnic group: ethnic groups, nation state and democaracy in Madagascar', *Transformation*, vol 22, pp15–29.
Rakotondrasoa, L. M. and Evers, S. J. T. M. (2010) 'Objectifs du millénaire pour le développement: sécurité, conflits de lois et accès à la terre dans le contexte

malgache,' also published as 'Malagasy challenges in achieving the Millennium Development Goals: security, competing jurisdictions, land access and livelihoods', *Taloha*, vol 19, www.taloha.info/document.php?id=888.

Rakotoson, L. R. and Tanner, K. (2006) 'Community-based governance of coastal zone and marine resources in Madagascar', *Ocean and Coastal Management*, vol 49, pp855–872.

Ralalarimanga, H. C. S. (2010) 'De l'oralité à l'écrit: droit et gestion durable des ressources naturelles renouvelables de la forêt de Merikanjaka', *Taloha*, vol 19, www.taloha.info/document.php?id=789.

Ramarolahy (1972) *Rakitry ny elan' ny Ntaolo Malagasy*, Imprimerie Catholique, Tananarive.

Randrianarison, M. and Karpe, P. (2010) 'Le contrat comme outil de gestion des ressources forestières', *Taloha*, vol 19, www.taloha.info/document.php?id=887.

Randrianja, S. and Ellis, S. (2009) *Madagascar: A Short History*, Hurst, London.

Randriatavy (1994) *L'Occupation de Éspace et l'Organisation Sociale à Bara Manamboay et à Andranomaitso*, World Wide Fund for Nature, Sakaraha.

Rarivoson, C. (2007) 'The Mandena *dina*: a potential tool at the local level for sustainable management of renewable natural resources', *SI/MAB Series 11*, pp309–315.

Ratsimbazafy, J. and Kaufmann, J. C (2008) 'An experiment in lessening cultural distance', in J. C. Kaufmann (ed.) *Greening the Great Red Island: Madagascar in Nature and Culture*, Africa Institute of South Africa, Pretoria.

Ratsimbazafy, J., Rakotoniaina, L. J. and Durbin, J. (2008) 'Cultural anthropologists and conservationists: can we learn from each other to conserve the diversity of Malagasy species and cultures?', in J. C. Kaufmann (ed.) *Greening the Great Red Island: Madagascar in Nature and Culture*, Africa Institute of South Africa, Pretoria.

Razafindrabe, M. and Thompson, J. (1994) *Local Governance in Madagascar*, KEPEM/USAID, Antananarivo.

Razanaka, S. (2000) 'Le dina: un mode de gestion communautaire moderne (Madagascar)', in D. Compagnon and F. Constantin (eds) *Administrer l'Environnement en Afrique*, Karthala, Paris.

Rey, J.-M. (1968) *La Part de l'Autre*, PUF, Paris.

Roberts, M., Normann, W., Minhinnick, N., Wihongi, D. and Kirkwood, C. (1995) 'Kaitiakitangata: Maori perspectives on conservation', *Pacific Conservation Biology*, vol 2, pp7–20.

Roberts, M., Haami, B., Benton, R., Satterfield, T., Finucane, M. and Henare, M. (2004) 'Whakapapa as a Maori mental construct: some implications for the debate over genetic modification of organisms', *The Contemporary Pacific*, vol 16, pp1–28.

Scales, I. R. (2012) 'Lost in translation: conflicting views of deforestation, land use and identity in western Madagascar', *The Geographical Journal*, vol 178, pp67–79.

Soulé, M. (1985) 'What is conservation biology?', *BioScience*, vol 35, pp727–734.

Tucker, B., Huff, A., Tsiazonera, Tombo, J., Hajasoa, P. and Nagnisaha, C. (2011) 'When the wealthy are poor: poverty explanations and local perspectives in southwestern Madagascar', *American Anthropologist*, vol 113, pp291–305.

Walsh, A. (2001) 'When origins matter: the politics of commemoration in northern Madagascar', *Ethnohistory*, vol 48, pp237–256.

Walsh, A. (2002) 'Responsibility, taboos and "the freedom to do otherwise" in Ankarana, northern Madagascar', *Journal of the Royal Anthropological Institute*, vol 8, pp451–468.
West, P. (2005) 'Translation, value, and space: theorizing an ethnographic and engaged environmental anthropology', *American Anthropologist*, vol 107, pp632–642.

15 Conclusion

The future of biodiversity conservation and environmental management in Madagascar: lessons from the past and challenges ahead

Ivan R. Scales

In the introduction to this volume, I set out the central challenge of conservation and environmental management in Madagascar: how to protect biodiversity at the same time as delivering economic growth and creating alternative livelihoods that place less pressure on ecosystems. The chapters that make up this book reveal the considerable size of this task. There is no doubt that Madagascar has experienced significant habitat loss and species extinction, with important implications not only for the island's biodiversity but also for the people who rely directly on its ecosystems for their livelihoods.

As well as offering an overview of the key issues, the chapters have provided a review and analysis of how policy has tried to deal with the conservation of biodiversity, the management of natural resources and poverty alleviation. They have shown how Madagascar has been a laboratory for policymakers, with some of the earliest examples in sub-Saharan Africa of protected areas (see Chapter 5 by Scales and Chapter 10 by Virah-Sawmy *et al.*), national environmental action plans (see Chapter 7 by Kull) and community management schemes (see Chapter 8 by Pollini *et al.*). As Ferguson and Gardner (2010, p75) remind us, progress has been made:

> Conservation in Madagascar has seen some notable advances over the last two decades; massive policy reform and the launch of a large number of field-level initiatives have resulted in a range of new policy frameworks and institutions, a reduction in deforestation rates in many regions, the creation of numerous new protected areas and the participation of local stakeholders in new forms of natural resource governance.

However, threats to biodiversity remain high and conservation policy has imposed significant costs on rural households. So what are the lessons that can be taken away from over a century of environmental management? What of the future?

Biological diversity and setting conservation priorities

Flagships, forests and furry animals: the narrow vision of environmental management in Madagascar

The past 30 years have seen huge efforts to document Madagascar's biological diversity and new species are discovered on a regular basis. Looking at lemurs, for example, only 32 species had been described by the early 1990s (Mittermeier *et al.*, 1992). By 2008, the number had gone up to 97 (Mittermeier *et al.*, 2008) and by 2013 there were over 100 known species (Rasoloarison *et al.*, 2013). The recent explosion in species discovery and description is the result of both huge efforts to collect specimens in remote parts of the island and developments in taxonomic methods, particularly the use of genetics to split existing groups (Thalmann, 2006).

While the diversity of life on Madagascar captivates biologists and tourists alike, research and conservation have tended to be rather lemur centric. Thalmann (2006, p6) has labelled the island's primates 'Ambassadors for Madagascar':

> As a primate group endemic to Madagascar they constitute a unique part of the world's natural heritage and a unique part of humankind's natural history. Being mostly forest dwelling animals they may serve as ambassadors for the forests of Madagascar and the whole wildlife in these forests all over the island where it remains. Lemur conservation equals forest conservation ... Because lemur conservation is forest conservation, the protection of lemurs also helps to grant important services by forests, such as reduced erosion, clear and sustainable water proliferation – *a better life for humans.*
>
> (emphasis added)

The last part of this quotation would be contested by many Malagasy living at the forest frontier. Lemur centric conservation has often imposed significant costs on forest-dependent households, having led to displacement, severe restrictions on natural resource use and the disruption of household economies (Ghimire, 1994; Peters, 1999).

From a conservation perspective, lemurs clearly fit the description of flagship species – charismatic species that help draw attention and funding for conservation (Walpole and Leader-Williams, 2002). They also fit the description of umbrella species (Walpole and Leader-Williams, 2002). To conserve lemurs means preserving forest ecosystems and all the species that contribute to ecosystem function. Most lemurs need large areas of relatively undisturbed forest and have played a key role in the establishment of protected areas. For example, the discovery of the golden bamboo lemur (*Hapalemur aureus*) led to discussions between American conservationists and Malagasy officials and precipitated the establishment of Ranomafana National Park. Under an agreement with the Malagasy government in 1990, biologists from Duke University negotiated a US$3.2 million grant from the United States Agency for International Development (USAID) for an integrated conservation and development project (Peters, 1999).

However, there are possible downsides to a flagship approach to conservation. Lemur focused conservation leads to a large amount of attention, funding, research and effort going on a specific group of mammals and the ecosystems that support them. Biodiversity conservation in Madagascar has tended to focus on the eastern rainforests at the expense of other biomes, especially the dry-deciduous and spiny forests of the west and south (Hannah *et al.*, 1998; Bollen and Donati, 2006). In focusing attention in this way, conservation policy helps to feed the 'fetishisation' of Madagascar's forests, especially its rainforests. I use the word 'fetishise' here both in its contemporary usage (i.e. to be excessively devoted to something) and in the anthropological sense, which refers to the process whereby immaterial objects are attributed magical powers. Policy expects the island's forests to do a lot – not only provide habitats for lemurs but also deliver a wide range of ecosystem services and create new livelihoods for rural households. Under Reducing Emissions from Deforestation and Forest Degradation (REDD+) and other carbon-offset schemes (see Chapter 13 by Brimont and Bidaud), the fetishisation of trees is only likely to increase, as forests are not only expected to solve the problems of biodiversity conservation and poverty alleviation but also climate change.

I do not wish to downplay the importance of forests and their significant role in terms of biodiversity conservation and ecosystem services. However, a narrow focus on forests means that other biomes and landscapes are relatively poorly understood. Madagascar's diverse ecosystems have followed different trajectories in different areas, yet little is known about the specifics of vegetation change and the role of human action (Kull, 2000; see Chapter 3 by Dewar). The danger in such fetishisation is that other important ecosystems are forgotten. For example, recent research on the island's grasslands has revealed that they are as diverse as anywhere in continental Africa, both in terms of flora and fauna (Bond *et al.*, 2008, p1753):

> We suggest that biologists should take a fresh look at Madagascan grasslands, not least because the grassland biota has been largely neglected in biological inventories for conservation in a country characterized by almost unparalleled levels of endemism. Grassy ecosystems in general have been viewed as an alien, rather than intrinsic, component of this extraordinary island.

The marine environment: out of sight, out of mind?

Readers might have noticed the relatively small number of marine case studies in this book. This is not through any intentional bias on the part of the authors. Unfortunately, our understanding of Madagascar's marine realm, particularly in terms of its management, lags behind considerably and 'environmental challenges below the waves have been eclipsed by the many threats facing the island's terrestrial biodiversity' (Harris, 2011, p8). While levels of biological endemism do not match those on land, with over 5,600 km of coastline Madagascar boasts a huge diversity of marine and coastal habitats (FAO, 2008).

Madagascar's marine resources are important as a source of both major export commodities for the national economy and subsistence for Malagasy households. Traditional fisheries, carried out from dugout canoes, exploit a wide range of species from fish to marine mammals, sea turtles, sharks, octopus, crustaceans and sea cucumbers (FAO, 2008). They are especially important in western and south-western Madagascar where semi-arid and arid conditions can make agriculture difficult (Le Manach *et al.*, 2012). Not only do they provide the basis of subsistence and livelihoods for fishing communities, it has been estimated that fish and fish products contribute about 20 per cent of animal protein consumption of Madagascar's population (FAO, 2008). They are therefore a crucial part of food security on the island. As well as these local, regional and national dimensions, local fisheries are connected to international markets, especially in China and Southeast Asia. There has been a significant increase in demand for products such as sea cucumber and shark fin, with major questions about the impact that this trade is having on species (McVean *et al.*, 2006; Le Manach *et al.*, 2012; Purcell *et al.*, 2013).

With regards to industrial fisheries, near shore shrimp fisheries constitute one of the island's biggest exports. This is carried out principally through trawling, with industrial fleets operating 1–4 km off the western, north-western and eastern coasts (FAO, 2008). Shrimp are exported to Japan, Europe, USA, Reunion and Mauritius. Catches have decreased over the past few years, probably due to a combination of over-exploitation and climate change (FAO, 2008). Several species of tuna (for example yellowfin, *Thunnus albacares*) are heavily targeted, both by legal European fleets and illegal Asian fleets (Le Manach *et al.*, 2012). Research shows that small-scale fisheries are probably reaching a plateau in terms of total catches despite growing numbers of fishers, and are projected to start declining within the next decade (Le Manach *et al.*, 2012). In addition, there are increasing tensions between commercial shrimp fleets and artisanal fishers.

Given the importance of Madagascar's fisheries for both economic growth and food security, marine environmental management is in need of much greater research and policy attention (Harris, 2011; Le Manach *et al.*, 2012). In terms of protection, strategies have focused primarily on protected areas, although there have been notable attempts to involve communities in resource management. The coverage of Marine Protected Areas (MPAs) has grown more than 50-fold in the past decade and the Durban Vision saw a large increase from three MPAs in 2002 to more than 15 under temporary or permanent protection in 2010, although these have tended to focus on coral reefs at the expense of other ecosystems, in particular seagrass beds and mangroves (Harris, 2011).

Environmental discourse and received wisdom: time for a new vocabulary?

The focus of policy and research on a narrow set of ecosystems and species is symptomatic of a much broader environmental discourse. I use the term discourse to mean a way of thinking and communicating about a set of issues that is underpinned by a key set of ideas, norms and values (see Chapter 4 by McConnell and Kull and Box 6.1 on page 131). Another way of putting it is that there is a distinct received wisdom that pervades the way a certain group of stakeholders (primarily conservation organisations and government ministries) think about human–environment interactions on the island, based on a set of recurring themes and informed by a key set of (often untested) assumptions. Over the past 30 years, with the greater influence of conservation organisations on environmental management, this received wisdom has become more prevalent and policy has seen a shift of emphasis from the sustainable management of resources to a much narrower focus on biodiversity first and foremost (see Chapter 7 by Kull).

The strongest theme at the heart of environmental discourse is that of 'Eden-like' nature. This is most clearly seen in the island-forest myth, which paints an image of a biological paradise devastated by human action (see Chapter 3 by Dewar and Chapter 4 by McConnell and Kull). Reading about Madagascar, one often comes across words such as 'forest', 'pristine', 'primary' and 'virgin' used without being defined and in binary opposition to 'grassland', 'secondary' and 'degraded'. This thinking has had a very strong influence on environmental management, from the French colonial policy of establishing vast plantations of exotic tree species to the anti-fire rhetoric and legislation that has stigmatised and criminalised the activities of pastoralists and swidden agriculturalists (Kull, 2004). The perception that Madagascar's grasslands are degraded landscapes has led to policies that 'recreate' Madagascar's forests through afforestation programmes, even though the history of land cover change in Madagascar is poorly understood (see Chapter 3 by Dewar and Chapter 4 by McConnell and Kull).

The second component of the received wisdom focuses on the drivers of degradation. If Madagascar is being rapidly degraded, the blame is usually placed on rural households who are painted as being too poor, backwards and/or stubborn to do anything different (see Chapter 5 by Scales). The received wisdom is explicitly neo-Malthusian:

> [T]he poverty that afflicts Madagascar's people threatens to destroy what remains of this unique biology ... widespread poverty, increasing

population, and the absence of resources and techniques to improve
the productivity of agricultural and pasture lands have led to massive
deforestation.

(Sussman *et al.*, 1994, p334)

It is a grim Malthusian cycle that is currently being played out by
unprecedented numbers of fishers along all but the most inaccessible
of Madagascar's western coasts, and exacerbated by the rapid rates of
human population growth typical of many coastal regions.

(Harris, 2011, p8)

Clearly humans have played a major role in shaping landscapes and human actions have led to the extinction of a number of species (see Chapter 3 by Dewar). Rapid population growth and poverty do constrain the livelihood choices of many rural households (see Chapter 5 by Scales). Given the island's environmental problems, it is easy to see why people often resort to hyperbole and catastrophic narratives when talking or writing about Madagascar. However, the received wisdom is problematic for two reasons. First, in many cases it is inaccurate. For example, Madagascar was never entirely forested. As our understanding of Holocene environmental change improves (see Chapter 3 by Dewar), it is clear that narratives of an island-forest ruined by the arrival of humans simply will not do. Ideas of the 'original' vegetation of Madagascar are highly problematic, especially given the incompleteness of evidence. As McConnell and Kull argue (Chapter 4), 'without denying that native forests – in general – are shrinking, more caution about estimates of forest area and loss is warranted than is generally accorded'. Bob Dewar (Chapter 3) reminds us that while Madagascar's past has much to teach us in terms of evolution, environmental change and human impacts, this history is not simply a morality tale. This also applies for the island's more recent past. Areas thought of and talked about as 'virgin' have often had much more human influence than the received wisdom would suggest, with parts of protected areas such as Ranomafana National Park (Peters, 1999) and Menabe Antimena (Scales, 2011), for example, having long histories of commercial logging prior to their establishment.

As well as land cover change being more complex, the drivers of livelihood choices and natural resource use are also more diverse and nuanced than simply population growth and poverty. They often involve powerful political and economic forces outside Madagascar (see Chapter 5 by Scales). Environmental discourse thus oversimplifies complex realities. As Kull argues (2000, p441): 'Received wisdoms about the environmental history of Madagascar include much confusion, misunderstanding and misinterpretation.' While policy must of course simplify reality in order to understand it and manage it, there are important differences between simplification, sweeping generalisation and plain myth.

The second and closely related problem with the received wisdom is that in obscuring the complexities of resource use, it constrains policy and pushes it down certain routes. While policy has touched on rural development, the lion's share of attention and funding has gone to the conservation of biodiversity and the separation of people and 'wilderness' (see Chapter 7 by Kull). The received wisdom thus frames debates in certain ways and makes clear statements about which options are acceptable and which are not, more often according to subjective values than pragmatic realities. For example, when a forest is described as 'pristine', and its history of logging is hidden, it makes it much easier to ban all consumptive uses of the forest. To acknowledge a more complex history of human use is to open the door to a more diverse and dynamic set of management options. This is powerfully demonstrated in the example of the Baobab Alley (see Chapter 11 by Scales) which, despite being an anthropogenic landscape, is currently being 'preserved' as a tourist attraction and in doing so threatening the livelihoods of those living close to the site. It can also be seen in the emphasis placed on nature tourism, which promises (and mostly fails to deliver) income from forests and coral reefs through tourists only 'taking pictures', 'leaving only footprints' and foreign currency (see Chapter 11 by Scales). Once again, the values underpinning conservation policy and the desire to try and avoid consumptive uses of nature become clear. As Cronon (1996, p16) reminds us: 'The dream of an unworked natural landscape is very much the fantasy of people who have never themselves had to work the land to make a living.'

It is therefore vital for the future of conservation and environmental policy in Madagascar that it starts from the right foundations. The received wisdom is important because it influences the questions asked by researchers, the evidence that is seen, the stories that are told and the actions that are taken by policymakers (see Chapter 4 by McConnell and Kull). To see Madagascar for what it is (an island with a complex environmental history rather than some sort of paradise lost) would be a good start. Conservation should not be about preserving the past and trying to halt human actions but about managing change and negotiating impact (Adams, 2003). In the case of Madagascar, we need far more research on the influence of fire on vegetation and the circumstances under which burning can be an appropriate land management strategy (Bloesch, 1999; Kull, 2002). With regards to agricultural practices in particular, the socio-ecological dynamics of swidden systems, their responses to socio-environmental change and their sustainability are still poorly understood (Hume, 2006; Styger *et al.*, 2007). It is time for policy to move beyond an approach dominated by 'fortress conservation' to a greater consideration of multiple and consumptive uses of natural resources:

> Multiple-use forests will play a crucial role in this approach, because many high-biodiversity areas will continue to fall in non-park forests.

> Examples of the successful use of multiple-use forests in maintaining biodiversity are notably lacking in temperate and tropical systems ... yet multiple-use forests will continue to be managed for both production and biodiversity. Madagascar, with so much at stake, must be a leader in defining management systems that can meet national needs, benefit communities and maintain biodiversity.
>
> (Hannah *et al.*, 1998, p35)

Tension, conflict and 'participation' in environmental management: the need to negotiate and embrace trade-offs

A recurring theme throughout the book has been that tensions and conflicts are often at the heart of environmental management issues. For example, while outsiders may see swidden cultivation as irrational and destructive, farmers see it as a way of making the land productive, feeding their families and maybe even, in good years, making a profit (see Chapter 5 by Scales). Similarly, while biologists discover the latest species of lemur and describe the island as a naturalist's paradise, those living at the forest frontier are left to wonder whether conservationists care more about lemurs than they do about people (Peters, 1998; Harper, 2002).

The conservation of biological diversity and management of natural resources involves multiple stakeholders, often with contrasting and conflicting priorities. The first step is to acknowledge the diversity of such viewpoints and the complexity of environmental politics. The policy implications of managing a forest for maximum biodiversity, carbon sequestration, minimum ecosystem function, key ecosystem services, nature tourism or a particular natural resource (e.g. timber, non-timber forest resources, swidden, agroforestry) are not identical. Policymakers need to think through and make explicit the trade-offs that must be made, both between various conservation and environmental management goals and between conservation and other social goals, such as poverty alleviation and economic growth (Leader-Williams *et al.*, 2010; Hirsch *et al.*, 2011; McShane *et al.*, 2011).

Once such diverse priorities and possible options have been acknowledged, the second step is to re-align the power dynamics of environmental management. As many chapters in this book show, the politics of natural resource use in Madagascar have been highly uneven, with the goals of the state, international donors and conservation organisations prioritised over local needs. This will not do for two reasons, one ethical and the other pragmatic. Ethically, there is something troubling about some of the poorest households on Earth being made to pay the majority of the costs of biodiversity conservation. There is something even more troubling about conservation policy that seeks to undermine local social cohesion by asking people to report other members of their community, or even their relatives, for environmental 'crimes' defined largely by outsiders (see Chapter 14 by Kaufmann).

As well as the ethical arguments against 'top-down' conservation and environmental management, there are also more pragmatic arguments against policies that exclude people from decision-making and resource use. Poor households are directly dependent on ecosystems for their livelihoods and unless policy is able to create genuine viable and acceptable alternatives, consumptive uses will continue – illegally if necessary. Experiments to create incentives for conservation have had mixed results and significant challenges remain. For example, the impacts of nature tourism have been highly localised and limited (see Chapter 11 by Scales). At present, the cost of saving Madagascar's forests while reducing poverty is higher than the money that can be generated through carbon markets (see Chapter 13 by Brimont and Bidaud). There have been significant barriers to the distribution of financial benefits and the effectiveness of incentive-based schemes. With bioprospecting, it costs thousands of dollars in legal fees and significant time and knowledge of the legal process to draft mutually beneficial contracts (see Chapter 12 by Neimark and Tilghman). On a similar note, forest management contracts require the services of environmental mediators to manage negotiations over resource use between the state and communities (see Chapter 8 by Pollini *et al.*). Access to such resources is often not possible for small communities and illiterate rural households.

While conservation policy has attempted to decentralise resource management, meaningful and substantial participation in decision-making has been limited. At first glance, the idea of involving local stakeholders in conservation to try and make it less 'top-down' seems unquestionably 'a good thing'. There are, however, important questions about *who* participates and *how* they participate in resource management. The danger is that in their attempts towards greater community participation, conservationists 'play with anthropology', i.e. attempt to 'involve' people from very different cultures and with very different perspectives and beliefs, without a sufficiently sensitive understanding of different worldviews and institutions. This has often resulted in clumsy attempts to blend local cultural values and institutions with a Western conservation ethic, with the weaker side being asked to give in to the more powerful side (see Chapter 14 by Kaufmann). For example, while *Gestion Locale Sécurisée* (GELOSE) started off with a bold vision for the local management of natural resources, in practice it has tended to follow familiar 'top-down' politics, with local decision-making power heavily constrained (see Chapter 8 by Pollini *et al.*).

As well as the 'top-down' nature of many attempts to be more 'bottom-up', projects have often played into local politics, leading to an uneven distribution of costs and benefits and sometimes increasing inequality within local communities (see Chapter 8 by Pollini *et al.* and Chapter 13 by Brimont and Bidaud). 'Communities' are rarely homogenous, representing a diverse set of priorities that cut across lines of class, kinship, gender, ethnicity and age (Agrawal and Gibson, 2001). As well as recognising the difference *between* and *within* communities, participatory processes must also recognise existing institutions and power structures rather than try and bypass them, and must also consider the likely legitimacy and meaningfulness of any newly created institutions (see Chapter 8 by Pollini *et al.* and Chapter 14 by Kaufmann).

The rush to implement the Durban Vision has been particularly damaging to community involvement in environmental management. Chapter 9 (Corson) shows that it was impossible to consult all the rural villages potentially affected by the expansion of protected areas and that this led to considerable debate between policymakers over how much consultation was really necessary. While some stressed the importance of a thorough consultation process and called for the expansion to be slowed down, others argued that it was critical that the programme be implemented rapidly, with Conservation International (CI) in particular pushing for the deadline to be met (see Chapter 9 by Corson). In practice, 'consultation' is in fact more an awareness raising exercise designed to convince and educate, usually after decisions have already been made.

More than ten years ago, Marshall Murphree (2000, p12) argued that in sub-Saharan Africa, 'CBC [community-based conservation] has to date not been tried and found wanting; it has been found difficult and rarely tried!' The history of conservation policy in Madagascar supports this view. While it is unlikely that community involvement in biodiversity conservation and natural resource management will ever become easier, policymakers should at least fully engage with it before 'throwing out the baby with the bathwater'. Ultimately, conservation and environmental management should involve a two-way conversation with the potential for negotiation, rather than a one-way imposition of external ideas (Richard and Dewar, 2001). This would require conservation organisations and government ministries to abandon their mission of educating people about the importance of conservation and be more receptive to other views and priorities (see Chapter 9 by Corson).

Instead of conservation policy continuing to impose a vision of Eden on Madagascar, the management of natural resources must ultimately be about trade-offs – between global visions and local priorities; between preservationist and utilitarian views of nature; and between biodiversity conservation and economic growth. Acknowledging trade-offs requires 'resisting the temptation to obscure political realities, flatten multiple dimensions of value into a single term, or ignore marginalized interests or ways of knowing' (Hirsch *et al.*, 2011, p263).

Conservation and environmental management in an uncertain world

Madagascar's political and economic history has been tumultuous – the arrival of French colonialism; independence from France in 1960;

experiments with socialism; the economic crises of the 1970s and 1980s; dramatic 'structural adjustment' of the economy; and repeated *coups d'états* have had huge impacts, both on natural resource use and on legislation and policy (see Chapter 6 by Scales and Chapter 7 by Kull). Radical political and economic change is often accompanied by dramatic land use and land cover change, with Madagascar's colonial history providing ample evidence (see Chapters 5 and 6 by Scales).

Since the debt crisis of the 1980s and the International Monetary Fund's (IMF) structural adjustment policies that were the conditions of the bailout (see Chapter 7 by Kull), the island has undergone rapid trade liberalisation (Barrett, 1994). The World Bank has repeatedly encouraged Madagascar to do more to attract Foreign Direct Investment (FDI) in order to promote economic growth and reduce debt (Sarrasin, 2006). The last decade has seen an increase in foreign land acquisitions, particularly for large-scale agricultural projects (Neimark, 2013). A 2011 survey carried out by the World Wide Fund for Nature (WWF) found 56 biofuel projects in various stages of preparation or operation (WWF, 2011).

Madagascar is an attractive proposition for foreign investors, possessing abundant cheap land and labour and a favourable climate for agriculture. These are the very factors that convinced the French colonial government that the island could become hugely profitable and led it to award large concessions to foreign companies (see Chapter 6 by Scales). However, such large-scale projects tend to trample over local resource management institutions. Most rural Malagasy do not hold official land titles, relying instead on customary claims based on kinship, historical lineage and other local institutions (Healy and Ratsimbarison, 1998; McConnell, 2002; Neimark, 2013; see also Chapter 14 by Kaufmann). Customary land claims tend to be ignored by the state and investors and seen as unused and vacant, so that many land acquisitions are in fact 'land grabs' (Neimark, 2013). The rush for Madagascar's land risks undermining precarious livelihoods. This is likely to be bad news, both for biodiversity and poverty alleviation.

As well as the rush for agricultural land, Madagascar has also seen an increase in large-scale mineral extraction, most notably the Rio Tinto/QMM ilmenite mine near Tolagnaro in south-eastern Madagascar (Seagle, 2012).[1] Mining operations began in 2005 and will eventually involve stripping a total of 6,000 hectares of rare littoral forest, although Rio Tinto/QMM have set aside 620 hectares as conservation zones and pledged to 'restore' 25 per cent of the area with endemic species and reforest 75 per cent with non-native *Eucalyptus* spp. (Seagle, 2012). Mining companies in Madagascar are increasingly involved in such 'biodiversity offsets' to compensate for habitat loss, but there are significant questions about just how much lands needs to be offset, especially in cases where there are high or even unique conservation and social values (Virah-Sawmy, 2009).

Such projects raise questions about the power dynamics of environmental management and the relationship between big business and conservation in Madagascar. CI, Flora and Fauna International, BirdLife International and the WWF have formed partnerships with Rio Tinto/QMM, not only providing advice on the environmental aspects of the project but also receiving funding and helping to legitimise such large-scale extractive activities under the banner of 'sustainable development' (Seagle, 2012). For example, the BirdLife website states that:

> In 2001, BirdLife International and Rio Tinto formed a partnership to achieve mutually held goals of biodiversity conservation within the context of the global transition to sustainable development. Since its establishment, the partnership has helped the two organisations deliver sustainable and far-reaching outcomes, where their shared objectives are more effectively fulfilled by working together than by acting alone.[2]

Together with the Malagasy state and non-governmental organizations (NGOs), mining companies have negotiated the boundaries and rights associated with new protected areas, helping to reduce the constraints that they might place on the rapidly expanding mining industry at the same time as helping to limit local natural resource use rights (see Chapter 9 by Corson). These developments have helped to reinforce the high levels of political influence that international conservation organisations have in Madagascar (Duffy, 2006; Corson, 2010; see also Chapter 7 by Kull).

The latest political instability has also had significant consequences for the island's flora and fauna, with reports of increases in illegal logging in national parks such as Masoala. Rosewood (*Dalbergia* spp.) in particular has been targeted, with unprecedented levels of highly organised illegal timber harvesting and trade on international markets, very little of which benefits local communities (Schuurman and Lowry II, 2009). Illegal logging is often accompanied by bushmeat hunting (Ormsby and Kaplin, 2005). Lemur hunting has also been found to increase following an influx of migrants to gold mining areas (Jenkins *et al.*, 2011). This should remind us of the importance of global political and economic forces. Policy needs to tackle not only local drivers of land use (for example poverty and a lack of livelihood options) but also consider other factors such as global commodity chains, trade agreements and the certification of products.

Climate change and environmental management

The future of environmental management in Madagascar will be made more difficult because of climate change. The island's climate has been described as hyper-variable, with rainfall that is highly spatially and temporally unpredictable (Dewar and Richard, 2007). Climate variability and dramatic events such as cyclones have been shown to influence population dynamics of species such a lemurs (Dunham *et al.*, 2011). Furthermore, Madagascar's vegetative cover has been shown to vary significantly over time, both annually and seasonally, with a strong correlation with the El Niño Southern Oscillation (ENSO) and corresponding drought events and associated wildfires (Ingram and Dawson, 2005). At sea, climate impacts range from coral reef bleaching to coastal erosion and increased cyclonic activity (Harris, 2011).

Over thousands of years Madagascar's flora and fauna has responded to changes in climate (see Chapter 2 by Ganzhorn *et al.* and Chapter 3 by Dewar). However, with more rapid climate change, there are important questions about the ability of species to migrate and adapt quickly enough. Predictions of what might happen to Madagascar, like anywhere else, are difficult. According to models, temperatures are likely to increase most in the arid south, rainfall should increase in summer (January to April), while winters (July to September) should be drier along the south-east coast but wetter elsewhere (Tadross *et al.*, 2008; Hannah *et al.*, 2008). If these predictions are correct, although some species will benefit, more will suffer (Hannah *et al.*, 2008). It is likely that cyclones and other extreme events will become more frequent (Dunham *et al.*, 2011). Global climate change is predicted to increase the frequency of the ENSO phenomenon, which might result in an increase in droughts and wildfires and corresponding changes in land cover (Ingram and Dawson, 2005). Changes in temperature, rainfall, drought frequency and the occurrence of wildfire will have major implications for the island's flora and fauna.

In the past, forest refugia and corridors have played a key role in allowing species to adapt to climate change, both through migration in the short-to-medium term and evolutionary adaptation in the longer term. The design of protected areas needs to take these facts into consideration, with corridors to maintain gene flow and exchange of species between existing protected areas (Hannah *et al.*, 1998). The design of protected areas in Madagascar has often been based on historical precedent, but many of these protected areas were established under different circumstances and with different goals in mind.

In terms of environmental management and natural resource use, there are also major implications. Changes in temperature and rainfall will affect the viability of agriculture and pastoralism. While an increase in temperature in the central highlands would probably lead to an increase in the yields of irrigated rice, changes in rainfall would have negative consequences for agricultural production, particularly in the semi-arid and arid west and south where temperatures are predicted to rise the most and where rainfall is likely to become less predictable (Hannah *et al.*, 2008; Vololona *et al.*, 2012). Without significant adaptation, innovation and investment, food insecurity is likely to increase. Similarly at sea, Madagascar is ranked among tropical coastal countries as having the lowest adaptive capacity to climate change (Harris, 2011), and there are already signs that changes in sea temperature are affecting shrimp fisheries (FAO, 2008).

Conclusion 355

Meeting the challenge: making a case for interdisciplinarity, crossing boundaries and moving beyond panaceas

At this stage, the challenge of successfully combining conservation and poverty alleviation seems both urgent and daunting. Little wonder then that policy often relies on overly simplistic narratives to make reality manageable. Perhaps it is not surprising that in the face of such socio-ecological complexities, policy looks for cure-all solutions. However, the history of environmental policy shows us that complex and multi-scalar, socio-ecological problems defy 'magic bullets' (Ostrom et al., 2007). Ultimately, the conservation of biodiversity is as much about people as it is about ecosystems or endangered species and no single perspective can hope to recognise all the dimensions of complex socio-ecological issues (Mascia et al., 2003; Sanderson, 2005; Hirsch et al., 2011). In the words of Bill Adams (2003, p209): 'There is no right way to do conservation. There are only choices.'

To help make these choices, research and policy in Madagascar desperately need more conversations – between biologists, anthropologists, archaeologists, economists, environmental historians and geographers; between researchers and practitioners; and between 'experts' and the individuals, households and communities directly dependent on the island's natural resources for their livelihoods. Such dialogue is never easy, having to cross barriers not only of vocabulary and language but also of contrasting and sometimes conflicting worldviews. Furthermore, it is crucial that well-intentioned ideas such as 'interdisciplinarity', 'participation' and 'dialogue' do not become overly prescriptive and simply a box-ticking exercise. Rather, what is needed is a change in attitude:

> The new face of science will be more public spirited and be characterized by its problem solving capability. It will draw upon the disciplines and not compete with them. It will seek models in which there is no longer the separation of the human from the rest of the biosphere, and in which uncertainty, surprise and incompleteness are not taken as signs of failure but rather as better approximations to the real world.
> (Rapport, 1997, p289)

Summary of key research and policy priorities

The following list sets out the key priorities for research and policy in Madagascar. It reflects the issues that have emerged most strongly from the chapters, as well as some of the topics that have received less attention in this book.

- Research and policy need to pay more attention to understudied species and ecosystems, both terrestrial (especially grasslands) and marine (especially coral reefs, mangroves and sea grass).

- Researchers and policymakers should adopt greater clarity with regards to categories used in describing vegetation and ecological change. Terms such as 'pristine' tend to reflect desires and preconceived ideas rather than reality. Overly simplistic narratives and myths (for example the idea of an Eden-like island-forest) hinder progress and limit options.
- Policymakers must acknowledge the different perceptions and priorities of stakeholders and be ready to make trade-offs between various environmental goals (e.g. maximum biodiversity, ecosystem function, the maintenance of resource stocks) and between environmental and social goals.
- Policy needs to move beyond panaceas, 'win–wins' and cure-all solutions. Environmental policy will invariably be messy, complicated and contingent on local realities.
- Policies must be able to deliver economic benefits and/or genuine viable livelihood alternatives for low-income rural households.
- There should be a re-alignment in the power dynamics of natural resource management, with greater involvement of rural households and communities in environmental research, planning and practice.
- Policy needs to consider the implications of climate change, both for the island's flora and fauna (particularly with regards to the design of protected areas) and for rural livelihoods.
- The key to all the above will be moving beyond traditional boundaries: between academic disciplines; between research and policy; and between 'experts' and 'laypeople'.

Notes

1 Rio-Tinto is a UK–Australian mining conglomerate. The mineral ilmenite is a source of titanium dioxide, which is used as a white pigment in paints, paper and cosmetics (including toothpaste). QIT Madagascar Minerals (QMM) is a Malagasy subsidiary of Rio Tinto and the Government of Madagascar.
2 www.birdlife.org/action/business/rio_tinto/index.html, accessed 3 May 2013.

References

Adams, W. M. (2003) *Future Nature: A Vision for Conservation*, Earthscan, Oxford.
Agrawal, A. and Gibson, C. C. (2001) *Communities and the Environment: Ethnicity, Gender, and the State in Community-based Conservation*, Rutgers University Press, New York.
Barrett, C. B. (1994) 'Understanding uneven agricultural liberalisation in Madagascar', *The Journal of Modern African Studies*, vol 32, pp449–476.
Bloesch, U. (1999) 'Fire as a tool in the management of a savanna/dry forest reserve in Madagascar', *Applied Vegetation Science*, vol 2, pp117–124.
Bollen, A. and Donati, G. (2006) 'Conservation status of the littoral forest of southeastern Madagascar: a review', *Oryx*, vol 40, pp57–66.

Bond, W. J., Silander, J. A., Ranaivonasy, J. and Ratsirarson, J. (2008) 'The antiquity of Madagascar's grasslands and the rise of C4 grassy biomes', *Journal of Biogeography*, vol 35, pp1743–1758.
Corson, C. (2010) 'Shifting environmental governance in a neoliberal world: USAID for conservation', *Antipode*, vol 42, pp576–602.
Cronon, W. (1996) 'The trouble with wilderness or, getting back to the wrong nature', *Environmental History*, vol 1, pp7–28.
Dewar, R. E. and Richard, A. F. (2007) 'Evolution in the hypervariable environment of Madagascar', *Proceedings of the National Academy of Sciences*, vol 104, pp13723–13727.
Duffy, R. (2006) 'Non-governmental organisations and governance states: the impact of transnational environmental management networks in Madagascar', *Environmental Politics*, vol 15, pp731–749.
Dunham, A. E., Erhart, E. M. and Wright, P. C. (2011) 'Global climate cycles and cyclones: consequences for rainfall patterns and lemur reproduction in southeastern Madagascar', *Global Change Biology*, vol 17, pp219–227.
FAO (2008) *Fishery Country Profile, The Republic of Madagascar*, Food and Agriculture Organisation of the United Nations, Rome.
Ferguson, B. and Gardner, C. J. (2010) 'Looking back and thinking ahead: where next for conservation in Madagascar?', *Madagascar Conservation and Development*, vol 5, pp75–76.
Ghimire, K. B. (1994) 'Parks and people: livelihood issues in national parks management in Thailand and Madagascar', *Development and Change*, vol 25, pp195–229.
Hannah, L., Dave, R., Lowry, P. P., Andelman, S., Andrianarisata, M., Andriamaro, L., Cameron, A., Hijmans, R., Kremen, C., MacKinnon, J., Randrianasolo, H. H., Andriambololonera, S., Razafimpahanana, A., Randriamahazo, H., Randrianarisoa, J., Razafinjatovo, P., Raxworthy, C., Schatz, G. E., Tadross, M. and Wilmee, L. (2008) 'Climate change adaptation for conservation in Madagascar', *Biology Letters*, vol 4, pp590–594.
Hannah, L., Rakotosamimanana, B., Ganzhorn, J. U., Mittermeier, R. A., Olivieri, S., Iyer, L., Rajaobelina, S., Hough, J., Andriamialisoa, F., Bowles, I. and Tilkin, G. (1998) 'Participatory planning, scientific priorities, and landscape conservation in Madagascar', *Environmental Conservation*, vol 25, pp30–36.
Harper, J. (2002) *Endangered Species: Health, Illness and Death Among Madagascar's People of the Forest*, Carolina Academic Press, Durham.
Harris, A. R. (2011) 'Out of sight but no longer out of mind: a climate of change for marine conservation in Madagascar', *Madagascar Conservation and Development*, vol 6, pp7–14.
Healy, T. M. and Ratsimbarison, R. (1998) 'Historical influences and the role of traditional land rights in Madagascar: legality versus legitimity', in M. Barry (ed.) *Proceedings of the International Conference on Land Tenure in the Developing World with a focus on Southern Africa*, University of Cape Town, pp365–377.
Hirsch, P. D., Adams, W. M., Brosius, J. P., Zia, A., Bariola, N. and Dammert, J. L. (2011) 'Acknowledging conservation trade-offs and embracing complexity', *Conservation Biology*, vol 25, pp259–264.
Hume, D. W. (2006) 'Swidden agriculture and conservation in eastern Madagascar: stakeholder perspectives and cultural belief systems', *Conservation and Society*, vol 4, pp287–303.

Ingram, J. C. and Dawson, T. P. (2005) 'Climate change impacts and vegetation response on the island of Madagascar', *Philosophical Transactions of the Royal Society of London Series a-Mathematical Physical and Engineering Sciences*, vol 363, pp55–59.

Jenkins, R. K. B., Keane, A., Rakotoarivelo, A. R., Rakotomboavonjy, V., Randrianandrianina, F. H., Razafimanahaka, H. J., Ralaiarimalala, S. R. and Jones, J. P. G. (2011) 'Analysis of patterns of bushmeat consumption reveals extensive exploitation of protected species in eastern Madagascar', *Plos One*, vol 6.

Kull, C. A. (2000) 'Deforestation, erosion, and fire: degradation myths in the environmental history of Madagascar', *Environment and History*, vol 6, pp423–450.

Kull, C. A. (2002) 'Madagascar aflame: landscape burning as peasant protest, resistance, or a resource management tool?', *Political Geography*, vol 21, pp927–953.

Kull, C. A. (2004) *Isle of Fire: The Political Ecology of Landscape Burning in Madagascar*, University of Chicago Press, Chicago.

Le Manach, F., Gough, C., Harris, A. R., Humber, F., Harper, S. and Zeller, D. (2012) 'Unreported fishing, hungry people and political turmoil: the recipe for a food security crisis in Madagascar', *Marine Policy*, vol 36, pp218–225.

Leader-Williams, N., Adams, W. M. and Smith, R. J. (2010) *Trade-offs in Conservation: Deciding What to Save*, Wiley-Blackwell, Oxford.

Mascia, M. B., Brosius, P. J., Dobson, T. A., Forbes, B. C., Horowitz, L., McKean, M. A. and Turner, N. J. (2003) 'Conservation and the social sciences', *Conservation Biology*, vol 17, pp649–650.

McConnell, W. J. (2002) 'Misconstrued land use in Vohibazaha: participatory planning in the periphery of Madagascar's Mantadia National Park', *Land Use Policy*, vol 19, pp217–230.

McShane, T. O., Hirsch, P. D., Trung, T. C., Songorwa, A. N., Kinzig, A., Monteferri, B., Mutekanga, D., Thang, H. V., Dammert, J. L., Pulgar-Vidal, M., Welch-Devine, M., Brosius, J. P., Coppolillo, P. and O'Connor, S. (2011) 'Hard choices: making trade-offs between biodiversity conservation and human well-being', *Biological Conservation*, vol 144, pp966–972.

McVean, A. R., Walker, R. C. J. and Fanning, E. (2006) 'The traditional shark fisheries of southwest Madagascar: a study in the Toliara region', *Fisheries Research*, vol 82, pp280–289.

Mittermeier, R., Ganzhorn, J., Konstant, W., Glander, K., Tattersall, I., Groves, C., Rylands, A., Hapke, A., Ratsimbazafy, J., Mayor, M., Louis, E. E., Rumpler, Y., Schwitzer, C. and Rasoloarison, R. (2008) 'Lemur diversity in Madagascar', *International Journal of Primatology*, vol 29, pp1607–1656.

Mittermeier, R. A., Konstant, W. R., Nicoll, M. E. and Langrand, O. (1992) *Lemurs of Madagascar: An Action Plan for their Conservation 1993–1999*, International Union for Conservation of Nature, Gland.

Murphree, M. (2000) 'Community-based conservation: old ways, new myths and enduring challenges', in R. D. Baldus and L. S. Siege (eds) *Tanzania Wildlife Discussion Paper No. 29*, Deutsche Gesellschaft für Technische Zusammenarbeit, Dar er Salaam.

Neimark, B. D. (2013) *The Land of our Ancestors: Property Rights, Social Resistance, and Alternatives to Land Grabbing in Madagascar*, LDPI Working Paper 26, The Land Deal Politics Initiative., International Institute of Social Studies, The Hague.

Ormsby, A. and Kaplin, B. A. (2005) 'A framework for understanding community resident perceptions of Masoala National Park, Madagascar', *Environmental Conservation*, vol 32, pp156–164.

Ostrom, E., Janssen, M. A. and Anderies, J. M. (2007) 'Going beyond panaceas', *Proceedings of the National Academy of Sciences of The United States of America*, vol 104, pp15176–15178.

Peters, J. (1998) 'Transforming the integrated conservation and development project (ICDP) approach: observations from the Ranomafana National Park project, Madagascar', *Journal of Agricultural and Environmental Ethics*, vol 11, pp17–47.

Peters, J. (1999) 'Understanding conflicts between people and parks at Ranomafana, Madagascar', *Agriculture and Human Values*, vol 16, pp65–74.

Purcell, S. W., Mercier, A., Conand, C., Hamel, J. F., Toral-Granda, M. V., Lovatelli, A. and Uthicke, S. (2013) 'Sea cucumber fisheries: global analysis of stocks, management measures and drivers of overfishing', *Fish and Fisheries*, vol 14, pp34–59.

Rapport, D. J. (1997) 'Transdisciplinarity: transcending the disciplines', *Trends in Ecology and Evolution*, vol 12, p289.

Rasoloarison, R., Weisrock, D. W., Yoder, A. D., Rakotondravony, D. and Kappeler, P. (2013) 'Two new species of mouse lemurs (Cheirogaleidae: Microcebus) from eastern Madagascar', *International Journal of Primatology*, vol 34, pp455–469.

Richard, A. F. and Dewar, R. E. (2001) 'Politics, negotiation and conservation: a view from Madagascar', in W. Weber, L. J. T. White, A. Vedder and L. Naughton-Treves (eds) *African Rain Forest Ecology and Conservation: An Interdisciplinary Perspective*, Yale University Press, New Haven.

Sanderson, S. (2005) 'Poverty and conservation: the new century's "peasant question"?', *World Development*, vol 33, pp323–332.

Sarrasin, B. (2006) 'The mining industry and the regulatory framework in Madagascar: some developmental and environmental issues', *Journal of Cleaner Production*, vol 14, pp388–396.

Scales, I. R. (2011) 'Farming at the forest frontier: land use and landscape change in western Madagascar, 1896 to 2005', *Environment and History*, vol 17, pp499–524.

Schuurman, D. and Lowry II, P. P. (2009) 'The Madagascar rosewood massacre', *Madagascar Conservation and Development*, vol 4, pp98–102.

Seagle, C. (2012) 'Inverting the impacts: mining, conservation and sustainability claims near the Rio Tinto/QMM ilmenite mine in southeast Madagascar', *Journal of Peasant Studies*, vol 39, pp447–477.

Styger, E., Rakotondramasy, H. M., Pfeffer, M. J., Fernandes, E. C. M. and Bates, D. M. (2007) 'Influence of slash-and-burn farming practices on fallow succession and land degradation in the rainforest region of Madagascar', *Agriculture, Ecosystems and Environment*, vol 119, pp257–269.

Sussman, R. W., Green, G. M. and Sussman, L. K. (1994) 'Satellite imagery, human ecology, anthropology, and deforestation in Madagascar', *Human Ecology*, vol 22, pp333–354.

Tadross, M., Randriamarolaza, L., Rabefitia, Z. and Zheng, K. Y. (2008) *Climate Change in Madagascar: Recent Past and Future*, World Bank, Washington, DC.

Thalmann, U. (2006) 'Lemurs: ambassadors for Madagascar', *Madagascar Conservation and Development*, vol 1, pp4–8.

Virah-Sawmy, M. (2009) 'Ecosystem management in Madagascar during global change', *Conservation Letters*, vol 2, pp163–170.

Vololona, M., Kyotalimye, M., Thomas, T. S. and Waithaka, M. (2012) *East African Agriculture and Climate Change: A Comprehensive Analysis – Madagascar*, International Food Policy Research Institute, Washington, DC.

Walpole, M. J. and Leader-Williams, N. (2002) 'Tourism and flagship species in conservation', *Biodiversity and Conservation*, vol 11, pp543–547.

WWF (2011) *Première Phase de l'Etude Stratégique du Développement du Secteur Agrocarburant à Madagascar: Etat des Lieux de la Situation Actuelle du Secteur*, World Wide Fund for Nature, Antananarivo.

Index

Locators in **bold** refer to figures/tables/boxes

access and benefit-sharing (ABS) agreements 274, 279–80
accountability, governance principle 9, 225, 242–3
acculturation 321, 329, 331–2 *see also* culture
Advanced Very High Resolution Radiometry *see* AVHRR
aerial photography, deforestation 71, **74–5**, 83, 84, 85, 95
agendas *see* political agendas
agriculture: animal farming 112–13, 117; archaeological evidence 51–3; intensification 183–4, 186, 314; and vegetation change 55–7, 67, 93, 105 *see also* cash crops; land use; subsistence agriculture; swidden farming
aid programs 163, 164–5, 166
Allnutt, T. F. **88**
allopatric speciation 20, 30
alternatives to slash-and-burn (ASB) case study **5**, 307, 313–14
Ambohiposa, archaeological evidence 49, 50
Amelot, X. 92–3
amphibians **18**, **19**, 26, 28
ancestors, cultural importance 320, 323–5, 328, 329
Andasibe-Mantadia National Park 227, 228, 263
Andreone, F. **89**
Andriamarovololona, M. M. 180, 183, 186
Andrianampoinimerina, King 133, 134, 149
ANGAP *see* Madagascar National Parks
Anjohibe, fossils 51
Ankarana National Park **32**, 58, **218**, 228, 264
Ankeniheny-Zahamena wildlife corridor 186–7, 195, 201–4, **206**, 208, **224**; case study **5**, 311–13, **312**
Ankodida case study 225, 229–34, 246–7; governance principles 236, 238, 240–1, 243–5; governance structure **232**; governance types **222**; maps **5**, **231**; protected area categories **219**, **220**; sacred forest **230**, 238; supporting institutions **224**
anthropogenic drivers of change *see* human impacts
anthropological perspectives 3, 172, 245; conflicting visions of nature 320–2, 325, 326, 331, 333, 335–7; future of conservation 344, 350, 355 *see also* ethnographic research
archaeological evidence, human settlements 49–53
assessment of deforestation *see* scientific assessment
Association Nationale d'Actions Environnementales (ANAE) 157, 158
Association Nationale pour la Gestion des Aires Protégées see Madagascar National Parks
assumptions, untested *see* received wisdom
attitude changes, future of conservation 355 *see also* stakeholder perceptions/perspectives
AVHRR (Advanced Very High Resolution Radiometry) 76, **80**, **81**, 83, **91**
aye-aye (*Daubentonia madagascariensis*) 33, 142, 151

Bakoariniaina, L. N. **87**
Baobabs 27; Baobab Alley/Avenue **5**, 260–2, **261**, 348
Barrett, M. A. **89**
bathymetric profile, Antarctica and Madagascar **25**
bats, speciation 19, **19**
beliefs *see* cultural beliefs and practices
benefits distribution: bioprospecting 288; PES 304–5, **309**; tourism 263
Benstead, J. **87**
Bérard, M. H. 180, 188, 189
bet-hedging life-history trait 47
Betsileo 109, 132, 137, 323
Betsimisaraka 106, 113, 178, 324; pre-colonial 132
biodiversity 1, 4, 17, **25**, 34–5; bats **19**, 19; biogeography 4, **18**, 18–19, 27–33, **29**; bioprospecting 272; climatic factors 28–33, **32**, 35; conservation politics 196–8; environmental discourse 348–9; future of conservation 11, 342, 343–6, 349; geological history 22–7, **23**, **25**; grasslands 344; number of species described 19–20, **20**, 343; and tourism 258 *see also* endemic species
biofuels 120, 352
biogeography, island 4, **18**, 18–19, 22, 27–33, **29**
biological endemism *see* endemic species
biomes 4, 17, 55, **69**, 70, 71; Ice Age 44–5
biopiracy 275-276, **277**, 279, 280
bioprospecting 10, 271–2; access to biogenetic resources 280–2; case study **5**, 282–9, **284**, **285**; and conservation 290–3; ethical/legal debates 275–80; future of conservation 350; historical contexts 272–5
BirdLife International 353
birds, extinct 34, 57, 58, **59**
black rat (*Rattus rattus*) 52, 266
Blanc-Pamard, C. 178, 182, 184
Blasco, F. **75**
Blue Ventures non-governmental organisation 184–5
Bollen, A. **88**, 344
bottom-up initiatives 177, 350 *see also* community-based conservation; community-based natural resource management; participatory governance/paradigm
burning *see* fire
butter beans (*Phaseolus lunatus*) 109

carbon offset and trading schemes 120 *see also* Reducing Emissions from Deforestation and Forest Degradation
cash crops 142; colonial period 136; and deforestation 108–10, 115–16, 119, 120, 259 *see also* butter beans; maize; rice
cattle *see* zebu cattle
CAZ *see* Ankeniheny-Zahamena wildlife corridor
CBC *see* community-based conservation
CBD (Convention on Biological Diversity) 274
CBNRM *see* community-based natural resource management
Centre de Coopération Internationale en Recherche Agronomique pour le Développement (CIRAD) 119
change, quantitative assessment 78–82, **80-2**, 95
charcoal 56, 112, 181, 187
charismatic species 9, 35, 258, 343
Charte de l'Environnement 155, 156
choices: future 355; politics of 182–3
CITES (Convention of International Trade in Endangered Species of Wild Fauna and Flora) 291
climate change 4; and biodiversity 28–33, **32**, 35; and deforestation 121; future of conservation 356; human impacts 55, 353–4; Ice Age 44–7; and species distribution 30–1, 45–7
climatic zones 28–30, **29**
Coase, R. 301
Code des Aires Protégées (COAP) **150**, 160, 161, 201, 206, 217
Code of 101 Articles 134
Code of 305 Articles 134
coercive legislation 3, 299, 302, 315
colonial period 135, 149; deforestation 68, 70, 71–3, **72**, **73**, 93, 108, 110; environmental management 129, 135–40, 173; protected areas/reserves 225, 227
colonization, species 4, 17, **18**, 22, **23**, 24–7, **25**
commodification of nature 10, 276 *see also* bioprospecting; incentive-based mechanisms; tourism

Index 363

Communautés de Bases (COBA) **175**, 176–80, 183, 186, 197, 203–4
community-based conservation (CBC) 3, 7, 9, 139, 158, 172; future of 351; incentive-based approaches 300
community-based natural resource management (CBNRM) 172; community regulations 330, 332–3; Durban Vision 197–8, 201, 209–10, 216, 225; GELOSE 174–7, **175**; historical/international contexts 173–4; participatory governance 232, 234, 236, 238, 241, 243; and voluntary servitude 334, 335 *see also* resource management transfers
community regulations *see dina*
complexity, ecosystem *see* biodiversity
compromises *see* trade-offs
conflagration, great 4, 56, 70 *see also* fire
conflict, environmental management 349–51 *see also* resistance
conflicting visions of nature *see* cultural beliefs and practices
conservation boom, history and politics 166–8; conservation funding 146, **147**, **148**; foreign input 164–5; global environmentalism 163; local/indigenous environmentalism 163–4; nineteen seventies/eighties developments 151–5; nineteen nineties to present 155–62; pre-nineteen seventies 149–51; real and imagined environment 162; rivalries 165–6; timeline of key events **150**; and tourism 255
conservation funding *see* funding conservation
conservation futures *see* future of conservation
Conservation International (CI) 92, 146, 177, 193, **224**, 258, 272, 307, 351
continental drift 4, **23**, 24–5
contracts, standardized 180, 181
conventions, community 322, 329–33
Convention of International Trade in Endangered Species of Wild Fauna and Flora (CITES) 291
Convention on Biological Diversity (CBD) 274
convergent evolution 33–4
Coopération Suisse see Swiss Aid
coral reefs, degradation 185
corridors, wildlife *see* wildlife corridors

corruption 10–11 *see also* incentive-based mechanisms; power dynamics
cost-benefits, tourism 263–6
Cours Darne, G. 71, 73, **74**, **75**, 79, 83, **90**
Craul, M. **89**
crisis narratives, environmental 6, 146, 163, 167
cultural beliefs and practices 114, 336–7; conservation and culture 320–2; *dina* 322, 329–33; *fady* 326–9; future of conservation 349; landscape mosaics 325–33; and participatory governance 238, 247; political ecology approach 258–63; trees, role of 322–5, 326; voluntary servitude 322, 333–6 *see also* social factors; stakeholder perceptions/perspectives
culture 3; and conservation 320–2; Durban Vision 217; and forest loss 7, 114–15, 116; reification of 327
cyclones 31, 33, 47, 353

Dawson, T. P. 84, **88**, **89**, **91**
de Flacourt, Etienne 131
debt crisis 152–4, 158, 164, 167, 352
Debt-for-Nature swaps 156
decentralization of government 173 *see also* community-based natural resource management; participatory governance/paradigm; resource management transfers
definitions: biodiversity **18**; community-based natural resource management 173–4; deforestation 107; discourse **131**; ecosystem services 301; original vegetation 55; Payments for Ecosystem Services 302–4; political ecology 258; politics **131**; protected areas/reserves **226**; state 129 *see also* terminology use
deforestation 70, 73, 79, 84; and biodiversity 34; definitions 107; and farming practices 67, 93, 105; human impacts 56; and poverty 105, 106, 115, 121, 242 *see also* island-wide forest hypothesis; land use and deforestation; reforestation; scientific assessment of deforestation
delegation, definition **226**
demographic changes 110, 116, 119 *see also* population growth
dina (community regulations) 180, **226**, 240, 322, 329–33

direction, governance principle 9, 225, 244–5
Direction Générale des Eaux et Forêts (DGEF) 154, 157, 199, 200, 202–3, 205–6
Direction du Système des Aires Protégées (DSAP) 217, 223, 241
discourse(s): definition **131**; deforestation 67, 85, 92, 93, 94, 95; environmental 6, 105, 106, 121, 346–9; pristine nature **131**, 176, 260, 346, 348, 356
diversity, species *see* biodiversity
Donati, G. **88**, 344
drugs, medicinal *see* bioprospecting
dry-deciduous forests **5**, 28, **29**, 140, 262, 344
DSAP *see Direction du Système des Aires Protégées*
Du Puy, D. J. 76, **87–9**, **91**, 91
Dufils, J.-M. 79, **88**
Durban Accord 216, 217
Durban Action Plan 216
Durban Vision 3, 9, 35, 159–61, 193–4, 207–8, 210; biodiversity priority setting 196–8; formulating the parks announcement 195–6; governance principles 216–17, 223, 225; governance types **221–2**; implementation 201–2, 351; list of protected areas/delegated institutions **223–4**; map of protected areas/reserves **206**; policy recommendations 209–10; protected area categories **218–20**; resource protection 198–9; resource use rights/debates 200–1; rural consultation process 202–7 *see also* participatory governance
Durban Vision Community 217
Durbin, J. C. 85, **86**, 257, 265
Durkin, L. **89**, 92

eastern region 79, 106, 113, 115, 120, 184-186, 202-203,
Eaux et Forêts see Forest Service
eco-certification schemes 307
ecology 163; deforestation/land use dynamics 113–14 *see also* political ecology
economic factors 1–2, 3, 10; and conservation 146, **147**, **148**, 300–6, 342; deforestation drivers 7, 106, 108, 110, 115, 116, 120; resource management transfers 186–7; self-sufficiency 136; timeline **8** *see also* incentive-based mechanisms
Eco-Regional Initiative (ERI) 186
eco-regions 158–9
ecosystem complexity *see* biodiversity
ecosystem goods/services 2, 3, 10, 301, 303, 349 *see also* incentive-based mechanisms; Payments for Ecosystem Services
ecosystems 28, **29**; lake 184–5; marine 184–5, 344–6, 355
ecotourism *see* tourism
ecotypes **29**, 29–30 *see also* vegetation types
Eden, Garden of 44, 346, 351, 356 *see also* pristine nature discourse
effectiveness, governance principles 9, 225, 244–5
eggshell fragments, ratites 58
El Niño Southern Oscillation (ENSO) 354
endemic species 1, 4, 9, 17, **18**, 20, 34, 35; bioprospecting 272, 282, 290; conservation book 151, 162; future of 346; geological history 22–7; macro-endemism/geological history 22–7, **23**, **25**; micro-endemism 30–3; protected area categories **219**; and tourism 258; tropical/subtropical habitats **21** *see also* biodiversity
Environmental Charter 110
environmental crisis discourse 6, 146, 163, 167
environmental discourse 346–9
environmental economics *see* economic factors
environmental management: and climate change 354; colonial period 135–40; pre-colonial history 132; resource management transfers 183–5; power dynamics 11; tourism 265–6 *see also* future of conservation; resource management transfers; state roles
environmental policy *see* policy frameworks
environmentalism, local/indigenous 163–4
equality, ICBG-Madagascar project case study **5**, 288–9 *see also* benefits distribution; fairness; justice; power dynamics
ERI (Eco-Regional Initiative) 186

ethics: bioprospecting 275–80; conservation policy 11, 350 see also cultural beliefs and practices
ethnic identity, tribal **138**
ethnographic research 207, 321, 335 see also anthropological perspectives
eucalyptus (*Eucalyptus* spp.) 139, 352
evidence-based policy 95, 96 see also scientific research
evolution, models 33–4
evolutionary history 4, 17, 22; climatic zones 33; conservation implications 34–5; endemic species **18**; molecular clocks 24, 33
explorers 131, 149, 271
extinctions 4, 17, **18**; human impacts 55, 57–8; mass 22, 24, 25; radiocarbon dating 57, 58, **59**

fady (forbidden behavior) 114, 238, 281, 326–9 see also taboos
fairness, governance principle 9, 225, 240–2 see also benefits distribution; justice
fallow periods, land use 106, 107, 113–14, 119
Fandriana-Vondrozo wildlife corridor 186–7, 195, 201–4, **206**, 208, **224**; mining 199
FAO (Food and Agricultural Organization) 78
Faramalala, M. H. 76, 77, 83, **86**, **90**, **91**, 91
farming practices see agriculture; land use and deforestation
FCPF (Forest Carbon Partnership Facility) 307
fire 6, 7; and deforestation 67, 107; environmental discourse 348; Holocene palaeo-environment 4; human impacts 54, 56; legislation 140, 142, 149, 151; post-colonial attitudes 141; species distribution 31 see also swidden farming practices
First Republic 141–2, 149–51, **150**
fish, freshwater 27
fishing 184, 330, 345, 347, 354 see also marine ecosystems
flagship species, lemurs as 343–4 see also charismatic species
fokonolona **175**, 176–7
food security 113, 345
Foreign Direct Investment (FDI) 352

Forest Carbon Partnership Facility (FCPF) 307
forest conservation see incentive-based mechanisms
Forest Resource Assessments, FAO 78
Forest Service **74**, 139, 154, 157 see also Direction Générale des Eaux et Forêts
forests: dry-deciduous **5**, 28, 29, 140, 262, 344; Pleistocene distribution 44–5; sacred 197, **219**, **220**, **230**, 229–31, 238 see also deforestation; island-wide forest hypothesis; spiny forest; timber resources; vegetation
fortress conservation approach 3, 7, 173, 187, 188, 300; conservation politics 196; environmental discourse 348; and tourism 255 see also protected areas
fossil record 22, 23, 24, 27, 47, 51 see also palaeoecology
France: colonization by see colonial period; modern links with 149, 151, 152, 166
freshwater fish 27
friendship with nature 333–4
frogs, convergent evolution 34
fruit trees **68**, 112, 120, 330
funding institutions, ICBG-Madagascar project **285**
future of conservation 11, 342; biodiversity/priorities 343–6; climate change/environmental management 353–4; environmental discourse 346–9; global political uncertainty 351–3; multi-disciplinary perspectives 355; priorities 355–6; socio-cultural beliefs and practices 336–7; trade-offs 349–51, 356

Gallieni, General Joseph Simon 136, 137, 139
Ganzhorn, J. U. **87**, 91
Garden of Eden 44, 346, 351, 356 see also pristine nature discourse
GCF (*Gestion Contractualisée des Forêts*) 159, 166, 174, 177–8, 180, 201
GELOSE see *Gestion Locale Sécurisée*
genetic (molecular) clocks 24, 33
Geographic Information Systems (GIS) 197, 198
geological isolation 4, 20, 22, 24, 35; and biodiversity 22–7, **23**, **25**; biogeography 27, 31, 33

Index

geopolitical rivalries/strategies 164, 165–6
Gestion Contractualisée des Forêts (GCF) 159, 177–8, 180
Gestion Locale Sécurisée (GELOSE) 7, 158, 159, 174–7, 236, 246; and *dina* 330; false assumptions 174–6; perversion of ideals of 176–7; stakeholders **175**, 176 *see also* resource management transfers
global contexts 7, 163; community-based natural resource management 173–4; conservation agendas 179–80; conservation boom 164–5; geopolitical rivalries/strategies 165–6; markets 133; political factors 351–3; scientific research 151–2
Global 200 Ecoregions (WWF) 34
Global Environment Facility (GEF) 197
Goedefroit, S. 179–80
golden-crowned sifaka (*Propithecus tattersalli*) 45
Goodman, S. **87**
governance, definition **226**
governance principles, Durban Vision 9, 216–17, 225, 229; accountability and transparency 9, 225, 242–3; direction and effectiveness 9, 225, 244–5; fairness 9, 240–2; inclusiveness 9, 225, 239–40; legitimacy 9, 225, 236–9 *see also* participatory governance; policy frameworks
Grandidier's baobab (*Adansonia grandidieri*) **5**, 260–2, **261**, 348
grasslands 28, 29; biodiversity 344; climate changes/variation 45; environmental discourse 346; human impacts 54–5; original vegetation 70
grazing, animal 112
great fire 4, 56, 70 *see also* fire
Green, G. M. 79, 82, **90**, 91
green economic developments 302, 310 *see also* sustainable development
Guichon, A. 71, **74**, 78, 79, 82
guidelines, governance principles 216–17 *see also* policy frameworks

Hannah, L. 85, **86**, **89**, 92
Hanski, I. **88**
Harper, G. J. 73, 76, 77, 79, 82, 84, **90**, 91–3
hatsake 106, 107 *see also* swidden farming practices, slash-and-burn

healers, traditional 281–2
herbal medicines *see* bioprospecting
Histoire de la Grande Isle Madagascar (de Flacourt) 131
historical contexts; bioprospecting 272–5; community-based natural resource management 173–4; deforestation and land use 107–10; oral history 51, 70, 134, 322–5; pre-colonial history 131–5; timelines **8**, **150** *see also* colonial period; conservation boom (history and politics)
A History of Madagascar (Brown) 130
Hockley, N. J. 180, 183, 186
Holocene palaeo-environment 4, 6, 45, 47; archaeological evidence 50; extinctions, animal 57; fire 56 *see also* human impacts
Horning, N. 76, 77, 78, 83, **90**, 91
hotspots, biodiversity 1, 17, 20, **21**, 85, 96, 162; bioprospecting 272; conflicting visions of nature 260; conservation interests 93
house mouse (*Mus musculus*) 52–3
households, rural 110–15
human impacts 1–2, 4, 6, 11, 44, 58, 60; animal extinctions 57–8, **59**; deforestation 67; empirical data 54–8; protected areas/reserves 228; settlers, origins and dating 47–54; sustainable 229; vegetation changes 55–7
human welfare, and conservation 153–4 *see also* poverty
Humbert, H. 71, **74**, **75**, 79, 83, 84, **90**, 93
Hume, D. W. **88**

ICBG-Madagascar project case study **5**, 282–9; financial structure 283–6; funding institutions/research organizations **285**; justice/equality 288–9; local perceptions/benefits **287**, 287–8; maps **5**, **284**; sustainable development 292–3
ICDPs (Integrated Conservation and Development Projects) 154, 157, 172, 228, 301
Ice Age 44–7, 55
identity: and place 325, 326, 327; tribal **138**
ideologies: geopolitical rivalries/strategies 165–6; neoliberal 153, 157, 161

Index 367

IEFN see *Inventaire Ecologique Forestier National*
imagined: communities 180, 181–2, 183; environments 162 see also stakeholder perceptions/perspectives
INBio (National Biodiversity Institute) 271, 274, 275
incentive-based mechanisms 10–11, 299–300, 314–15; case studies **5**, 306–14, **309**, **312**; and deforestation 3, 108, 109, 117; *and dina* 332; economics and environment 300–6; future of conservation 350 see also bioprospecting; economic factors; Payments for Ecosystem Services; Reducing Emissions from Deforestation and Forest Degradation; tourism
inclusiveness: governance principles 9, 225, 239–40
independence, post-colonial 141–2
indigenous environmentalism 163–4
indigenous people: conservation and culture 322; settlement origins/dating 48–9; rights 216, 228 see also cultural beliefs and practices
infrastructure improvements, colonial period 137
Ingram, J. C. 84, **88, 89, 91**
Integrated Conservation and Development Projects (ICDPs) 154, 157, 172, 228, 301
intellectual property rights (IPRs) 278–9
intensification, agricultural 183–4, 186, 314
interdisciplinary perspectives 3, 11, 355, 356
international perspectives see global contexts
International Conference on Conservation and Development 153–4
International Conference on the Conservation of Natural Resources 142, 151
International Cooperative Biodiversity Groups 271–2, 274–6 see also ICBG-Madagascar project case study
international debt 152–4, 158, 164, 167, 352
International Monetary Fund (IMF) 118, 152–3, 352
international trade see trade

International Union for Conservation of Nature (IUCN) 9, 141–2, 151, 160, 193; World Parks Congress 216–17; protected area categories **218–20** see also Durban Vision
Inventaire Ecologique Forestier National (IEFN) 71, 76, 78, 82
invertebrates: colonization 27; convergent evolution 34
IPRs (intellectual property rights) 278–9
Isalo National Park 228
island biogeography 4, **18**, 18–19, 22, 27–33, **29**
island hopping, colonization 24, **25**
island-wide forest hypothesis 6, 55, **68**, 346; conservation interests 93–4; discourse 67, **131**; origins of claim 85, 91–3; scientific assessment 67–8, 70, 73, 84–94, **86–9, 90–1**
Isle of Fire: The Political Ecology of Landscape Burning in Madagascar (Kull) 130
isolation, geological see geological isolation
IUCN see International Union for Conservation of Nature

jinja 106, 107 see also swidden farming practices
Johnson, S. E. **89**
justice, ICBG-Madagascar project case study 288–9 see also benefits distribution; fairness

kaitiakitanga (resource management terminology), New Zealand 320–1
kingdoms, pre-colonial history 131–5
Kirindy Forest 263
knowledge, traditional 279, 281–2, 321
K-T (mass extinction) event 22, 24, 25
Kull, Christian: *Isle of Fire* 130

La Boétie, Étienne 322, 333–5
Lakaton'i Anja, archaeological evidence 50
lake ecosystems 184–5
Lake Tritrivakely 45, **46,** 56
land tenure/land registration 108, 281, 305, 330–1, 352
land use and deforestation 4, 6–7, 28, 105–8, 121–2; aerial photography 95; complexity of land use decisions 110–15, 122, 346, 347, 352; cultural

dimensions 114–15; drivers 115–18; emotive terminology 106–7, 110; historical contexts 107–10; remote sensing **80–2**; social and ecological dynamics 113–14; swidden farming practices **111**, 111–13, **112**, 118–21; trade 117–18, 120 *see also* deforestation; vegetation
Landsat TM remote sensing 76, 77, 79
landscape mosaics, conflicting visions of nature 325–33
language, Malagasy people 48–9, 51, 52, 134, 327
Le Discours sur la Servitude Volontaire (La Boétie) 333
Le Territoire de Conservation et Développement (Development and Conservation Territory) 200
legal dualism 174–6
legal frameworks 7; bioprospecting 275–80, 281, 282; burning/forest fires 140, 142, 149, 151, 160; COAP **150**, 160, 161, 201, 206, 217; Codes of Articles 134; coercive legislation 3, 299, 302, 315; PEI 157; post-colonial attitudes 141; species distribution 31 *see also* policy frameworks
legitimacy: *and dina* 330; governance principles 9, 225, 236–9
Lehman, S. M. **87**, **88**, 91
Lehtinen, R. M. **87**
lemur species: climate changes 47; conservation 35; endemism 17; evolution models 33; extinctions 57; as flagship species 343–4; fossils 51; future of conservation 349; protectionist view of nature 139; radiocarbon dating **59**; speciation 24; and tourism **257**, 258, 263, 266
life-history traits, endemic mammals 47
lima beans (*Phaseolus lunatus*) 109
livelihood strategies 113, 116, 184
local environmentalism 163–4
logging *see* timber resources
Lowry, P. P. I. 78, **86**, 91

macro-endemism 22–7
Madagascar: A Short History (Randrianja and Ellis) 130
Madagascar Action Plan (2007) 1
Madagascar Biodiversity Fund 161
Madagascar National Parks (MNP - formerly ANGAP) **150**, 157, 158, 161, 195, 199, 202–3, 209; Durban Vision 217, 227–8, 229; nature tourism 259
Mahafale/Mahafaly people 187, 220, 234, 323, 326-329,
Mahilaka, archaeological evidence 52
maize: conservation politics **206**; and deforestation 67, 106, 109–10, 117, 118, 120; Ranobe PK32 case study **5**, **234**, 234, **235**; resource management transfers 187; swidden farming practices 111, 112–13
Makira REDD+ project case study **5**, 304, 307–11, **309** *see also* Reducing Emissions from Deforestation and Forest Degradation
Malagasy people *see* indigenous people
mammals, large: colonization **23**, 26; speciation **18**
Management Effectiveness Tracking Tool (METT) 245
management transfers *see* resource management transfers
Manambolomaty Lakes complex 184
manioc (*Manihot esculenta*) 54, 111, 113
maps: case studies **5**, **231**, **233**, **284**; community-based natural resource management 197–8; palaeoecological sites **46**; vegetation zones **69**
marine ecosystems 184–5, 344–6, 355
Marine Protected Areas (MPAs) 345
market-based instruments (MBIs) 301–2 *see also* incentive-based mechanisms
market environmentalism 336
Masoala National Park 244, 265
mass extinction (K-T event) 22, 24, 25
Mayaux, P. 76, 77, 79, **81**, 83
Mayeur, Nicolas 56
measurement of deforestation *see* scientific assessment
mediation, environmental 179–80, 350
medicines *see* bioprospecting
Menabe region 108, **224**, 347
Méral, P. 116, 117, 118
Merina 109, 178, 323; colonial relations 137, 173; pre-colonial 132-134, 149
Mesozoic geological era 22
micro-endemism 30–3
mid-domain-effect hypothesis, species distribution 30
migrant populations 184, 188; deforestation drivers 110, 115–16; Ranobe PK32 case study 233–4, 240
Mikea Forest 265, 266

milestone payments, bioprospecting 280
Millennium Development Goals 196
mineral resources/mining 164, 209; and bioprospecting 290–1; future of conservation 352–3; policy recommendations 209; resource protection 198–9; resource use rights/debates 200
Minten, B. 116, 117, 118
mission civilisatrice (civilizing mission) 137
missionaries 133–4, 149, 322
MITOIMAFI inter-communal association 234, **235**, 239
MNP *see* Madagascar National Parks
Moat, J. 76, **87–9**, **91**, 91
molecular clocks 24, 33
Montagne d'Ambre National Park 56, 228
Montagne des Français 50, 58, **224**, 271, 281, 289
MPAs (Marine Protected Areas) 345
MSS (Multi-Spectral Scanner)-based remote sensing 76, 79, 83
multi-disciplinary perspectives 3, 11, 355, 356
Muttenzer, F. 181, 183
Myers, N. **86**, **87**, **88**, **89**

narratives: island-wide forest **131**; simplistic 4, 44, 122, 176, 355, 356
National Biodiversity Institute (INBio) 271, 274, 275
National Environmental Action Plan (NEAP) **148**, 149, **150**, 154–6, 162, 193; bioprospecting 272; community-based natural resource management 174, 181; conservation funding 146; conservation politics 196–7; environment programs 156–61; nature tourism 259; participatory governance 225–7
national parks *see* protected areas; Andasibe-Mantadia; Ankarana; Isalo; Madagascar National Parks; Masoala; Montagne d'Ambre; protected areas; Ranomafana; Tsingy de Bemaraha
National Protected Area Management Plan (*PlanGrap*) 197
National Strategy for Conservation and Development 153, 154
Natural Change and Human Impact in Madagascar (Goodman and Patterson) 3

The Natural History of Madagascar (Goodman et al.) 3
nature, friendship with 333–4
nature tourism *see* tourism
NEAP *see* National Environmental Action Plan
Nelson, R. 76, 77, 78, 83, **90**, 91
neoliberal/neoliberalism 153, 157, 161, 163,
non-governmental organisations (NGOs) 130, 157; Blue Venture 184–5; COBA 177, 178–9; conflicting visions of nature 260; Durban Vision 193–203, 205, 207, 210, 246; mining 353; REDD+ mechanism 306–7; resource management transfers 184–5, 189; swidden farming practices 119
Norris, S. **88**
no-take zones, marine 184–5, 330

ocean currents, colonization 25
offset schemes, mining 352
open access 174–6, 185
oral history 51, 70, 134, 322–5
original vegetation *see* island-wide forest hypothesis
Ostrom, Elinor 245–7
overfishing 184–5

palaeoecology 3, 44; human impacts 55, 56; Lake Tritrivakely 45; site locations 45, **46** *see also* fossil record
Paleotropis 22
panaceas: bioprospecting 275; future of conservation 355–6; tourism 255, 259, 264, 266
paper parks 195, 209, 227
Participatory Ecological Monitoring and Community Based Competition 307
participatory governance, Durban Vision 216–17, 225, 245–7; case studies 5, 229–36, **230–5**; definitions **226**; governance structures **232**, **235**, **237**; governance types **221–2**; list of protected areas/delegated institutions **223–4**; maps, reserves **231**, **233**; protected area categories **218–20**; protected areas/reserves, expansion 225–9 *see also* governance principles; policy frameworks
participatory paradigm 9, 161, 166, 174, 176; future of conservation 349–51
parks *see* protected areas

pastoralism: archaeological evidence 51–3; *dina* 330; environmental discourse 346; vegetation changes 55–7 *see also* subsistence agriculture
patents on life 276
Payments for Ecosystem Services (PES) 10, 161, 299, 302–7; land tenure and land registration 305; socio-cultural beliefs and practices 336; unequal distribution of benefits 304–5, **309** *see also* incentive-based mechanisms
PDFIV (*Projet de Développement Intégré Forestier Villageois*) 178
PEI environment programs, NEAP 156–61
people, indigenous *see* indigenous people
perceptions *see* stakeholder perceptions/perspectives
performance based market-based instruments 301
Perrier de la Bâthie, H. 70, 71, 84, 85, 93, 94
pharmaceutical medicines *see* bioprospecting
PHCF (*Projet Holistique de Conservation de la Forêt*) 304
phylogenetic distinctness 17
phytogeography **29**, 29–30, **32** *see also* vegetation
Pigou, Alfred 300–1
PlanGrap (National Protected Area Management Plan) 197
plant species, colonization 27 *see also* vegetation
plantations, trees 139 *see also* reforestation
Pleistocene 24, 31, 44–7, 55
policy frameworks 2–3, 6, 7, 9, 11; deforestation and land use 109; Durban Vision 209–10; European Union 117; evidence-based 95, 96; priorities 355–6 *see also* legal frameworks
political agendas: conservation interests 93–4; resource management transfers 179–80, 187
political ecology 7, 9, 10, 255–8; bioprospecting 272; conflicting visions of nature 258–63; and tourism 255–7
political factors 2, 3, 7; choice and recognition model 182–3; environmental management 129–30,

131–2, 133, 142; conservation interests 94; deforestation drivers 106, 108, 110, 116, 120; future of conservation 351–3; international 164; timelines **8**, **150**; resource protection 198–9 *see also* conservation boom; power dynamics; resistance; state roles; top down initiatives
pollen analyses 45, 54–5, 56
pollution 300–1
population growth: case study 314; deforestation drivers 105, 115, 121; environmental discourse 347; post-sixteenth century 53–4; and poverty 259; received wisdom **131**; resource management transfers 185, 186 *see also* demographic changes
poverty 2, 6, 9; deforestation drivers 105, 106, 115, 121, 242; environmental discourse 346–7; future of conservation 349–50, 356; and population growth 259; and protected areas/reserves 216; received wisdom **131**; and swidden farming practices 259–60; and tourism 258
power dynamics 9, 11, 152; COBA 186; conservation and culture 321, 322, 328, 331–2; conservation boom 164–5; future of conservation 349, 352–3; geopolitical rivalries/strategies 165–6; ICBG-Madagascar project case study 288–9; state roles in environmental management **131–2** *see also* resource management transfers; voluntary servitude
preservationist perspectives 3, 160, 163, 166, 351
Prince Philip 2, 154, 193
pristine nature discourse **131**, 176, 260, 346, 348, 356
Programmes d'Option Spécifiques à l'Eloignement et à l'Insularité (POSEI) 117
Projet de Développement Intégré Forestier Villageois (PDFIV) 178
Projet Holistique de Conservation de la Forêt (PHCF) 304
protected areas: biodiversity 34–5; bioprospecting 281; colonial period 139–40; conservation boom 167; conservation politics 193–5, 198–9; definitions **226**; expansion 225–9;

PEI 157; nineteen seventies/eighties developments 154; post-colonial 141; and poverty 216; and tourism 262 see also Durban Vision
Protection de la Nature (MEEF) 142

quantitative change assessment, deforestation 78–82, **80–2**, 95
QMM (Qit Madagascar Minerals) 198, 223, 353

Radama I, King 133–4
Radama II, King 134
radiated tortoise (*Geochelone radiata*) 328–9
radiocarbon dating 50, 51, 57, 58, **59**
rafting, colonization of Madagascar through 25–6, 27
rainfall 28, 30–1, 45–7, 121 see also climate change
Rajoelina, President 2, **8**, **150**, 162
Rakoto Ramiarantsoa, H. 178, 182, 184
Rakotomaria, E. 151, 152
Ramamonjisoa, B. 187
Ranobe PK32 case study **5**, **224**, 229, 232–4, 239–47; governance structures **235**; maize cultivation **234**, **235**; maps **5**, **233**
Ranomafana National Park **5**, 156–7, 178, 228, 263–4, 343, 347
ratites, subfossils 58
Ratrimoarisaona, S. **86**, 91, 257, 265
Ratsiraka, President 2, **8**, **150**, 152–4, 156, 158–9
Ravalomanana, President 2, **8**, 9, 35, **150**, 159–61, 162; conservation politics 193, 198, 217
Rabemananjara, Z. 187
received wisdom 6; deforestation and land use 110, 115; environmental discourse 346–9; nature tourism 259–60; population growth/poverty **131**, 346–7; real and imagined environment 162
Reducing Emissions from Deforestation and Forest Degradation (REDD+) mechanism 10, 11, 302–4, 306–7, 314, 344; Makira project **5**, 307–11, **309** see also incentive-based mechanisms
reforestation: colonial period 139; post-colonial 141
refugia 31, 35
regional approaches 158–9

regulations, community (*dina*) 329–33 see also policy frameworks
reification: culture 327; imagined communities 180–3
remote sensing 71, 92; satellite-based 73, 76–9, **80–2**, 83–5, 95, 107
reptiles: colonization **23**, 26–7; extinct **59**; radiate tortoise 328–9
Réseau de la Biodiversité de Madagascar (REBIOMA) 197
reserves see protected areas
resistance, passive 334
resistance, political 135–6, 140, 143; and dina 332; Penan people 320–1 see also conflict
resource management see environmental management
resource management transfers 172, 187–9, 239; economic effects 186–7; environmental impacts 183–5; social impacts 177–83 see also community-based natural resource management; *Gestion Locale Sécurisée*
resource protection, conservation politics 198–9
Réunion 117–18, 133
Ribot, J. 182
rice 106, 109, 111, 119
rights, resource use 200–1, 216, 228
ring-tailed lemurs (*Lemur catta*) **257**, 258, 263, 266
Rio Tinto 353
rituals, cultural 114–15, 322–5 see also cultural beliefs and practices
river hypothesis, species distribution 30
roads: colonial period 149; deforestation drivers 116; and tourism 257
rosy periwinkle (*Catharanthus roseus*) 10, 276, **277**
royalties, bioprospecting 280
rural consultation process, Durban Vision 202–7
rural households, land use decisions 110–15

sacred 321, 329; forest 197, **219**, **220**, **230**, 229–31, 238 see also cultural beliefs and practices
Sakalava 109, 114, 323; colonial relations 136, 137; kingdom, pre-colonial history 132
Sandy, C. **88**
SAPM see Durban Vision

satellite-based remote sensing 73, 76–9, 80–2, 83–5, 95, 107
scientific assessment of deforestation 67–70, 94–6; colonial period 71–3, **72**, **73**, 93; forest patches in the highlands landscape **68**; island-wide forest hypothesis 6, 70, 84–94, **86–9**, **90–1**; methodology issues/accuracy assessment 83–4; original vegetation 67–8, 70, 73; recent forest cover 73–7, **74–5**; trends, quantitative change assessment 78–82, **80–2**, 95; vegetation zones **69**, 71
scientific research, conservation 3–4, 6, 11, 137; biodiversity 35; evidence-based: policy 95, 96; foreign input 151–2; local/indigenous 163–4; multi-disciplinary perspectives 355; overfishing 184; priorities 355–6
Second Republic **8**, 152, 153
sediment sampling, palaeoecology 45
sensibilisation (persuasive education, outreach and awareness raising) 204, 208
Service de la Main d'Oeuvre des Travaux d'Interet Général (SMOTIG) 137
Service des Eaux et Forêts see Forest Service
services, ecosystem *see* ecosystem goods/services
sharecropping 109, 110 *see also* cash crops
shifting cultivation 84, 106, 119 *see also* swidden farming practices
Sibree, James 71, **72**
simplistic narratives 4, 44, 122, 176, 355, 356
slash-and-burn agriculture 93, 94; ASB case study **5**, 307, 313–14; emotive terminology 106–7; legislation 160 *see also* swidden farming practices
slave trade 53, 54, 56, 132, 133
slavery, and voluntary servitude 333, 334
social cohesion 11, 332, 333, 334
social factors 6, 60; deforestation 96, 113–14, 121; protected areas 216, 217, 228, 240; resource management transfers 177–83; swidden farming practices 348 *see also* cultural beliefs and practices
Société de Production de Stockage et de Manutention des Produits Agricoles (SOPAGRI) 117
socio-political organizations, pre-colonial history 131–2

speciation **18**, 19, **19**, 20
species: colonization 4, 17, **18**, 22, **23**, 24–7, **25**; distribution **29**, 29–30, 44; total numbers 19–20, **20**, 343 *see also* biodiversity; endemic species
spiny forest 5, **5**, 28, **29**, 76, 78, 117, 120, 344; Durban Vision **220**, 229
SPOT-4 VEGETATION remote sensing 76–7, 84
stakeholder perceptions/perspectives 11, 356; ICBG-Madagascar project case study **287**, 287–8; local/indigenous environmentalism 163–4; real and imagined environment 162; reification of imagined communities 180–3; tension/conflict 349–51 *see also* cultural beliefs and practices
stakeholders: future of conservation 349, 350, 356; GELOSE **175**, 176 *see also* political agendas
standardized contracts 180, 181
state roles in environmental management 7, 129–30, 142–3; colonial period 129, 135–40; environmental politics/power **131–2**; independence 141–2; pre-colonial history 131–5; tribal ethnic identity **138**
structural adjustment 152–3
subfossils 57–8 *see also* fossils
subsistence agriculture 2; colonial period 136; deforestation and land use 108, 109, 119; historical contexts 173 *see also* pastoralism; swidden farming practices
supernatural beliefs *see* cultural beliefs and practices
Sussman, R. W. 79, 82, **90**, 91
sustainable development 154, 163, 168; and bioprospecting 290–2; conflicting visions of nature 260; deforestation and land use 105, 107, 118; Durban Vision 217; environmental discourse 348; forest management 3; future of conservation 353; geopolitical rivalries/strategies 165; Reducing Emissions from Deforestation and Forest Degradation mechanism 302
swidden farming practices 6, 52, 93, 94; ASB case study **5**, 307, 313–14; colonial period 137, 139, 140; community-based

conservation 181; cultural rituals 115; definition of 107; deforestation and land use 84, 106–13, **111**, **112**, 118–21; description of basic practice 111; Durban Vision 238; environmental discourse 346; future of conservation 349; legislation 160; pre-colonial history 134; socio-ecological dynamics 348; and tourism 259–60
Swiss Aid 142, 146, **148**, 151, 158, 166, 263
Système des Aires Protégées de Madagascar see Durban Vision

taboos 11, 114, 121, **138**, 238, 281, 320, 322 see also *fady*
tavy 106, 107, 119 see also swidden farming practices
taxes, colonial period 137
tenure, land 108, 281, 305, 330–1, 352
terminology use: emotive 106–7, 110; environmental discourse 346–9; future of conservation 356; ICBG-Madagascar project case study 287–8, **287** see also definitions
Third Republic 156
timber resources 93; and bioprospecting 290–1; colonial period 149; environmental discourse 347, 348; future of conservation 353; political resistance 320–1; pre-colonial history 134–5; resource protection 198–9; resource use rights/debates 200; state roles in management 133; and tourism 263
Toillier, A. 184, 187
top down initiatives 7, 9, 350; Durban Vision 159, 160, 206; ethics 350; forest management 256; rules and regulations 331; tourism 266, 267 see also resource management transfers
topography of Madagascar 27–8, 47, 71
tortoise, radiated (*Geochelone radiata*) 328–9
tourism 9–10, 85, 255–7, 266–7; Baobab Alley **5**, 260–2, **261**, 348; and biodiversity 258; conflicting visions of nature 258–63; cost-benefits/ politics 263–6; environmental discourse 348; environmental politics/power **131**; future of conservation 349, 350; political ecology approach 255–7; and

poverty 258; ring-tailed lemurs **257**, 258, 263, 266; visitor numbers 256–7
trade: archaeological evidence 52; colonial period 136; deforestation and land use 117–18, 120; global markets 133; post-sixteenth century 53–4; pre-colonial history 132, 133; slave trade 53, 54, 56, 132, 133
Trade Related Aspects of Intellectual Property Rights (TRIPS) 278–9
trade-offs 11, 256, 267; future of conservation 349–51, 356
traditional beliefs see cultural beliefs and practices
traditional knowledge 279, 281–2, 321
trans-oceanic dispersal, colonization 25–7 see also rafting
transparency governance principle 9, 225, 242–3
trees, meaning for indigenous people 322–5, 326 see also cultural beliefs and practices; forests; timber resources
tribal ethnic identity **138**
Tritrivakely 45, **46**, 56
Tsiebo, Calvin (Vice President of the Republic) 151
Tsingy de Bemaraha National Park 228
Tsiranana, President Philibert 141, 149, **150**, 152
turtles, freshwater, colonization **23**, 26–7

unequal distribution of benefits see benefits distribution
United Nations Development Programme (UNDP) 155
United Nations Educational, Scientific and Cultural Organization (UNESCO) 154, 155
United States: conservation funding 146; geopolitical rivalries/strategies 166
United States Agency for International Development (USAID) 77, 142, 156, 157–8
utilitarian view of nature 3, 139, 301, 351 see also commodification of nature

Vazimba people 51
vegetation: climatic conditions 28–33, **29**, **32**; colonization 27; ecotypes **29**, 29–30; human impacts 55–7; phytogeography **29**, 29–30, **32**; zones **69** see also deforestation; forests

Velondriake Locally Managed Marine Area (LMMA) 185
virgin forest, environmental discourse 85, 346, 347
visions of nature, conflicting *see* cultural beliefs and practices
Vohémar region 49, 50, 53–4
Volampeno, N. S. M. **89**
voluntary servitude 322, 333–6

Washington Consensus 153
water availability, deforestation and land use 120
watershed chorological hypothesis, species distribution 31
weather phenomena 31, 33, 47, 353 *see also* climate change
western region 108, 114, 345
Whitehurst, A. S. **89**
wilderness ideology 142, 149, 166, 260
wildlife corridors 35, 158–9, 186–7 *see also* Ankeniheny-Zahamena; Fandriana-Vondrozo
wildlife tourism *see* tourism
win–win solutions 10, 356; bioprospecting 274; Durban Vision 239; future of conservation 356; incentivising forest conservation 299, 302; Reducing Emissions from Deforestation and Forest Degradation mechanism 302; resource management transfers 182, 185, 186; tourism 256, 267
World Bank 2, 105, 152–7, 160–2, 164, 306; conservation politics 206, 207, 229; and tourism 258
World Bank Forest Carbon Partnership Facility (FCPF) 307
World Conservation Strategy 153, 154
World Trade Organization (WTO) 278–9
World Wide Fund for Nature (WWF) 92, 141–2, 151, 153, 156; conservation funding 146, **147**; *and dina* 331; Durban Vision 246; Global 200 Ecoregions 34

Yellowstone Park model 196 *see also* fortress conservation

Zafy, President 156–8
zebu cattle 113, 132, 241, 312, 327
zoning, protected areas **226**, **231**